STRUCTURE OF THE NUCLEON AT LARGE BJORKEN x

Related Titles from AIP Conference Proceedings

744 Latin-American School of Physics, XXXV ELAF Supersymmetries in Physics and Its Applications
Edited by R. Bijker, O. Castaños, D. Fernández, H. Morales-Técotl, L. Urrutia, and C. Villarreal, January 2005, 0-7354-0228-0

739 IX Hadron Physics and VII Relativistic Aspects of Nuclear Physics: A Joint Meeting on QCD and QGP
Edited by M. E. Bracco, M. Chiapparini, E. M. Ferreira, and T. Kodama, November 2004, 0-7354-0223-X

737 Advanced Accelerator Concepts: Eleventh Advanced Accelerator Concepts Workshop
Edited by Vitaly Yakimenko, December 2004, CD-ROM included, 0-7354-0220-5

717 Hadron Spectroscopy: Tenth International Conference
Edited by Eberhard Klempt, Helmut Koch, and Herbert Orth, August 2004, 0-7354-0197-7

698 Intersections of Particle and Nuclear Physics, 8th Conference; CIPANP 2003
Edited by Zohreh Parsa, February 2004, CD-ROM included, 0-7354-0169-1

688 Scalar Mesons: An Interesting Puzzle for QCD
Edited by Amir H. Fariborz, November 2003, 0-7354-0159-4

687 High Energy Physics: The 25th Annual Montreal-Rochester-Syracuse-Toronto Conference on High Energy Physics, MRST 2003: A Tribute to Joe Schechter
Edited by Amir H. Fariborz, November 2003, 0-7354-0161-6

675 Spin 2002: 15th International Spin Physics Symposium and Workshop on Polarized Electron Sources and Polarimeters
Edited by Y. Makdisi, A. Luccio, and W. MacKay, July 2003, CD-ROM included, 0-7354-0136-5

672 Short Distance Behavior of Fundamental Interactions: 31st Coral Gables Conference on High Energy Physics and Cosmology
Edited by Behram N. Kursunoglu, Metin Camcigil, Stephan L. Mintz, and Arnold Perlmutter, June 2003, 0-7354-0139-X

670 Particles and Fields: Tenth Mexican School
Edited by U. Cotti, M. Mondragón, G. Tavares-Velasco, June 2003, 0-7354-0135-7

667 Increasing the AGS Polarization
Edited by A. D. Krisch, A. M. T. Lin, and T. Roser, May 2003, 0-7354-0130-6

660 Hadron Physics: Effective Theories of Low Energy QCD;
Second International Workshop on Hadron Physics
Edited by A. H. Blin, B. Hiller, A. A. Osiplov, M. C. Ruivo, and E. van Beveren, March 2003, 0-7354-0120-9

To learn more about these titles, or the AIP Conference Proceedings Series, please visit the webpage **http://proceedings.aip.org/proceedings**

STRUCTURE OF THE NUCLEON AT LARGE BJORKEN *x*

2nd International Workshop on the Structure of the Nucleon at Large Bjorken *x*

HiX2004

Marseille, France 26 - 28 July 2004

EDITORS

Claude Bourrely

Centre de Physique Théorique-CNRS
Marseille, France

Zein-Eddine Meziani

Temple University
Philadelphia, Pennsylvania

Jacques Soffer

Centre de Physique Théorique-CNRS
Marseille, France

SPONSORING ORGANIZATIONS

Centre de Physique Théorique de Marseille, France
Centre National de la Recherche Scientifique - CNRS, France
IN2P3, France
Université de Provence, Marseille, France
Jefferson Laboratory, Newport News, Virginia
Temple University, Philadelphia, Pennsylvania

AMERICAN INSTITUTE OF PHYSICS

Melville, New York, 2005

AIP CONFERENCE PROCEEDINGS ■ VOLUME 747

Editors:

Claude Bourrely
Jacques Soffer

Centre de Physique Théorique-CNRS
Luminy Case 907
13288 Marseille Cedex 9
FRANCE

E-mail : bourrely@cpt.univ-mrs.fr
soffer@cpt.univ-mrs.fr

Zein-Eddine Meziani

Temple University
Barton Hall A116
1900 N. 13th Street
Philadelphia, PA 19122-6082
U.S.A.

E-mail: meziani@temple.edu

L.C. Catalog Card No. 2005920502
ISBN 0-7354-0232-9
ISSN 0094-243X
Printed in the United States of America

CONTENTS

Preface..ix
Local Organizing and International Advisory Committees....................xi
Old Port of Marseille Photograph xiii

OVERVIEW

QCD Theory at Large x..3
X. Ji
Large-x Resummations in QCD 9
W. Vogelsang
MRST Parton Distributions 16
W. J. Stirling, A. D. Martin, R. G. Roberts, and R. S. Thorne

UNPOLARIZED STRUCTURE FUNCTIONS

Global QCD Analyses of Parton Distributions at Large x and Small...........25
W.-K. Tung
Measuring F_2^n/F_2^p and d/u: Past and Future...............................32
G. G. Petratos
Prospects for and Status of the Measurement of High x
Parton Distributions at HERA 40
Z. Zhang
The BoNuS Experiment at Jefferson Lab......................................46
S. Bültmann
Non-Singlet QCD Analysis of F_2 in 3-Loops...................................50
J. Blümlein, H. Böttcher, and A. Guffanti

POLARIZED STRUCTURE FUNCTIONS—I

Statistical Polarized Parton Distributions at Large-x 57
C. Bourrely
Status of the RHIC Spin Program..63
L. C. Bland
The Compass Experiment at CERN 72
A. Magnon *(on behalf of the COMPASS Collaboration)*

POLARIZED STRUCTURE FUNCTIONS—II

Neutron Spin Structure Measurements in JLab Hall A 83
J.-P. Chen *(on behalf of the JLab Hall A and E94-010 Collaborations)*

Spin Structure of the Proton and Deuteron with CLAS at Jefferson Lab 89
P. E. Bosted *(on behalf of the CLAS Collaboration)*
Polarized Structure Functions: Proton/Deuteron Measurements in Hall C 95
O. A. Rondon
How High is High x? 101
J. Franklin

DUALITY AND HIGHER TWISTS

Quark-Hadron Duality: Higher Twists and Large-x Structure Functions 107
W. Melnitchouk
Quark-Hadron Duality: Experimental Status and Prospects 114
M. E. Christy
Spin Duality in the Nucleon: Measurements at Jefferson Lab Hall A 118
N. Liyanage
Higher Twists and the Measurements of g_2. 122
W. Korsch
Quark-Hadron Duality and Higher Twist Contributions in Structure Functions 126
A. Fantoni, N. Bianchi, and S. Liuti

SEMI-INCLUSIVE DEEP INELASTIC SCATTERING

Semi-Inclusive DIS and Transversity 133
P. G. Ratcliffe
Transversity and SIDIS at Jefferson Lab 141
J.-C. Peng, J.-P. Chen, X. Jiang, and L. Zhu
Semi-Inclusive DIS in NLO QCD 147
M. Stratmann
Helicity and Transversity Distributions at HERMES 153
L. De Nardo *(on behalf of the HERMES Collaboration)*
Novel Azimuthal Asymmetries in Drell Yan and Semi-inclusive Deep Inelastic Scattering 159
L. P. Gamberg, G. R. Goldstein, and K. A. Oganessyan

EXCLUSIVE REACTIONS

GPDs from Small to Large x 167
A. Schäfer
Generalized Parton Distributions and Deep Exclusive Reactions: Present Program at JLab 173
M. Garçon

Pion Photoproduction on the Nucleon 179
H. Gao and L. Zhu
Generalized Parton Distributions and Constituent Quark Models 185
M. Traini, B. Pasquini, and S. Boffi
Nucleon DIS Structure Functions from Semi-inclusive *ed→enX* Reactions 191
M. Sargsian

PARITY VIOLATING EXPERIMENTS AND DRELL-YAN REACTIONS

Parity at High *x* 199
P. A. Souder
Current Status of the NuTeV Anomaly 205
J. T. Londergan
The Past and Future of Drell-Yan Physics at Fermilab 211
R. S. Towell *(on behalf of the FNAL E866/NuSea and E906 Collaborations)*
Q^2 Dependence of the Azimuthal Asymmetry in Unpolarized Drell-Yan 215
E. Di Salvo

LATTICE QCD AND NUCLEAR DEEP INELASTIC SCATTERING

Large *x* Behaviour and the Non-Perturbative Structure of Hadronic Systems 221
A. W. Thomas
Generalized Parton Distributions and Nucleon Structure at Large x from Lattice QCD 227
G. Schierholz
Nucleon Structure Functions and Nuclear DIS 233
A. Molochkov
Mapping High *x* Structure of the Nucleon: Which Data is Missing? 237
G. I. Smirnov

FUTURE AND SUMMARY

Probing Correlations in the Large x Proton Wave Function at *pp* and *pA* Colliders 243
M. Strikman
Research Perspectives at Jefferson Lab: 12 GeV and Beyond 248
K. de Jager
Overview on the Future Electron-Ion Collider Projects 254
E. W. Hughes

APPENDIX

HiX 2004 Workshop Agenda 263
HiX 2004 List of Participants 267

Author Index 271

Preface

This workshop, held on July 26-28 2004, at the Centre de Physique des Particules de Marseille (CPPM) Luminy, in Marseille, France was a follow up meeting, but with a wider scope, to a first workshop (HiX 2000) held in March, 2000 in Philadelphia,PA, USA. In HiX 2000, the goal was to identify opportunities for the 12 GeV upgrade program at Jefferson Lab in order to contribute to our overall understanding of the electromagnetic and spin structure of the nucleon in the valence quark region. However, due to the success of that workshop, we recognized that the interest of the subject of nucleon structure in the valence region was shared by a broader hadronic physics community, working at different laboratories, like FermiLab, JLab, RHIC, CERN and DESY. Therefore, we decided to broaden the program reach and invited speakers for a wider community.

At Marseille, the emphasis was not only to discuss the present theoretical and experimental status of our understanding of the nucleon valence quark region, but to keep seeking for new opportunities at existing facilities as well as planned ones. The subjects ranged from how to best access the helicity-independent distributions with flavor decomposition to identifiying new key observables which need to be measured in order to advance our grasp of the nucleon structure, like, generalized parton distributions (GPD's) and transverse momentum quark distributions. Many data taken at existing facilities were presented showing the pace of progress and the need for more precise data in the region of interest, namely the valence quark region.

Among the identified issues let us mention the determination of helicity-independent and helicity-dependent quark distributions, at very large values of Bjorken x ($x > 0.6$), which remain poorly known. There was a clear complementarity in the proposals put forward at different laboratories involved in the planned measurements of relevant observables using neutral and charged currents. The attendees were eager to see these measurements carried out and checked against our understanding of the structure of the nucleon in the valence region. It was important also to witness the theoretical progress made in preparation of the physics analysis of present and future data. We recall, for example, the degree of sophistication in evaluating the corrections of the data from soft gluons radiation and higher twists effects, in order to unravel the genuine quark distributions in the very high x region. Although it was not discussed extensively the physics presented in this workshop clearly finds its natural extension in the future construction of a high luminosity polarized Electron Ion Collider (EIC) wherever it will be located. This new facility will be crucial for sustaining our present progress into the future in the area of hadronic physics. This area of physics has been recently rewarded by the 2004 Nobel price award to David J. Gross, H. David Politzer and Frank Wilczek *for the discovery of asymptotic freedom in the theory of strong interaction.*

The meeting was well attended and revealed considerable enthusiam in the community for carrying the planned measurements and continuing to explore QCD through the study of the structure of the nucleon, seen as its best laboratory. The meeting was organized with technical and financial support from the Centre de Physique Théorique de Marseille, the Centre National de la Recherche Scientifique, the IN2P3, the Université de Provence, Jefferson Laboratory and Temple University.

We thank Roy Aleksan, Director of CPPM, for his kind hospitality.

Claude BOURRELY
Zein-Eddine MEZIANI
Jacques SOFFER

Local Organizing Committee

C. Bourrely (Marseille)
V. Burkert (JLab)
J.-P. Chen (JLab) co-chair
W. Melnitchouk (JLab)
Z.-E. Meziani (Temple) co-chair
G. G. Petratos (Kent State)
O. Rondon-Aramayo (Virginia)
J. Soffer (Marseille) chair

International Advisory Committee

S. Brodsky (SLAC)
F. Close (Oxford)
K. de Jager (JLab)
R. Jaffe (MIT)
X. Ji (Maryland)
M. Klein (DESY)
P. Mulders (Amsterdam)
J.-C. Peng (Illinois)
P. A. Souder (Syracuse)
W. Stirling (Durham)
W. K. Tung (Michigan State)
A. W. Thomas (JLab)
W. Vogelsang (BNL)

Entrance of the Old Port of Marseille. Photo by Jacques Soffer

OVERVIEW

QCD Theory at Large x

Xiangdong Ji

Department of Physics, University of Maryland, College Park, MD 20742, USA

Abstract. In this introductory talk, I discuss what we know about the large-x parton distributions and related hard processes in quantum chromodynamics. I use lepton-nucleon deep-inelastic scattering as a concrete example to illustrate the physics.

GENERAL REMARKS ON DIS AT LARGE X

This workshop is about parton distributions at large x. In this introductory talk, I will concentrate on what do we really know about the large-x parton distributions in the fundamental theory—quantum chromodynamics (QCD). Since I will mostly talk about deep-inelastic scattering (DIS), it is useful to start with a few remarks about the process. In the ordinary inclusive DIS, there are two mass scales, $\Lambda_{\rm QCD}$ and Q, involved. However, at large x, there is a new scale $(1-x)Q^2$ emerging. For example, the hadron invariant mass $W^2 = M^2 + Q^2(1-x)/x$ is of that order.

We will be interested in the regime $(1-x)Q^2 \gg \Lambda^2_{\rm QCD}$, where the standard QCD factorization is still valid. However, we now have an hierarchy of mass scales

$$Q^2 \gg (1-x)Q^2 \gg \Lambda^2_{\rm QCD} \, . \tag{1}$$

Because of that, the standard perturbative QCD (pQCD) series must be re-organized. The invariant mass of the hadron final state is so small that DIS is close to an exclusive scattering (for example, the so-called Sudakov form factor starts to creep in). Scale evolution of parton distributions can potentially be complicated.

DGLAP EVOLUTION AT LARGE X

DGLAP (Dokshitzer-Gribov-Lipatov-Altarelli-Parisi) evolution equation determines how the parton densities $f_{i/H}(x)$ evolve with probing scale μ,

$$\mu \frac{d}{d\mu} f_{i/H}(x) = \sum_j \int_x^1 \frac{dy}{y} P_{ij}\left(\frac{x}{y}\right) f_{j/H}(y) \, , \tag{2}$$

where P_{ij} are called parton splitting function. The large-x behavior of $P(x)$ determines the evolution of the parton densities at large x.

CP747, *Structure of the Nucleon at Large Bjorken x,* edited by C. Bourrely, Z.-E. Meziani, and J. Soffer

It is known that in the first two (and now three) orders of pQCD calculation, the splitting functions have the following leading large-x behavior

$$P_{ij}(x) = \left[2A_i \frac{1}{(1-x)_+} - 2B_i \delta(1-x)\right] \delta_{ij} + \mathcal{O}((1-x)^0) , \tag{3}$$

where A_i and B_i are perturbation series in α_s.

Define as usual the moments,

$$M_i(N) = \int_0^1 dx x^{N-1} f_i(x); \qquad \gamma_{ij}(N) = \int_0^1 dx x^{N-1} P_{ij}(x) . \tag{4}$$

The evolution equation becomes,

$$\mu \frac{d}{d\mu} M_i(N) = \sum_j \gamma_{ij}(N) M_j(N) . \tag{5}$$

The large-x behavior of P_{ij} corresponds to large-N behavior of γ_{ij}. The above large-x result for the splitting functions corresponds to

$$\gamma_{ij}(N) = [-2A_i \ln(Ne^C) - 2B_i]\delta_{ij} + \mathcal{O}(1/N) . \tag{6}$$

where C is the Euler constant.

What is the behavior of the evolution kernels to all orders in perturbation theory? This question has been answered by G. P. Korchemsky in 1989 [1]. He found that the structure of the leading orders persists to all orders in pQCD,

$$\begin{aligned} P_{ij}(x) &= \left[2A_i \frac{1}{(1-x)_+} - 2B_i \delta(1-x)\right] \delta_{ij} + \mathcal{O}((1-x)^0) , \\ \gamma_{ij}(N) &= [-2A_i \ln(Ne^C) - 2B_i]\delta_{ij} + \mathcal{O}(1/N) \end{aligned} \tag{7}$$

where A_i and B_i are perturbation series in α_s. He showed also that A_i is related to the cusp anomalous dimension of the two straight Wilson lines for which $A_q = (C_F/C_A)A_g$, where C_F and C_A are group theoretical factors.

Here is a brief summary of the proof. At large x, the quark distribution in the axial gauge can be re-factorized into hard, soft, and collinear parts. The x-dependence appears only in the soft part,

$$\phi(x) = (HJ)^2 \frac{p^+}{2\pi} \int dy e^{iy^- p^+(1-x)} S(x) + \mathcal{O}((1-x)^0) , \tag{8}$$

where the soft function is a vacuum expectation value of the Wilson lines, renormalized according to the cusp anomalous dimension. By taking derivative of the above with respect to the renormalization scale, one arrives at Korchemsky's conclusion.

What is the implication of the above result? Let us suppose the parton distribution at large x has the form,

$$f(x) = C(1-x)^{N(\mu)} . \tag{9}$$

The exponent depends on the renormalization scale. The evolution equation for the exponent goes like

$$\frac{d}{d\mu}N(\mu) = 2A \, . \tag{10}$$

Since A is positive definite, the exponent grows with μ and the large-x parton distributions vanish faster as the probing scale increases. This is the direct consequence of parton radiation.

Similar conclusion holds for polarized parton distribution. In particular, the soft part is independent of the quark spin and therefore is universal.

COEFFICIENT FUNCTIONS AT LARGE X AND RE-SUMMATION OF DOUBLE LOGARITHMS

Let us now consider the large-x behavior of the DIS structure function. The standard QCD factorization yields,

$$F_2(x,Q^2) = \sum_i \int_0^1 \frac{dy}{y} C_i\left(\frac{x}{y}, \alpha_s\right) q_i(y,Q^2) + \dots \, , \tag{11}$$

where the coefficient function C is a perturbation series in α_s and q_i is a quark distribution.

Leading-order calculations indicate that at order α_s^n, the x-dependence of the coefficient function goes like

$$\left[\frac{1}{1-x}\sum_{i=1}^{n} c_i^n \ln^{2i-1}(1-x)\right]_+ \, , \tag{12}$$

where $+$ is a plus function which subtracts off the singularity at $x = 1$. Although regular, the function is very large near $x = 1$. Hence the perturbative expansion is not convergent and must be re-summed.

In the moment space, define

$$F_2(N,Q^2) = \int_0^1 dx x^{N-1} F_2(x,Q^2) \, . \tag{13}$$

The factorization theorem takes the form

$$F_2(N,Q^2) = \sum_i C_i(N, \alpha_s(Q^2)) M_i(N,Q^2) \, . \tag{14}$$

The double logarithms are now in the form of $\ln N$—the coefficient of the α_s^n term in $C(N)$ is

$$\sum_{i=1}^{n} c_i^n \ln^{2i} N \, . \tag{15}$$

What is the physics of the double logarithms? The final-state hadrons have a small invariant mass compared to the virtuality of the photon, and therefore act like a resonance of mass $\sim (1-x)Q^2$. Thus the radiative corrections to the structure function are dominated by the form-factor type of diagrams. There is an incomplete cancellation of soft divergences from virtual and real emissions of gluons, resulting in large perturbative corrections. The Sudakov double logarithms in $\ln(1-x)$ can be re-summed using the formalism developed by A. Mueller, J. Collins, and A. Sen, etc. This has been done in the literature by G. Sterman [2], S. Catani and L. Trantadue [3].

To calculate the coefficient functions, one may consider DIS on a single quark at large x. In this case, the most general reduced diagrams consist of jets with a quark quantum number scattering with a photon, producing a single jet in the final state. The soft gluons carry away the energy momentum of order $(1-x)p$.

The factorization in the moment space reads,

$$F_2(N,Q^2) = |H(Q^2)|^2 S(N) J(N,Q^2) \phi(N,Q^2) \,, \tag{16}$$

where H is a perturbative function, S is a soft factor, J is the final-state jet factor, and ϕ is an energy-dependent parton distribution function in a parton. The jet function and ϕ contain double logarithms in N and obey an evolution equation in energy. By solving the evolution equation, the DIS moment can be shown to exponentiate.

A similar factorization works for the $\overline{\rm MS}$ parton distribution. Combining the two, one can extract the coefficient function as

$$C(N,\alpha) = \exp\left[-\int_0^1 dz \frac{z^{N-1}-1}{1-z}\left\{\int_{1-z}^1 \frac{d\lambda}{\lambda} g_1(\alpha(\lambda Q^2)) + g_1(\alpha)\right\}\right] \,, \tag{17}$$

where $g_i = \sum_{n=1} \alpha_s^n g_i^{(n)}$ is just twice of the cusp anomalous dimension. This resummation formula has not yet been, as far as I know, implemented systematically in the parton distribution fitting programs.

Recently, there has been much progress in developing the so-called soft-collinear effective theory (SCET), in which the soft and collinear gluons are separated at the lagrangian level. In this theory, Manohar has calculated the large N behavior of the factorized structure function [4]. The result has been achieved in two steps. First, the quark currents are matched to SCET currents at scale Q^2. Then, operator production expansion is applied to SCET currents at the scale $Q^2(1-x)$. Manohar obtained a result similar to the above equation, but the so-called infrared renormalon problem does not exist in this approach.

A NEW FACTORIZATION THEOREM FOR PARTON DISTRIBUTION AT LARGE X

In this last part of the talk, I discuss the question: Can one make a sensible pQCD prediction for the large-x behavior of a parton distribution? There is a famous classical result by Farrar and Jackson that the up and down quark distributions in the nucleon go like [5]

$$q(x) \sim (1-x)^3 \,, \tag{18}$$

at large x. The coefficient of this scaling, however, is infrared divergent, as shown below. On the other hand, lattice QCD calculates the moments of the parton distributions,

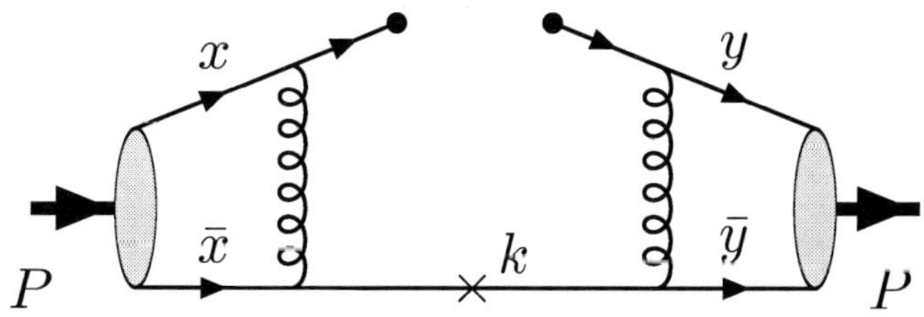

FIGURE 1. Quark distribution at large x in the pion.

and high moments are hard to obtain for technical reasons. There are some model predictions for the large-x properties of the parton distributions. However, to relate them to experimental data, one in principle has to know the relationship between constituent and current quarks.

More theoretical efforts are needed to understand QCD factorizations at large x, in particular, the theoretical foundation of the Farrar-Jackson result (X. Ji, J.P. Ma, and F. Yuan, to be published). In the following, we consider the example of pion. The leading-order diagram contributing to the parton distribution at large x is shown in Fig. 1. As $x \to 1$, the on-shell quark has a soft longitudinal momentum $1-x$. On the other hand, the virtuality of other quark and gluon line become very large.

A straightforward calculation yields,

$$q(x) \sim (1-x)^2 \int \frac{dk_\perp^2}{(k_\perp^2)^2} . \tag{19}$$

Therefore, the quark distribution in the pion goes like $(1-x)^2$ at large x. The coefficient, however, diverges at small transverse momentum. One might argue that the $k_\perp$-integral is regulated by quark confinement. However, without an explicit recipe to handle the divergence, one cannot take the scaling prediction seriously. In other words, one needs a factorization theorem for the large-x parton distribution.

To see the form of factorization, let us consider the above calculation in the soft approximation. Since the $+$ component of the spectator momentum is very soft, one can make soft approximation when its momentum is flowing into the gluon and quark lines,

$$\frac{1}{(xp+k)^2+i\varepsilon} \approx \frac{1}{2x(p\cdot k)+i\varepsilon} , \tag{20}$$

After the approximation, the internal quark and gluon lines are the same as the eikonal lines along the direction of the pion momentum.

One can collect together the eikonal lines and the soft-quark propagator to define a soft function. Then the diagram can be factorized into a soft factor and jet wave function. Of course, the soft factor cannot be calculated in perturbative QCD; neither can the wave function amplitude. If one insists on calculating the soft factor in perturbation theory,

$$S \sim \int d^4k \frac{k_\perp^2}{(k \cdot p)^4} \delta(k^2) \mathrm{Tr}(k_\mu \gamma^\mu \gamma^-) \,. \tag{21}$$

The first factor here is a part of the eikonal lines, and the second factor comes from the soft-quark final state. Carrying out the integral,

$$S \sim (1-x)^2 \int \frac{d^2 k_\perp}{k_\perp^4} \tag{22}$$

This is exactly what we had before.

Finally, let us generalize the above result to all orders in perturbation theory. The most general reduced diagrams for parton distributions at large x consist of jet funtions before and after the hard scattering, and a soft blob connecting the jets. First of all, the jets can be factorized from the hard part using Ward identities. Next the soft lines can be factorized from jets by using soft approximation and Ward identities. The final product is a factorization theorem of the following type,

$$f(x) = H(p,\mu)^2 J_L(p,\mu) J_R(p,\mu) S((1-x)p,\mu) \,. \tag{23}$$

The scaling rule now comes from the power counting of the soft function which itself, however, is nonperturbative.

We finally summarize the main results of this talk. There is a factorization theorem for parton distributions at large x, which confirms the long-established power counting. The evolution of the parton distribution at large x follows the anomalous dimension of the Wilson-line cusp. Finally, the DIS structure function at large x contains large Sudakov double logarithms which, when summed, could modify the behavior of the extracted parton distributions in an essential way.

ACKNOWLEDGMENTS

This work is supported partly by the DOE grant DE-FG-02-93ER-40762.

REFERENCES

1. Korchemsky, G. P., *Mod. Phys. Lett.*, **A4**, 1257 (1989).
2. Sterman, G., *Nucl. Phys.*, **B281**, 310 (1987).
3. Catani, S., and Trentadue, L., *Nucl. Phys.*, **B327**, 323 (1989).
4. Manohar, A., *Phys. Rev.*, **D68**, 114019 (2003).
5. Farrar, G. R., and Jackson, D. R., *Phys. Rev. Lett.*, **35**, 1416 (1975).

Large-x Resummations in QCD

Werner Vogelsang

Physics Department and RIKEN-BNL Research Center,
Brookhaven National Laboratory, Upton, New York 11973, U.S.A.

Abstract. We give a brief introduction to the resummation of a class of large logarithmic perturbative corrections to partonic hard-scattering cross sections. These corrections occur in deeply-inelastic structure functions at high Bjorken-x, and near partonic threshold in cross sections for large produced invariant mass or transverse momentum, such as the Drell-Yan process or hadronic prompt-photon production. They are associated with soft and/or collinear gluon emission.

INTRODUCTION

Hadronic cross sections at large momentum transfer factorize to leading power in the hard scale into process-independent long-distance pieces (usually parton distributions) and short-distance parts that describe the hard interactions of the partons and are amenable to QCD perturbation theory. For example, for the Drell-Yan reaction $pp \to \mu^+\mu^- X$ at high invariant muon pair mass Q one has:

$$\frac{Q^4 d\sigma_{\rm DY}(\tau)}{dQ^2} = \sum_{a,b=q,\bar{q},g} f_a(x_a,\mu) \otimes f_b(x_b,\mu) \otimes \frac{Q^4 d\hat{\sigma}^{ab}_{\rm DY}\left(\frac{\tau}{x_a x_b}, \frac{Q}{\mu}, \alpha_s(\mu)\right)}{dQ^2}, \quad (1)$$

where $\tau = Q^2/S$, f_a, f_b are the parton densities, $\otimes$ denotes an appropriate convolution, and where the sum is over all contributing partonic channels $a+b \to \mu^+\mu^- + x$, with $d\hat{\sigma}^{ab}_{\rm DY}$ the associated partonic cross section. The latter has the perturbative expansion $d\hat{\sigma}^{ab}_{\rm DY} = d\hat{\sigma}^{ab,(0)}_{\rm DY} + \frac{\alpha_s}{\pi} d\hat{\sigma}^{ab,(1)}_{\rm DY} + \ldots$, corresponding to lowest order (LO) and next-to-leading order (NLO) etc. of perturbation theory. μ is the renormalization/factorization scale. Corrections to the right-hand-side of Eq. (1) are down by inverse powers of Q.

When probed near an exclusive boundary of phase space, the perturbative partonic hard-scattering cross sections acquire large logarithmic corrections arising from incomplete cancellations of soft-gluon effects between virtual and real diagrams. A prominent example are threshold corrections. For the Drell-Yan case above, these are of the form $\alpha_s^n \ln^m(1-z)/(1-z)$, where $m \leq 2n-1$, and become large when $z = Q^2/\hat{s} \to 1$, with $\hat{s}$ the partonic c.m. energy. Sufficiently close to the phase-space boundary, i.e. in the limit of soft and/or collinear radiation, fixed-order perturbation theory is bound to fail. A proper treatment of the cross section requires resummation of the logarithmic corrections to all orders. The techniques for this are well established for many reactions of interest, starting with the Drell-Yan process [1, 2]. Fig. 1 compares the fixed-order (LO, NLO, ...) and the resummed (leading logarithms (LL), next-to-leading logarithms (NLL), and so forth) approaches qualitatively.

CP747, *Structure of the Nucleon at Large Bjorken x,* edited by C. Bourrely, Z.-E. Meziani, and J. Soffer

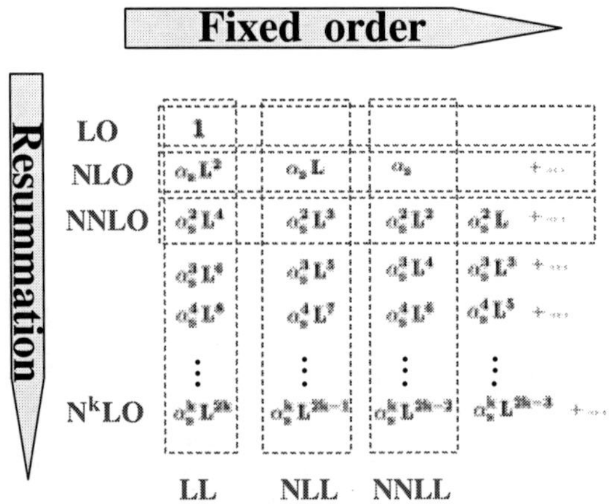

FIGURE 1. Fixed orders in perturbation theory, and resummation. L is a large logarithm near threshold in the appropriate Mellin moment variable; see below.

BASICS OF THRESHOLD RESUMMATION

For illustration, let us sketch the derivation of the LL threshold resummation for the Drell-Yan process. For rigorous and alternative derivations, see [1, 2, 3, 4]. We start with the first-order corrections $q(p)\bar{q}(\bar{p}) \to \gamma^* g(k)$ when the gluon is soft:

$$|M(p,\bar{p};k)|^2 \approx_{k\to 0} |M_{\rm Born}|^2 \cdot (4\pi\alpha_s)\, C_F \frac{2(p\cdot\bar{p})}{(p\cdot k)(\bar{p}\cdot k)} \,. \tag{2}$$

We write the gluon momentum $k = xp + y\bar{p} + k_\perp$ with $x = k\cdot\bar{p}/p\cdot\bar{p} \approx 2k\cdot\bar{p}/Q^2$, $y = k\cdot p/p\cdot\bar{p} \approx 2k\cdot p/Q^2$. One has $k_\perp^2 \approx xyQ^2$. Inserting the gluon phase space yields

$$\int d\Phi_g\, |M(p,\bar{p};k)|^2 \approx 8(4\pi\alpha_s)\, C_F \int \frac{d^{n-1}k}{(2\pi)^3 2k_0} \frac{1}{k_\perp^2} \delta\left(1-z-z_k\right) , \tag{3}$$

where $z = Q^2/\hat{s}$ and $z_k = 2k_0/\sqrt{\hat{s}} = x+y$ is the energy fraction of the emitted gluon. We use dimensional regularization for the infrared and collinear divergencies present in Eq. (3). We now write the δ-function in (3) as an inverse Mellin (or Laplace) transform,

$$\delta\left(1-z-z_k\right) \approx \frac{1}{2\pi i}\int_C dN\, \mathrm{e}^{N\left(1-z-z_k\right)} , \tag{4}$$

which implies that soft emission (large z / small z_k) corresponds to large N. Then,

$$\int d\Phi_g\, |M(p,\bar{p};k)|^2 \approx \frac{1}{2\pi i}\int_C dN\, \mathrm{e}^{N(1-z)}\, \frac{\alpha_s C_F}{\pi} \int \frac{dx}{x} \int \frac{dk_\perp^2}{\left(k_\perp^2\right)^{1+\varepsilon}}\; \mathrm{e}^{-N\left(x+\frac{k_\perp^2}{Q^2 x}\right)} \,. \tag{5}$$

There are two regions that contribute, $x \leq y \leq 1$ and $y \leq x \leq 1$, where $y = k_\perp^2/Q^2 x$. Thus we will have integrations over $\int_0^1 \frac{dx}{x} \int_0^{x^2 Q^2} dk_\perp^2$ and $\int_0^1 \frac{dx}{x} \int_{x^2 Q^2}^{x Q^2} dk_\perp^2$. Substituting x by y in the second integral shows that the two regions give identical contributions, since the integrand is symmetric in x and y. Hence, adding the contributions from virtual diagrams, the right-hand-side of Eq. (5) becomes

$$\frac{2\,\alpha_s C_F}{\pi} \int_0^1 \frac{dx}{1-x} \int_0^{Q^2(1-x)^2} \frac{dk_\perp^2}{\left(k_\perp^2\right)^{1+\varepsilon}} \left[\mathrm{e}^{-N\left(1-x+\frac{k_\perp^2}{Q^2(1-x)}\right)} - 1 \right], \tag{6}$$

where we have substituted $x \to 1-x$ and suppressed the term $\frac{1}{2\pi i}\int_C dN\, \mathrm{e}^{N(1-z)}$ which just gives the Mellin-inverse. The expression in (6) still has a collinear singularity for $k_\perp \to 0$, which is removed by collinear factorization, i.e., by applying the $\overline{\mathrm{MS}}$ subtraction

$$\frac{2\,\alpha_s C_F}{\pi} \int_0^1 \frac{dx}{1-x} \left[x^N - 1 \right] \int_0^{\mu^2} \frac{dk_\perp^2}{\left(k_\perp^2\right)^{1+\varepsilon}}, \tag{7}$$

where μ is the factorization scale, chosen here for simplicity as Q. Combining Eqs. (6) and (7) one finds that the dominant behavior at large N is contained in the expression

$$\frac{2\,\alpha_s C_F}{\pi} \int_0^1 \frac{dx}{1-x} \int_{Q^2}^{Q^2(1-x)^2} \frac{dk_\perp^2}{k_\perp^2} \left[\mathrm{e}^{-N(1-x)} - 1 \right] = \frac{2\,\alpha_s C_F}{\pi} \ln^2(\bar{N}) + \mathcal{O}\left(\frac{1}{N}\right), \tag{8}$$

where $\bar{N} = N \mathrm{e}^{\gamma_E}$ with γ_E the Euler constant. When the Mellin inverse is taken, this result yields the term $\alpha_s C_F/\pi \ln(1-z)/(1-z)$ that dominates the first-order Drell-Yan coefficient function at large z. Obviously we could have derived this result directly in "z-space", without using Mellin moments. However, the use of moments is crucial for deriving the resummed cross section. This is because the phase space for multi-gluon emission contains energy-momentum conserving δ functions connecting components of the gluon momenta. In our case at hand, for emission of n gluons, Eq. (4) turns into

$$\delta\left(1 - z - \sum_{i=1}^{n} z_i\right) \approx \frac{1}{2\pi i} \int_C dN\, \mathrm{e}^{N\left(1-z-\sum_{i=1}^{n} z_i\right)} = \frac{1}{2\pi i} \int_C dN\, \mathrm{e}^{N(1-z)} \prod_{i=1}^{n} \mathrm{e}^{-N z_i}, \tag{9}$$

leading to a factorization of the n-gluon phase space under Mellin moments. At the same time, the squared n-soft-gluon matrix element has QED-like factorization properties and is essentially a product of n single-gluon emission factors, times the Born matrix element. As a result, *exponentiation* of the soft-gluon terms occurs in Mellin-moment space, with an exponent that to LL is just the one-loop expression derived in Eq. (8):

$$\hat{\sigma}_{\mathrm{DY}}^{q\bar{q}}(N) \propto \exp\left[2 \int_0^1 dx \frac{x^N - 1}{1-x} \int_{Q^2}^{Q^2(1-x)^2} \frac{dk_\perp^2}{k_\perp^2} A(\alpha_s(k_\perp^2)) \right], \tag{10}$$

where

$$A(\alpha_s) = \frac{\alpha_s}{\pi} C_F + \left(\frac{\alpha_s}{\pi}\right)^2 \frac{C_F}{2} \left[C_A \left(\frac{67}{18} - \frac{\pi^2}{6} \right) - \frac{10}{9} T_R N_F \right]. \tag{11}$$

Here, the first term in A leads to resummation of the LL terms $\alpha_s^k \ln^{2k} N$ in moment space. The second term gives next-to-leading logarithmic contributions of the form $\alpha_s^k \ln^{2k-1} N$ to the cross section. It is evident that the exponent is *positive*, so that resummation will *enhance* the cross section. This is related to the fact that soft-gluon effects are partly already contained in the parton distributions, as shown by the subtraction in (7).

We note that in QCD the scale $k_\perp$ appears in the strong coupling constant in the exponent in Eq. (10). Because of the singularity of the perturbative coupling at $\Lambda_{\rm QCD}$, the perturbative expansion of the expression shows factorial divergence, which in QCD corresponds to a power-like ambiguity of the series. A closer look, however, shows that the resummed logarithmic terms in N have the form (we define $\alpha_s(Q^2)\ln(N) \equiv \alpha_s L$)

$$\hat{\sigma}_{\rm DY}^{q\bar{q}}(N) \propto \exp\left[\sum_{k=1}^{\infty} \alpha_s^k L^k \left(a_k L + b_k\right) + \mathcal{O}(\alpha_s^{k+1} L^k)\right] , \tag{12}$$

in which the coefficients a_k, b_k have no factorial behavior [5]. The LL terms in the exponent are *single* logarithms of the form $\alpha_s^k L^{k+1}$, which after expansion of the exponential generate the leading *double* logarithms $\alpha_s^k L^{2k}$ in the cross section. The factorial divergence thus appears only at nonleading powers of momentum transfer. This implies that perturbative resummation can suggest [6] the form of nonperturbative, power-suppressed, dynamics. Expanding the full resummed exponential at small $k_\perp$, one finds

$$\exp\left[\frac{2N^2}{Q^2}\int_0^{\lambda^2} dk_T^2 A_a(\alpha_s(k_T)) \ln\left(\frac{Q}{\bar{N}k_T}\right)\right] \to \exp\left[\frac{2C_F}{\pi}\frac{N^2}{Q^2}\left\{g_1 + g_2 \ln\left(\frac{Q}{NQ_0}\right)\right\}\right] \tag{13}$$

with g_1 and g_2 nonperturbative parameters to be determined by comparison to data.

Threshold resummation has been worked out to NLL accuracy for most reactions of current interest; see for example [1, 2, 7, 8, 9, 10], and in some cases to NNLL [11, 12]. If more than two partons are involved in the hard-scattering, exponentiation still occurs but, because of color interferences and correlations, usually comes in the form of sums of exponentials, rather than a single exponential [8, 9]. However, at the LL level, the resummation is always very simple. Each initial parton receives a factor (see Eq. (10))

$$\exp\left[\int_0^1 dx \frac{x^N - 1}{1-x}\int_{Q^2}^{Q^2(1-x)^2} \frac{dk_\perp^2}{k_\perp^2}\frac{C_i \alpha_s(k_\perp^2)}{\pi} + \ldots\right] \to \exp\left[\frac{\alpha_s C_i}{\pi}\ln^2(\bar{N}) + \ldots\right] , \tag{14}$$

where $C_i = C_F$ for (anti)quarks and $C_i = C_A$ for gluons, and where on the right we have for illustration given the simple large-N result obtained for fixed coupling. An "observed" final-state parton (that is, a parton fragmenting into an observed hadron) has the same factor [10]. For an "unobserved" final-state parton, one finds

$$\exp\left[\int_0^1 dx \frac{x^N - 1}{1-x}\int_{Q^2(1-x)^2}^{Q^2(1-x)} \frac{dk_\perp^2}{k_\perp^2}\frac{C_i \alpha_s(k_\perp^2)}{\pi} + \ldots\right] \to \exp\left[-\frac{\alpha_s C_i}{2\pi}\ln^2(\bar{N}) + \ldots\right] . \tag{15}$$

As one can see, this exponent is negative, corresponding to Sudakov-*suppression*, as expected, since for an "unobserved" final-state parton collinear singularities cancel. Also note that for fixed coupling the exponent in Eq. (15) has half the size of that in (14).

RESUMMATIONS FOR VARIOUS PROCESSES

It is instructive to compare the resummed LL exponents for various processes of interest. These will not be sufficient for quantitative calculations (and in fact in the numerical applications discussed below the NLL terms will be included); however, they will indicate where threshold resummation effects will potentially be most important.

- **Drell-Yan $\mathbf{q\bar{q}} \to \gamma^*$:** As shown above,

$$\exp\left[(C_F + C_F)\frac{\alpha_s}{\pi}\ln^2(\bar{N}) + \ldots\right] . \tag{16}$$

  Higgs production via gluon-gluon fusion has the same exponent with $C_F \to C_A$.

- **DIS $\gamma^*\mathbf{q} \to \mathbf{q}$:** Mellin moments are taken in light-cone momentum fractions x.

$$\exp\left[\left(C_F - \frac{C_F}{2}\right)\frac{\alpha_s}{\pi}\ln^2(\bar{N}) + \ldots\right] . \tag{17}$$

- **Prompt-photon production :** for single-inclusive reactions such as prompt-photon production $pp \to \gamma X$, integrated for simplicity over all photon rapidities, a partonic threshold is reached when the center-of-mass energy of the incoming partons is just large enough to produce the photon and the recoiling parton, that is, when $\hat{x}_T \equiv 2p_T^\gamma/\sqrt{\hat{s}} \to 1$. Threshold logarithms occur as $\alpha_s^k \ln^{2k}(1-\hat{x}_T^2)$. Taking Mellin moments in $\hat{x}_T^2$, one obtains the resummed exponents for the two partonic channels:

$$q\bar{q} \to \gamma g : \qquad \exp\left[\left(C_F + C_F - \frac{C_A}{2}\right)\frac{\alpha_s}{\pi}\ln^2(\bar{N}) + \ldots\right] ,$$

$$qg \to \gamma q : \qquad \exp\left[\left(C_F + C_A - \frac{C_F}{2}\right)\frac{\alpha_s}{\pi}\ln^2(\bar{N}) + \ldots\right] . \tag{18}$$

- **Single-inclusive hadron production .** Kinematics for the process $pp \to hX$ are similar to the prompt photon case. A major difference is that one final-state parton is "observed" since it fragments into the observed hadron h. One therefore finds extra enhancement for the cross section, due to the factor in Eq. (14). In addition, all $2 \to 2$ QCD subprocesses contribute at Born level, among them $gg \to gg$. Because of the large gluon color charge $C_A = 3$, resummation effects are expected to be large. We only give examples for two subprocesses here; the others follow directly.

$$gg \to gg : \qquad \exp\left[\left(C_A + C_A + C_A - \frac{C_A}{2}\right)\frac{\alpha_s}{\pi}\ln^2(\bar{N}) + \ldots\right] ,$$

$$qg \to qg : \qquad \exp\left[\left(C_F + C_A + C_F - \frac{C_A}{2}\right)\frac{\alpha_s}{\pi}\ln^2(\bar{N}) + \ldots\right] . \tag{19}$$

- **Semi-inclusive DIS $\gamma^*\mathbf{q} \to \mathbf{q}$:** here one has two separate Mellin moments, for the light-cone momentum fractions associated with the initial state and with the fragmentation process. One finds to LL:

$$\exp\left[\frac{C_F\alpha_s}{2\pi}\left(\ln(\bar{N}) + \ln(\bar{M})\right)^2 + \ldots\right] . \tag{20}$$

PHENOMENOLOGICAL EXAMPLES

Among the many processes to which threshold resummation has been applied, we choose three examples.

Higgs production at the LHC. Here the full NNLO fixed-order corrections have been calculated [13], which allow to improve the resummation to NNLL accuracy. Fig. 2 shows results from the study in [12] for the Higgs K-factor. On the left, resummation is done to NLL and compared to the NLO prediction. On the right, one can see the NNLL resummation and the NNLO result. It becomes evident that the NLL resummation yields a fairly good prediction of the NNLO result. NNLL resummation predicts a further, moderate, increase of the cross section over NNLO.

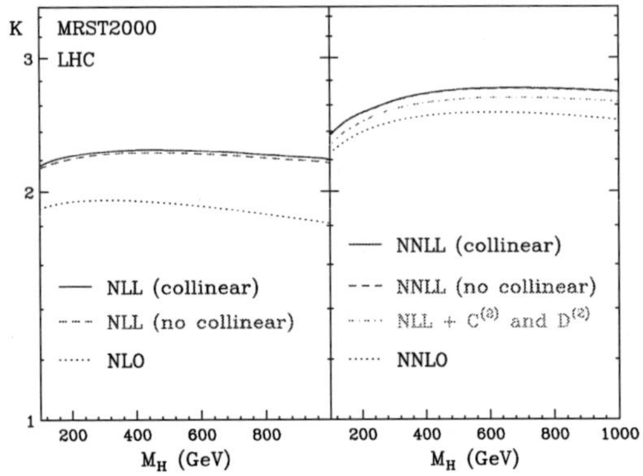

FIGURE 2. Fixed-order (NLO, NNLO) and resummed (NLL, NNLL) predictions for Higgs production at the LHC. Taken from [12].

Semi-inclusive DIS. Here we focus on the "multiplicities" $\frac{d\sigma^{\pi^+ + \pi^-}/dx_B dz}{d\sigma/dx_B}$, where $z = E^\pi/\nu$. Fig. 3 (left) shows NLO and resummed results for the kinematic regime relevant to HERMES measurements [14]. Even though we are considering a ratio of cross sections here, resummation effects become important towards larger z and Bjorken-x_B.

Single-inclusive hadron production pp $\rightarrow \pi$X. NLO significantly underpredicts [15] data for this process in the fixed-target regime, as may be seen from the lower lines in Fig. (3) (right). On account of Eq. (19), we expect large effects from threshold resummation. Indeed, as the upper lines in the figure show, the agreement with data is substantially improved by NLL resummation, and the scale dependence is strongly reduced. Details will be given elsewhere [17].

ACKNOWLEDGMENTS

I am grateful to C. Bourrely and J. Soffer for their invitation, and to D. de Florian, S. Kretzer, G. Sterman, and M. Stratmann for collaboration on some of the topics presented here. I thank RIKEN, Brookhaven National Laboratory and the U.S. Department of Energy (contract number DE-AC02-98CH10886).

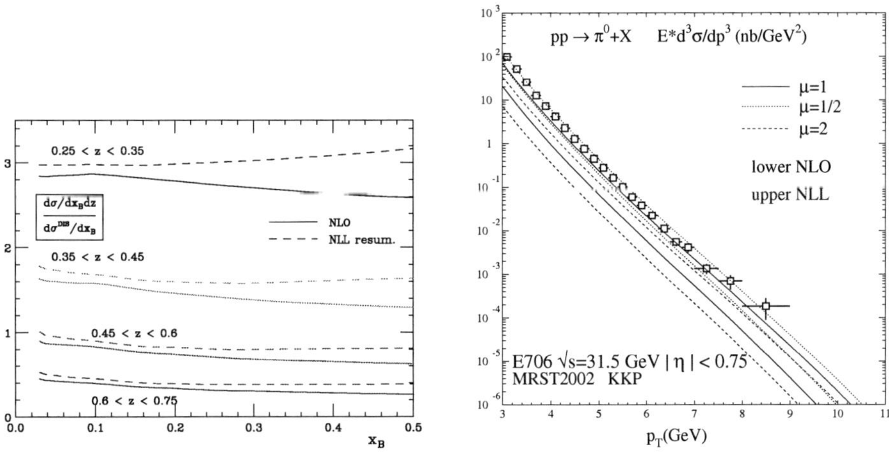

FIGURE 3. Left: Resummation for semi-inclusive DIS in the HERMES kinematic regime. Solid is NLO, dashed NLL resummed. Right: NLO and threshold-resummed results for $pp \to \pi X$ in the fixed-target regime. Data are from E706 [16].

REFERENCES

1. G. Sterman, Nucl. Phys. **B281**, 310 (1987).
2. S. Catani and L. Trentadue, Nucl. Phys. **B327**, 323 (1989); *ibid.* **353**, 183 (1991).
3. E. Laenen, G. Sterman, and W. Vogelsang, Phys. Rev. D **63**, 114018 (2001).
4. T. O. Eynck, E. Laenen and L. Magnea, JHEP **0306**, 057 (2003).
5. S. Catani et al., Nucl. Phys. B **478**, 273 (1996).
6. H. Contopanagos and G. Sterman, Nucl. Phys. **B419** (1994) 77; B.R. Webber, Phys. Lett. **B339** (1994) 148; G.P. Korchemsky, G. Sterman, Nucl. Phys. **B437** (1995) 415; see also: M. Beneke, Phys. Rept. **317** (1999) 1, G. Sterman, A. Kulesza, and W. Vogelsang, Phys. Rev. D **66**, 014011 (2002); G. Sterman and W. Vogelsang, hep-ph/9910371; hep-ph/0109234.
7. E. Laenen, G. Oderda, and G. Sterman, Phys. Lett. B **438**, 173 (1998); S. Catani, M. L. Mangano, and P. Nason, JHEP **9807**, 024 (1998); S. Catani et al., JHEP **9903**, 025 (1999); N. Kidonakis and J. F. Owens, Phys. Rev. D **61**, 094004 (2000); G. Sterman and W. Vogelsang, JHEP **0102**, 016 (2001).
8. N. Kidonakis, G. Oderda, and G. Sterman, Nucl. Phys. B **525**, 299 (1998); Nucl. Phys. B **531**, 365 (1998); N. Kidonakis and J. F. Owens, Phys. Rev. D **63**, 054019 (2001); R. Bonciani, et al., Phys. Lett. B **575**, 268 (2003).
9. N. Kidonakis and G. Sterman, Nucl. Phys. B **505**, 321 (1997); N. Kidonakis, J. Smith, and R. Vogt, Phys. Rev. D **56**, 1553 (1997); N. Kidonakis and R. Vogt, Phys. Rev. D **59**, 074014 (1999); R. Bonciani et al., Nucl. Phys. B **529**, 424 (1998); E. L. Berger and H. Contopanagos, Phys. Rev. D **57**, 253 (1998).
10. M. Cacciari and S. Catani, Nucl. Phys. B **617**, 253 (2001).
11. A. Vogt, Phys. Lett. B **497**, 228 (2001).
12. S. Catani, et al., JHEP **0307**, 028 (2003), and their contribution in hep-ph/0204316.
13. C. Anastasiou and K. Melnikov, Nucl. Phys. B **646**, 220 (2002); R. V. Harlander and W. B. Kilgore, Phys. Rev. Lett. **88** (2002) 201801.
14. A. Airapetian *et al.* [HERMES Collaboration], Eur. Phys. J. C **21**, 599 (2001).
15. P. Aurenche et al., Eur. Phys. J. C **13**, 347 (2000); U. Baur *et al.*, hep-ph/0005226; C. Bourrely and J. Soffer, Eur. Phys. J. C **36**, 371 (2004).
16. L. Apanasevich *et al.* [Fermilab E706 Collaboration], Phys. Rev. D **68**, 052001 (2003).
17. D. de Florian and W. Vogelsang, in preparation.

MRST Parton Distributions

W.J. Stirling and A.D. Martin*, R.G. Roberts† and R.S. Thorne**

*Institute for Particle Physics Phenomenology, Department of Physics, University of Durham, Durham, DH1 3LE, U.K.
†Rutherford Appleton Laboratory, Chilton, Didcot, Oxon, OX11 0QX, U.K.
**Cavendish Laboratory, University of Cambridge, Madingley Road, Cambridge, CB3 0HE, U.K.

Abstract. A brief summary of some of the recent developments in the MRST determinations of NLO and NNLO parton distribution functions is presented.

Keywords: Parton Distribution Functions, Deep Inelastic Scattering, Quantum Chromodynamics
PACS: 12.38.Bx,12.38.Qk, 13.60.Hb

INTRODUCTION

A knowledge of the partonic structure of the proton is an essential ingredient in the analysis of hard scattering data from hadron-hadron or lepton-hadron high energy collisions. Recently the focus has been on obtaining reliable *uncertainties* on the parton distributions obtained from global or semi-global fits. One obvious uncertainty comes from the systematic and statistical errors of the data, i.e. the *experimental* errors. They have been estimated by several groups [1]–[10], working within a NLO (and sometimes NNLO) framework using a variety of different procedures. The general conclusion is that in all approaches the uncertainties are of the order of $\pm\,(2-5)\%$, except for certain partons and regions of phase space (for example, gluons and sea quarks at very high x) where the uncertainty can be bigger. However, when the partons and predictions for physical quantities using different parton sets are compared, the deviations can be significantly greater than the quoted errors.[1] This is most likely due to different assumptions made by different groups when carrying out their fitting procedures, such as differences in (i) the data sets used in the fit, (ii) the functional forms used for the input distributions, (iii) the choice of heavy flavour prescriptions, etc. However, the different results for partons obtained when making different assumptions also suggest that the standard fixed order in the perturbative expansion is not completely adequate to describe the data, with different individual data sets making different and conflicting demands on the partons and on α_S. Indeed, in our (MRST) analysis we do see certain difficulties in obtaining the best possible fit to the data with the current theoretical treatment – the HERA and NMC DIS structure function data in the range $x \sim 0.01-0.001$ appear to increase too rapidly with Q^2; the gluon is negative (or at best valence-like) at small x and Q^2; and the Tevatron jet data prefer a different shaped gluon to the DIS data. Hence, in order to understand the uncertainties in parton distributions, it is vital to consider theoretical corrections. These

[1] A recent summary can be found in [11].

CP747, *Structure of the Nucleon at Large Bjorken x,* edited by C. Bourrely, Z.-E. Meziani, and J. Soffer

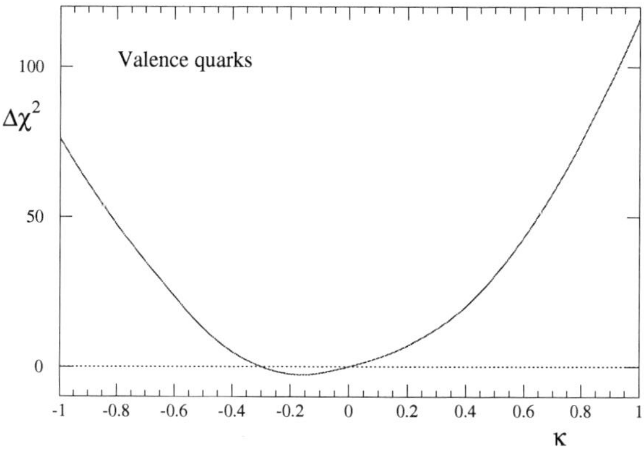

FIGURE 1. Variation of the global fit χ^2 with the value of κ from Ref. [12]

include the possibility of isospin violation; $s(x) \neq \bar{s}(x)$; higher orders in perturbation theory (e.g. NNLO); QED effects (which *a priori* could be comparable to NNLO pQCD corrections, since $\alpha_s^3 \sim \alpha$); large-x logarithmic corrections ($\sim \alpha_s^n \ln^{2n-1}(1-x)$); low-$Q^2$ power corrections (higher twist); small x logarithmic corrections ($\sim \alpha_s^n \ln^{n-1}(1/x)$). The uncertainties associated with many of these possible sources have recently been considered [12]. In this review we will highlight some of the main conclusions and some additional results which postdate this analysis.

ISOSPIN VIOLATION

One of the issues we have examined is the possibility of isospin violation in the parton distributions, i.e. $u^p \neq d^n$ etc. This is particularly relevant because of the so-called 'NuTeV $\sin^2\theta_W$ anomaly' [13]. NuTeV measure $\sin^2\theta_W$, with a 3σ discrepancy from the world average value, essentially from the Paschos-Wolfenstein ratio of charged and neutral current (anti-)neutrino-nucleus cross sections

$$R^- = \frac{\sigma_{\rm NC}^{\nu} - \sigma_{\rm NC}^{\bar{\nu}}}{\sigma_{\rm CC}^{\nu} - \sigma_{\rm CC}^{\bar{\nu}}} = \frac{1}{2} - \sin^2\theta_W + (1 - \frac{7}{3}\sin^2\theta_W)\frac{[\delta U_{\rm v}] - [\delta D_{\rm v}]}{2[V^-]}.$$

where $[\delta U_{\rm v}] = [U_{\rm v}^p] - [D_{\rm v}^n]$ and $[\delta D_{\rm v}] = [D_{\rm v}^p] - [U_{\rm v}^n]$, $[U_{\rm v}^p] = \int_0^1 x u_{\rm v}^p(x)\,dx$ etc.. If we parameterize possible isospin violation by

$$u_{\rm v}^p(x) = d_{\rm v}^n(x) + \kappa f(x), \qquad d_{\rm v}^p(x) = u_{\rm v}^n(x) - \kappa f(x),$$

where $f(x)$ is a number conserving function with valence-like behaviour at large and small x [12], then we obtain the variation of the χ^2 for the global fit as a function of κ shown in Fig. 1, and the best fit leads to $\kappa = -0.2$ (which agrees very well with a model-based prediction in Ref. [14]), and to a reduction of the NuTeV anomaly to $\sim 1.5\sigma$, i.e.

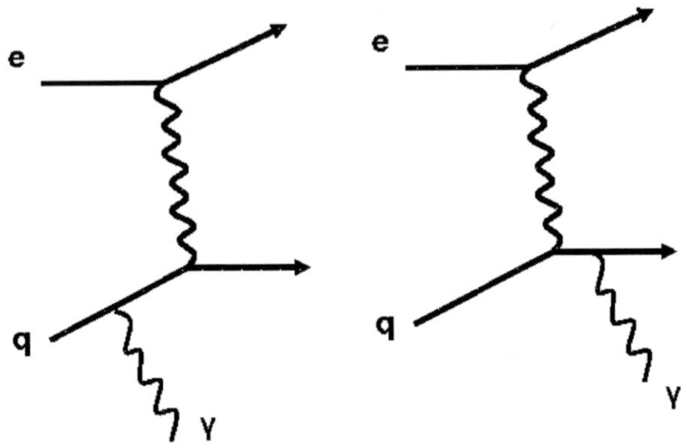

FIGURE 2. Diagrams for QED corrections to deep inelastic scattering associated with radiation off the quark lines

$\Delta \sin^2\theta_W \sim -0.002$. However, a wide variation in κ is possible, and the anomaly may be reduced further or enlarged. Other issues which affect the NuTeV $\sin^2\theta_W$ measurement are the strangeness asymmetry in the proton (i.e. $s \neq \bar{s}$) recently investigated in [15] and electroweak corrections [16]. In each case the uncertainties may be large as the anomaly, though there is some disagreement regarding the strangeness asymmetry.

QCD EFFECTS

We have also recently examined the effects on the parton distribution analysis due to QED, i.e. contributions from photon emissions from quark lines, see Fig. 2. These give rise to large logarithmic $\alpha\log(Q^2/m_q^2)$ contributions from photons emitted collinearly with the incoming and outgoing quarks, the analogue of the $\alpha\log(Q^2/m_e^2)$ radiative corrections familiar in e^+e^- collisions. One could take these explicitly into account, but this would require a consistent choice of input quark masses. Furthermore, at the very high Q^2 scales probed at hadron colliders, one should in principle resum these logarithms. Fortunately the QCD factorisation theorem applies also to QED corrections, and as a result such collinear (photon-induced) logarithms can be absorbed into the parton distributions functions, exactly as for the collinear $\alpha_S\log Q^2$ logarithms of perturbative QCD. There are two effects of this: the normal DGLAP evolution equations are slightly modified — the emitted photon carries away some of the quark's momentum — and a 'photon parton distribution' of the proton, $\gamma^p(x,Q^2)$, is generated. By correctly taking account of these QED effects through modified DGLAP evolution equations, we obtain a consistent procedure for dealing with this part of the overall electroweak correction in all hard-scattering processes involving initial-state hadrons.

We have performed new global NLO and NNLO fits taking these QED corrections into account in the DGLAP equations. In both cases the QED corrections do not alter the fit quality in any significant way. The effect on the gluon distribution is negligible, as is that on the quark distributions at small x where the gluon contribution dominates the evolution. Only at large x is the effect noticeable ($\mathcal{O}(1\%)$) as first noted in [17], and more recently verified in [18], and in fact is equivalent to a slight shift in α_S: $\Delta\alpha_S(M_Z^2) \simeq +0.0003$. This is much smaller than many other sources of uncertainty. However, these QED effects, although small, do lead to isospin violation. $u_V^p(x)$ quarks radiate more photons than $d_V^n(x)$ quarks. To a rough approximation, the photon distribution should be

$$\gamma(x,Q^2) = \sum_j e_j^2 \frac{\alpha}{2\pi} \ln(Q^2/m_q^2) \int_x^1 \frac{dy}{y} P_{\gamma q}(y)\; q_j(\frac{x}{y},Q^2)\,,$$

and so there is more photon momentum in the proton than in the neutron — high-x up quarks radiate more than high-x down quarks. Momentum conservation leads to $u_V^p(x) < d_V^n(x)$ at high x. Hence, $[\delta U_{\rm v}] < 0$ as required by the NuTeV anomaly and as found by our best fit described above. Our estimates imply that the QED effects lead to $\sim 3/4$ of the isospin violation observed by our best fit. Further details of this and other aspects of parton distributions involving QED corrections can be found in Ref. [19].

NNLO CORRECTIONS

NNLO perturbative QCD corrections are generally much more important in the context of the parton distribution global fit. The coefficient functions for $F_2(x,Q^2)$ are known at NNLO [20], and recently the full three-loop (NNLO) DGLAP have been calculated by Moch, Vermaseren and Vogt [21]. When the NNLO corrections are included, they improve the overall quality of the (NLO) fit very slightly [22], and reduce $\alpha_S(M_Z^2)$ to ~ 0.1155.

Fig. 3 shows theoretical predictions for the W and Z total cross sections at the Tevatron and LHC colliders, at LO, NLO, and NNLO, together with recent CDF and D0 Tevatron cross-section data. Taking all sources of uncertainties in the theoretical analysis into account [12], we quote a $\pm 4\%$ uncertainty in the predictions, shown as the horizontal dashed lines bracketing the NNLO cross sections in the figure. Evidently the perturbation series is well under control, and there is also good agreement between theory and experiment. However, this good convergence of the perturbation series for $\sigma(W,Z)$ is largely guaranteed because the quarks are fit directly to data, and it is certainly much worse for gluon-dominated quantities such as $F_L(x,Q^2)$ or the Higgs production cross section in hadron-hadron collisions.

The failure of the NNLO corrections to solve completely the problems with the quality of the fit leads us to try an alternative approach. In order to investigate the real quality of the fit and the regions that give rise to problems, we vary the kinematic cuts on the input data, i.e. we change W_{cut}^2, Q_{cut}^2 and x_{cut}, re-fit and see if the quality of the fit to the remaining data improves and/or the input parameters change dramatically. This is continued until the quality of the fit and the partons stabilize. This procedure is discussed in detail in Ref. [12]. To summarize: our W^2 cut of 12.5 GeV^2 seems reasonable, whereas

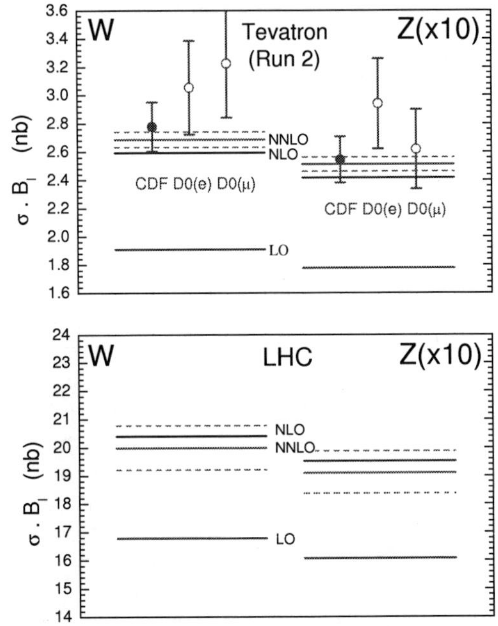

FIGURE 3. Predictions for the W and Z total cross sections at the Tevatron and LHC [22]

the fit improves gradually until $Q^2_{cut} \sim 10\ \mathrm{GeV}^2$ and $x_{cut} \sim 0.005$.[2] This is true at both NLO and NNLO, but the modification of the partons with changing cuts is much greater at NLO. For both NLO and NNLO we have defined 'conservative' partons [12] that are obtained using only the data within the conservative cuts.

In order to investigate the possible theoretical improvements which could modify the partons at small x and low Q^2 (particularly at high x) we have tried including known $\ln(1-x)$ resummations, saturation corrections, and purely phenomenological higher-twist and $\ln(1/x)$ corrections. At high x, higher orders in the coefficient functions and higher-twist contributions *can* improve the fit, but are difficult to distinguish. At low x, there is no evidence for phenomenological higher-twist contributions. Saturation corrections [24] do not seem to help the NLO or NNLO fits. There is a distinct improvement in the fit quality when (momentum conserving) $\ln(1/x)$ corrections are added to the splitting functions, increasing the small-x evolution. Hence, there is some reason for uncertainty about what is the correct theoretical approach at very small x.

In summary, as well as the uncertainties on partons due to experimental errors on the input data, there are uncertainties on partons, and related physical quantities, due to

---

[2] We note, however, that the CTEQ collaboration do not seem to reproduce these conclusions [23].

assumptions in the fit and having to work within an incomplete theoretical framework. Some of the uncertainties are small, but in some cases, e.g. QED corrections or $s(x) \neq \bar{s}(x)$, they can be important for specific quantities. Others, associated with low Q^2 and small x, may be potentially large, and more theoretical work is needed, along with specifically targeted data to guide and verify this work.

ACKNOWLEDGMENTS

WJS would like to thank the organisers of the HiX2004 Workshop for the invitation to present this talk, and for the efficient and friendly way in which the meeting was organised.

REFERENCES

1. M. Botje, Eur. Phys. J. **C14** (2000) 285.
2. W.T. Giele and S. Keller, Phys. Rev. **D58** (1998) 094023;
   W.T. Giele, S. Keller and D.A. Kosower, arXiv:hep-ph/0104052.
3. S.I. Alekhin, Phys. Rev. **D68** (2003) 014002.
4. CTEQ Collaboration: D. Stump *et al.*, Phys. Rev. **D65** (2002) 014012.
5. CTEQ Collaboration: J. Pumplin *et al.*, Phys. Rev. **D65** (2002) 014013.
6. CTEQ Collaboration: J. Pumplin *et al.*, JHEP 0207:012 (2002).
7. H1 Collaboration: C. Adloff *et al.*, Eur. Phys. J. **C21** (2001) 33.
8. H1 Collaboration: C. Adloff *et al.*, arXiv:hep-ex/0304003.
9. A.M. Cooper-Sarkar, arXiv:hep-ph/0205153, J. Phys. **G28** (2002) 2669;
   ZEUS Collaboration: S. Chekanov *et al.*, Phys. Rev. **D67** (2003) 012007.
10. A.D. Martin, R.G. Roberts, W.J. Stirling and R.S. Thorne, Eur. Phys. J. **C28** (2003) 455.
11. R.S. Thorne, arXiv:hep-ph/0309343, Invited talk at 21st International Symposium on Lepton and Photon Interactions at High Energies (LP03), Batavia, Illinois, 11-16 Aug. 2003, published in Int. J. Mod. Phys. **A19** (2004) 1074-1088.
12. A.D. Martin, R.G. Roberts, W.J. Stirling and R.S. Thorne, Eur. Phys. J. **C35** (2004) 325.
13. G.P. Zeller *et al.*, Phys. Rev. **88** (2002) 091802.
14. J.T. Londergan, G.T. Carvey, G.Q. Liu, E.N. Rodinov and A.W. Thomas, Phys. Lett. **B40** (1994) 115.
15. S. Kretzer *et al.*, arXiv:hep-ph/0312322; F. Olness *et al.*, arXiv:hep-ph/0312323;
    NuTeV collaboration arXiv:hep ex/0405037.
16. K.P.O. Diener, S. Dittmaier and W. Hollik, Phys. Rev. **D69** (2004) 073005;
    A.B. Arbuzov, D.Yu. Barin and L.V. Kalinovskaya, arXiv:hep-ph/0407203.
17. H. Spiesberger, Phys. Rev. **D52** (1995) 4936.
18. M. Roth and S. Weinzierl, Phys. Lett. **B90** (2004) 190.
19. A.D. Martin, R.G. Roberts, W.J. Stirling and R.S. Thorne, arXiv:hep-ph/0411040.
20. E.B. Zijlstra and W.L. van Neerven, Phys. Lett. **B272** (1991) 127; *ibid.*, **B273** (1991) 476; *ibid.*, **B297** (1992) 377; Nucl. Phys. **B383** (1992) 525;
    S. Moch and J.A.M. Vermaseren, Nucl. Phys. **B573** (2000) 853.
21. S. Moch, J.A.M. Vermaseren and A. Vogt, Nucl. Phys. **B688** (2004) 101); *ibid.*, arXiv:hep-ph/040411.
22. A.D. Martin, R.G. Roberts, W.J. Stirling and R.S. Thorne, Eur. Phys. J. **C18** (2000) 117; *ibid.*, Phys. Lett. **B531** (2002) 216.
23. W.K. Tung, these proceedings.
24. A.H. Mueller and J. Qiu, Nucl. Phys. **B268** (1986) 427;
    E.M. Levin and M.G. Ryskin, Phys. Rep. **189** (1990) 267.

UNPOLARIZED STRUCTURE FUNCTIONS

Global QCD Analyses of Parton Distributions at Large x and Small

Wu-Ki Tung[1]

Michigan State University, E. Lansing, MI, USA

Abstract. A brief overview of recent work of the CTEQ group on global QCD analysis of the parton structure of the nucleon is presented. A detailed study of the stability of predictions of parton distribution functions (PDFs) in NLO global analysis is investigated. Two topics on flavor dependence of PDFs are touched upon: (i) status of $d(x)/u(x)$ at large x; and (ii) charge asymmetry in the strangeness sector of nucleon structure ($s(x) \neq \bar{s}(x)$). All three issues involve interaction between large and small x physics.

We survey a few issues of current interest in the global QCD analysis of parton distribution functions (PDFs) carried out by members of CTEQ and collaborators. Within the global analysis context, "high x" physics and "low x" physics are inseparable, due to sum rules ... etc. Most of the topics discussed in the following will involve implications of data in the small x region on large x physics, and vice versa.

Stability of NLO global analysis

The vast majority of work on global QCD analysis of parton distribution functions and the application of PDFs to the calculation of high energy processes has been performed at NLO (1-loop hard cross sections and 2-loop evolution kernel). In recent years, some preliminary NNLO analyses have been carried out, even if not all necessary hard cross sections are yet available, and the evolution kernel was only known approximately (until very recently). Since errors on most of the experimental data used in global analysis are generally larger than the known NNLO corrections, the necessity to extend the analysis to NNLO, at the expense of vastly more complicated calculations, is not obvious.

A strong motivation to go to NNLO would exist, however, if the conventional NLO analysis does not yield stable PDFs and physical predictions. This possibility was indeed raised by a recent MRST study [1]. In particular, they found a 20% variation in the predicted cross section for W production at the LHC—a very important "standard candle" process for hadron colliders—in their NLO analysis, depending on kinematic

[1] The work reported in this talk is mostly based on collaborations with Joey Huston, Stefan Kretzer, Stephen Kuhlmann, H.L. Lai, Frederick Olness, Joseph Owens, Jonathan Pumplin, and Daniel Stump.

CP747, *Structure of the Nucleon at Large Bjorken x,* edited by C. Bourrely, Z.-E. Meziani, and J. Soffer

cuts placed on input data.[2] Their results on the total cross section and on the rapidity

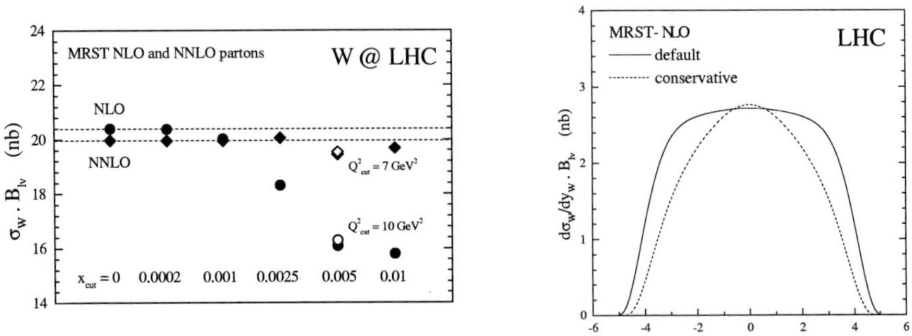

FIGURE 1. (a) Dependence of MRST predictions for W total cross section at LHC on kinematic cuts of input data; (b) W rapidity distribution according to the MRST default and the "conservative" (relatively high x and Q cuts) PDFs.

distribution of the W boson are shown in Fig. 1. For both the Tevatron Run II and LHC physics programs this is clearly a critical issue; it is important to investigate whether this result is confirmed by an independent study.

We have investigated this problem in considerable detail within the CTEQ global analysis framework [3]. Applying the same theoretical and experimental inputs to the global analysis, we systematically varied the kinematic cuts in x and Q, and generated new sets of best-fit PDFs. To explore the dependence of the results on assumptions made about the parametrization of PDFs at the starting scale Q_0 (1.3 GeV), we also studied cases where the gluon distribution function is allow to go negative at small x, a possibility favored by the MRST NLO analysis in the past few years. The main results of this study [4] are discussed below.

Fig. 2a shows the variation of the W total (W^++W^-) cross section at LHC with respect to a series of best fit PDFs obtained with increasing x-cuts comparable to those of the MRST study; the results are super-imposed on Fig. 1a. Also included is a point corresponding to a higher Q-cut of $Q^2 > 10$ GeV^2. These results show full stability of the predicted W cross section versus the kinematic cuts used in the global analysis of PDFs, in marked contrast to the results of [1]. (Results at Tevatron energies, not shown due to lack of space, are even more stable, as one would expect.) Fig. 2b shows a comparison of the predicted rapidity (y) distribution for (W^++W^-) at LHC. Instead of showing the predicted y-distribution from the fits with different kinematic cuts (which are all similar to one another), we show the range of variation of the y-distribution when the 2nd moment of the y-distribution is extremized using the Lagrange multiplier method of [5]. These "extreme" cases allowed by our global analysis are compared to

[2] In the absence of general methods of assessing theoretical uncertainties in the global analysis due to a variety of effects such as power-law corrections (low Q), parton saturation (low x), ... etc., raising kinematic cuts on input experimental data serves, in principle, as a poor-man's method of reducing the theoretical uncertainties. This is done, however, at the expense of leaving out large amounts of data that can otherwise provide valuable constraints on the PDFs.

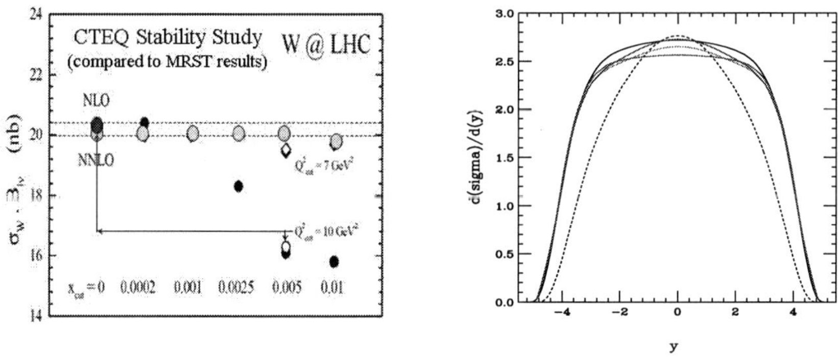

FIGURE 2. (a) Dependence of CTEQ predictions (light dots) on W total cross section at LHC on kinematic cuts of input data, super-imposed on Fig. 1a; (b) W rapidity distribution at LHC, CTEQ6 prediction plus those extremized w.r.t. the 2nd moment of y in solid color lines, compared to that of the MRST default and conservative PDF predictions (in black solid and dotted lines).

the MRST "default" and "conservative" predictions. The MRST conservative prediction, which corresponds to strong kinematic cuts, clearly stands out.

The stability of the prediction with respect to variations of kinematic cuts seen in analysis is reassuring. It indicates that NLO QCD should provide an adequate framework for studying high energy phenomenology at the Tevatron and LHC, except perhaps for processes that are known to have large corrections to the hard cross section beyond NLO. On the other hand, the difference between our results and those of MRST needs to be understood. The two global analysis efforts share many common theoretical and experimental inputs. Most of their predictions have been found to be compatible with each other. However, historically, they also have arrived at different conclusions on a number of specific issues, due to subtle (or not so subtle) differences in methodology and/or input. These were eventually resolved when the causes for the difference were identified. The case on the stability of NLO analysis may represent such a situation. In particular, the instability of the MRST NLO analyses appears to be associated with two other unique features of their recent analyses: (i) an apparent "tension" between inclusive jet production data at relatively large x and DIS data at small x from HERA; and (ii) a relatively strong preference for a negative gluon distribution $g(x,Q)$ at small x and small Q.

In order to look a little closer into these issues, we have investigated the χ^2 values of the jet data sets and the HERA data sets separately as we vary the kinematic cuts. No discernable improvement in the fit to the jet data sets is seen as the x-cut on input DIS data is raised.[3] The χ^2/N for the CDF and D0 experiments are $(1.47, 1.48, 1.47, 1.47, 1.48)$ and $(0.718, 0.715, 0.723, 0.726, 0.697)$ for x-cuts of $(0, .001, .0025, .005, .01)$ respectively. The quality of fit to the HERA data (as measured by χ^2/N) also remains stable

[3] If there were tension between the two, an improvement would result, because the pull of small-x HERA data would have been reduced.

with respect to the change in the kinematic cuts.

An important effect of raising the kinematic cuts in x and Q is that constraints provided by the precision HERA data in the small x and low Q region are removed from the global fit. Thus, the uncertainty of the resulting PDFs and their physical predictions will increase. We quantified this effect for the specific case of W production at LHC. Using the Lagrange Multiplier method [5], we examine the best χ^2 values obtained in constrained global fits, as a function of the W cross section, for several choices of kinematic cuts: (i) the *normal* CTEQ cuts ($Q^2 > 4$ GeV^2); (ii) *medium* cuts ($Q^2 > 6.25$ GeV^2, $x > 0.001$); (iii) *strong* cuts ($Q^2 > 10$ GeV^2, $x > 0.005$); and (iv) *very strong* cuts ($Q^2 > 100$ GeV^2).[4] Fig. 3a shows the results when the normal CTEQ parametrization, with positive PDFs at the input scale of $Q = 1.3$ GeV, is used. We see that, as stronger kinematic cuts are imposed, the range of uncertainty on the prediction for the W production cross section becomes progressively larger. Whereas the change from the normal to medium cuts is small, the range of uncertainty almost doubles from normal to strong cuts at any tolerance level of χ^2. The *very-strong-cuts* case removes most DIS data from the fit, thus heavily emphasizes collider data—it represents a step toward a hypothetical "collider-data-only prediction" from Tevatron to LHC. It results in a very large range of uncertainty for the LHC cross section, as one would have expected.

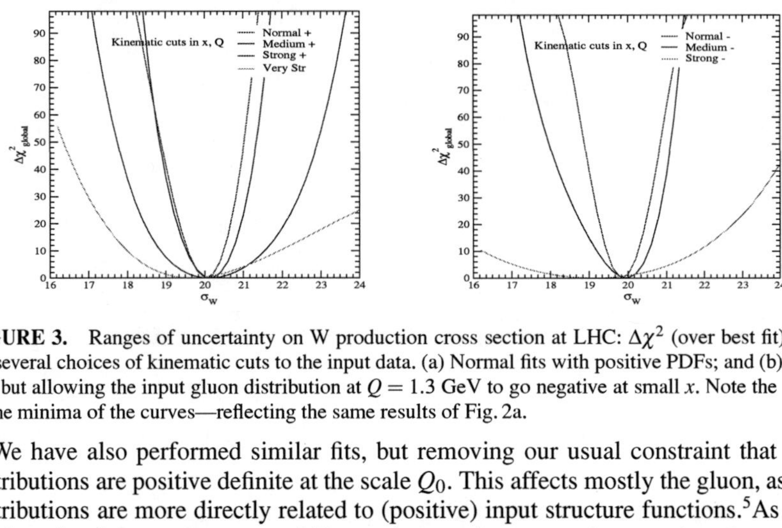

FIGURE 3. Ranges of uncertainty on W production cross section at LHC: $\Delta\chi^2$ (over best fit) vs. σ_W for several choices of kinematic cuts to the input data. (a) Normal fits with positive PDFs; and (b) Similar fits, but allowing the input gluon distribution at $Q = 1.3$ GeV to go negative at small x. Note the stability of the minima of the curves—reflecting the same results of Fig. 2a.

We have also performed similar fits, but removing our usual constraint that parton distributions are positive definite at the scale Q_0. This affects mostly the gluon, as quark distributions are more directly related to (positive) input structure functions.[5]As a rule, our best fits for any given sets of kinematic cuts do not require negative gluons at our input scale of $Q_0 = 1.3$ GeV. However, when σ_W^{LHC} is pulled away from the best-fit value by the constrained fits in the Lagrange Multiplier study, the additional freedom provided by a negative gluon widens the allowed range of variation of σ_W^{LHC}. The results

---

[4] A W-cut of 3.5 GeV is maintained in all cases. Note that the strong cuts are similar to those of the MRST conservative fit.

[5] A strong enough negative initial gluon at small x can induce negative quark distributions at small x and larger Q through QCD evolution. This happens to the MRST conservative PDF set, for instance, at $Q^2 = 5$ GeV^2 for $x < 0.00015$.

of this analysis (except that of the very strong cuts) are shown in Fig. 3b. Compared to the corresponding cases in Fig. 3a, we see that the medium-cuts curve has opened up noticeably at the lower end, and the strong-cuts case has widened at both ends.

Should the fits with negative gluons be taken seriously as candidate PDF sets, hence providing better determination of the true range of uncertainty? We think not. The reason is that whereas PDFs are not strictly forbidden to become negative, ***all*** *physical quantities calculated from them, in* ***all*** *parts of physical phase space, must remain positive definite*. This is a very strong condition that is extremely difficult to satisfy if some PDFs become negative in some region of the (x, Q) plane. For all the solutions involving negative gluons at small x and low Q (including the MRST ones), we found it is possible to identify ***some*** physical cross sections, at ***some*** high energies, near ***some*** boundary of phase space, that become negative. These negative gluon solutions must therefore be considered as unphysical (at this order of α_s).

Another possible source of discrepancy between the CTEQ and MRST results is the treatment of heavy quark mass effects in DIS. We primarily use the widely used zero-mass $\overline{MS}$ formalism, whereas MRST uses the Thorne-Roberts prescription for the non-zero-charm-mass variable flavor number scheme. In general, the difference is nominally small. However, since charm production accounts for nearly 25% of the DIS cross section at small x, and since the HERA data at small x have very small errors, the difference could have an effect on the issues discussed above, because the R-T prescription results in an unusual behavior of the charm structure function just above the threshold. Further study is clearly needed to settle fully the stability issue.

We should add that, independent of the stability of NLO global analysis, NNLO calculations are needed for processes that require a high level of accuracy, or that are known to have large corrections.

Ratio of d/u at large x

One of the open issues in quark distributions is the behavior of d/u at large x. (Cf. [6]) In most conventional global analyses, this ratio is constrained to approach zero as $x \to 1$ by the parametrization adopted for the non-perturbative input functional forms. Whereas the usual assumption $d/u \to (1-x)^{\alpha}$ with $\alpha \sim 1$ is suggested by quark counting rules, other quark model and related SU(6) considerations lead to constant ratios in the $x \to 1$ limit. Only detailed global analysis work based on solid experimental input can determine the true behavior.

Preliminary results from E866 (NuSea) on pp and pd lepton-pair production (Drell-Yan) cross sections raised the possibility that the separation of u and d quarks in existing global analyses may not be correct in the large x region. [7] This motivated us to study this issue in detail, including an assessment of the effects due to deuteron target correction, which can affect the outcome of the analysis. In the absence of a generally accepted reliable theoretical calculation, we have tried to use a phenomenological approach, and make a best estimate of, say, $R(x) = F_2^d(x)/2F_2^p(x)$ by fitting available high precision fixed-target DIS data on deuteron and on proton, with only general theoretical constraints. The plan is to incorporate this effect in a comprehensive global analysis, including the final E866 results, to determine both the best fit to $d(x)$ and $u(x)$ and the range of uncertainty of the d/u ratio at large x, without requiring it to approach 0 in the $x \to 1$ limit. This analysis will be completed as soon as the final E866 data become available [7].

Strangeness Asymmetry

An important, but so far poorly determined, frontier of parton structure of the nucleon is the strangeness sector. This can be separated into two parts: the total strangeness sea, $s_+(x) = s(x) + \bar{s}(x)$, and the strangeness charge asymmetry $s_-(x) = s(x) - \bar{s}(x)$. Advance on both fronts is expected by detailed study of the recent CCFR-NuTeV dimuon production data in neutrino and anti-neutrino scattering [8] in the global analysis context.[6]

The separate measurement of ν and $\bar{\nu}$ cross sections is crucial in order to separate the s and $\bar{s}$ distributions, and hence allow the measurement of $s_-(x)$ and determine whether the strange sea of the nucleon is charge asymmetric. Although the strangeness number sum rule requires $\int_0^1 s_-(x)\,dx = 0$, a charge asymmetric strange sea, manifested by, say, a non-zero first moment $[S_-] = \int_0^1 x\, s_-(x)\,dx$, has important physical consequences such as the NuTeV anomaly [9].

Our group has studied this issue in the past year [10], and is continuing the effort to achieve a full NLO analysis of the CCFR-NuTeV data in the global analysis context. Due to space limitations here, we can only mention the current status of this study, and leave the details to [10, 9]. Our study is currently limited by the fact that the CCFR-NuTeV measurement is not available as inclusive cross sections (that theory can calculate), but is in the form of cross sections with specific kinematic cuts. In order to perform a global fit, incorporating these new data, a LO model bridging the inclusive and the measured cross sections has to be invoked. With this caveat, we found that a positive value for $[S_-]$ on the order of 0.0017 is favored. We also found that the range of uncertainty, both theoretical and experimental, is rather large. We estimate that $-0.001 < [S_-] < 0.004$. This result will be significantly improved in the next round of analysis, in which we are collaborating with members of CCFR-NuTeV to incorporate the best NLO MC model in conjunction with the NLO global analysis.

REFERENCES

1. A. D. Martin, R. G. Roberts, W. J. Stirling and R. S. Thorne, Eur. Phys. J. C **35**, 325 (2004).
2. R. S. Thorne, Proceedings of DIS04, hep-ph/0407311.
3. CTEQ6: J. Pumplin, D. R. Stump *et al.*, JHEP **0207**, 012 (2002).
4. J. Huston, J. Pumplin, *et al*, under preparation.
5. D. Stump *et al.*, Phys. Rev. D **65**, 014012 (2002); J. Pumplin *et al.*, Phys. Rev. D **65**, 014013 (2002); and Phys. Rev. D **65**, 014011 (2002).
6. W.K. Tung, plenary talk, Proceedings of DIS04, hep-ph/0409145.
7. J. C. Webb *et al.* [NuSea Collaboration], hep-ex/0302019; and D. Isenhower, Proceedings of DIS04.

[6] PDFs in PQCD, by definition, are *universal*, i.e. they must be applicable to all processes. Specialized PDFs derived from analysis of very limited data sets (or processes) that do not fit a broad range of data may not reflect true behavior of PDFs of nature, hence can be misleading.

8. CCFR and NuTeV Collab. (M. Goncharov *et al.*), Phys. Rev. **D64**, 112006 (2001); NuTeV Collab. (M. Tzanov *et al.*), hep-ex/0306035; P. Spentzouris, Proceedings of DIS04.
9. NuTeV Collaboration, G.P. Zeller *et al.*, Phys. Rev.Lett. **88**, 091802 (2002); for physics consequences, see, e.g. S. Davidson, S. Forte, P. Gambino, N. Rius, A. Strumia, JHEP **0202**, 037, 2002.
10. F. Olness *et al.*, hep-ph/0312323; S. Kretzer *et al.*, Phys. Rev. Lett. **93**, 041802 (2004); S. Kretzer, Proceedings of DIS04, hep-ph/0408287.

Measuring F_2^n/F_2^p and d/u: Past and Future

Gerassimos G. Petratos

Department of Physics, Kent State University, Kent, OH 44242, USA

Abstract. The reliability of the existing extraction of the nucleon structure function and quark distribution ratios F_2^n/F_2^p and d/u from proton and deuteron deep inelastic measurements is discussed. Future measurements expected to provide reliable data for the two ratios, free of theoretical uncertainties, are presented. Particular emphasis is given in i) semi-inclusive inelastic electron-deuteron measurements with detection of backward spectator protons and ii) inelastic measurements of electron scattering from the ^{3}He and ^{3}H mirror nuclei. These data are expected to test perturbative and non-perturbative mechanisms of spin-flavor symmetry breaking in the nucleon, and constrain the structure function parametrizations needed for the interpretation of high energy electron-proton, proton-proton and proton-antiproton collider data.

INTRODUCTION

The proton and neutron structure functions, extracted from deep inelastic electron scattering off the proton and deuteron, have been of fundamental importance in establishing the internal quark structure of the nucleon [1]. The cross section for inelastic electron-nucleon scattering is given in terms of the structure functions $F_1(\nu,Q^2)$ and $F_2(\nu,Q^2)$ of the nucleon by:

$$\sigma \equiv \frac{d^2\sigma}{d\Omega dE'}(E,E',\theta) = \frac{4\alpha^2(E')^2}{Q^4}\cos^2\left(\frac{\theta}{2}\right)\left[\frac{F_2(\nu,Q^2)}{\nu} + \frac{2F_1(\nu,Q^2)}{M}\tan^2\left(\frac{\theta}{2}\right)\right], \tag{1}$$

where α is the fine-structure constant, E is the incident electron energy, E' and θ are the scattered electron energy and angle, $\nu = E - E'$ is the energy transfer, $Q^2 = 4EE'\sin^2(\theta/2)$ is the four-momentum transfer squared, and M is the nucleon mass.

The basic idea of the quark-parton model [2, 3] is to represent inelastic electron-nucleon scattering as quasi-free scattering from the partons/quarks in the nucleon, when viewed in a frame where the nucleon has infinite momentum (the center-of-mass frame is a very good approximation to such a frame). The fractional momentum of the nucleon carried by the struck quark is given by the Bjorken scaling variable, $x = Q^2/2M\nu$. In the limit where $\nu \to \infty$, $Q^2 \to \infty$ with x fixed, the nucleon structure functions become:

$$F_1 = \frac{1}{2}\sum_i e_i^2 f_i(x), \qquad F_2 = x\sum_i e_i^2 f_i(x). \tag{2}$$

Here, e_i is the fractional charge of quark type i, $f_i(x)dx$ is the probability that a quark of type i carries momentum in the range between x and $x+dx$, and the sum runs over all quark types.

CP747, *Structure of the Nucleon at Large Bjorken x*, edited by C. Bourrely, Z.-E. Meziani, and J. Soffer

Isospin symmetry and structure function positiviness arguments suggest that the ratio of the neutron to proton F_2 structure functions is bounded for all values of x:

$$\frac{1}{4} \leq \frac{F_2^n}{F_2^p} \leq 4\,, \tag{3}$$

a relationship known as the Nachtmann inequality. Dominance of the valence quarks at large x results in the simple relationship:

$$\frac{F_2^n}{F_2^p} = \frac{1+4(d/u)}{4+(d/u)}\,, \tag{4}$$

with u and d being proton quark probability distributions. Figure 1 (left) shows all the SLAC deep inelastic scattering (DIS) (large Q^2 and ν) data from the pioneering SLAC/MIT Collaboration experiments on the F_2^n/F_2^p ratio versus x. The ratio has been extracted from DIS measurements off the proton and deuteron [4], using a non-relativistic smearing model to account for the Fermi-motion of the nucleons in the deuteron [5]. The ratio data are within the bounds of the Nachtmann inequality. For large x, the ratio is about 1/4 which can only be reached if $d = \bar{d} = s = \bar{s} = 0$. This suggests a picture in which the high momentum partons in the proton (neutron) are mainly up (down) quarks. For small x the ratio is close to 1, suggesting little influence of valence quarks and dominance of the quark-antiquark "sea".

THEORY OVERVIEW

The F_2^n/F_2^p and subsequently the d/u ratio can be calculated in a number of models of the nucleon. In deep inelastic scattering, exact SU(6) symmetry would be manifested in equivalent shapes for the valence quark distributions of the proton, which would be related simply by $u(x) = 2d(x)$ for all x. For the neutron to proton structure function ratio this would imply: $F_2^n/F_2^p = 2/3$.

In nature, spin-flavor SU(6) symmetry is, of course, broken. The nucleon and Δ masses are split by some 300 MeV. In deep inelastic scattering off the nucleon, this symmetry breaking is reflected in the experimental observation that the d quark distribution is softer than the u quark distribution, with the F_2^n/F_2^p ratio deviating from the SU(6) expectation. The correlation between the mass splitting in the **56** baryons and the large-x behavior of F_2^n/F_2^p was observed some time ago by Close [6] and Carlitz [7]. Based on phenomenological [6] and Regge [7] considerations, and kinematic arguments by Close and Thomas [8], the breaking of the symmetry was argued to arise from suppression of "diquark" configurations having $S = 1$ relative to the $S = 0$ configuration as $x \to 1$. A dominant scalar valence diquark component of the proton suggests that in the $x \to 1$ limit, F_2^p is essentially given by a single quark distribution (i.e. the u), in which case: $F_2^n/F_2^p \to 1/4$ and $d/u \to 0$ This expectation has, in fact, been built into most phenomenological fits to the parton distribution data [9].

The phenomenological suppression of the d quark distribution at intermediate and large x can be understood in terms of the same hyperfine interaction responsible for the

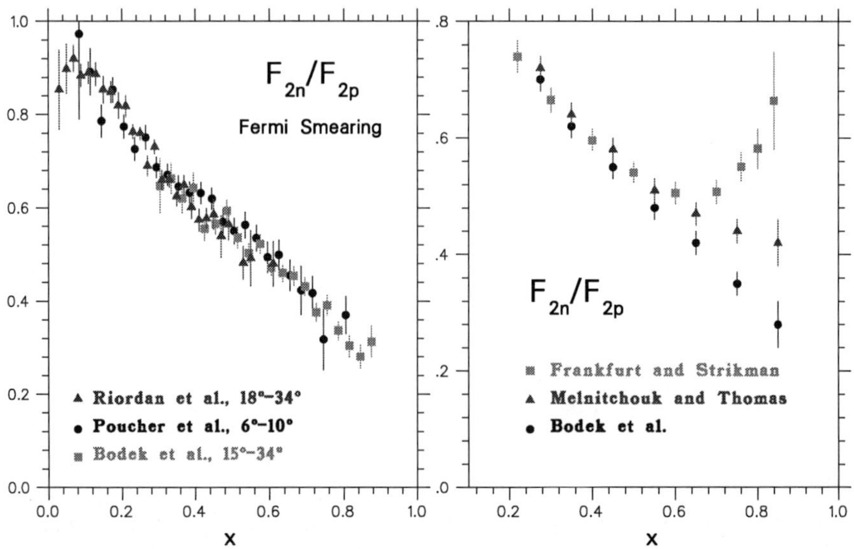

FIGURE 1. Left: SLAC data [4] on the F_2^n/F_2^p ratio extracted from proton and deuteron DIS measurements with a Fermi-smearing model [5]. Right: The F_2^n/F_2^p ratio extracted from proton and deuteron DIS measurements [4] with a) a Fermi-smearing model (Bodek *et al.* [5]), b) a covariant model that includes binding and off-shell effects (Melnitchouk and Thomas [15]), and c) the "nuclear density model" that also incorporates binding effects (Frankfurt and Strikman [16, 17, 19]).

$N-\Delta$ splitting [8, 10, 11]. This color hyperfine interaction is generated by one-gluon exchange between quarks in the core [10]. At lowest order, the Hamiltonian for the color-magnetic hyperfine interaction [10] between two quarks is proportional to $\vec{S}_i \cdot \vec{S}_j$, where $\vec{S}_i$ is the spin vector of quark i. Because this force is repulsive if the spins of the quarks are parallel and attractive if they are antiparallel, Isgur [11] argued that the SU(6) wave function naturally leads to an increase in the mass of the Δ and a lowering of the mass of the nucleon, and a softening of the d quark distribution relative to the u.

An alternative suggestion, based on a perturbative QCD argument, was originally formulated by Farrar and Jackson [12]. Their proposal might be expected to hold at very large x, where the struck quark has momentum approaching infinity, which goes beyond mean field theory (like the simple quark model). They showed that the exchange of longitudinal gluons, which are the only type permitted when the spins of the two quarks in $(qq)_S$ are aligned, would introduce a factor $(1-x)^{1/2}$ into the Compton amplitude – in comparison with the exchange of a transverse gluon between quarks with spins anti-aligned. In this approach the relevant component of the proton valence wave function at large x is that associated with states in which the total "diquark" spin projection, S_z, is zero as $x \to 1$. Consequently, scattering from a quark polarized in the opposite direction to the proton polarization is suppressed by a factor $(1-x)$ relative to the helicity-aligned configuration.

A similar result is also obtained in the treatment of Brodsky *et al.* [13] (based on quark-counting rules), where the large-x behavior of the parton distribution for a quark polarized parallel ($\Delta S_z = 1$) or antiparallel ($\Delta S_z = 0$) to the proton helicity is given by: $q^{\uparrow\downarrow}(x) = (1-x)^{2n-1+\Delta S_z}$, where n is the minimum number of non-interacting quarks (equal to 2 for the valence quark distributions). In this scheme, it is predicted, for the $x \to 1$ limit, that $F_2^n/F_2^p \to 3/7$ and that $d/u \to 1/5$. It is noted that in the latter two treatments, the d/u ratio does not vanish and the F_2^n/F_2^p ratio is 3/7 instead of 1/4.

MOTIVATION FOR NEW EXPERIMENTS

Although the problem of extracting neutron structure functions from deuteron data is rather old [14], the discussion has been recently revived [15, 16] with the realization [17] that F_2^n, extracted from F_2^d by taking into account Fermi-motion *and* binding effects, could be significantly larger [15, 16, 17] than that extracted in earlier analyses [5] in which only Fermi-motion corrections were applied.

Melnitchouk and Thomas [15] have incorporated binding and off-shell effects within a covariant framework in terms of relativistic deuteron wave functions. Neglecting the relativistic deuteron P-states and off-shell deformation of the bound nucleon structure function, which were found to contribute at the $\sim 1\%$ level, the deuteron structure function, F_2^d, can be written as a convolution of the free proton and neutron F_2 structure functions and a nucleon momentum distribution in the deuteron, $f_{N/d}$:

$$F_2^d(x,Q^2) = \int dy\, f_{N/d}(y)\, [F_2^p(x/y,Q^2) + F_2^n(x/y,Q^2)], \qquad (5)$$

where y is the fraction of the 'plus'-component of the nuclear momentum carried by the interacting nucleon, and $f_{N/d}(y)$ takes into account both Fermi-motion and binding effects. Their calculation results in larger F_2^n/F_2^p values compared with the Fermi-motion only extracted values. As can be seen in Figure 1 (right), the difference at $x = 0.85$ can be up to $\sim 50\%$.

Whitlow *et al.* [17] incorporated binding effects by estimating the EMC effect [18] of the deuteron itself (difference between the free and bound nucleon quark distributions) using the "nuclear density model" of Frankfurt and Strikman [19]. In this model, the EMC effect for the deuteron scales with nuclear density as for heavy nuclei:

$$\frac{F_2^d}{F_2^p + F_2^n} = 1 + \frac{\rho_d}{\rho_A - \rho_d}\left[\frac{F_2^A}{F_2^d} - 1\right], \qquad (6)$$

where ρ_d is the deuteron charge density, and ρ_A and F_2^A refer to a heavy nucleus with mass number A. This model predicts F_2^n/F_2^p values that are significantly higher (in excess of 100%) than the Fermi-motion only extracted ones at high x, as can be seen in Figure 1 (right).

It is evident from the above two models that neglecting nuclear binding effects in the deuteron can introduce, at large x, a significant uncertainty in the extraction of F_2^n/F_2^p and d/u. In the absence of experimental data or a unique theory for the magnitude of binding effects and the existence of the EMC effect in the deuteron, the question of the large-x behavior of F_2^n/F_2^p and d/u can only be settled by a measurement which does not rely on inclusive electron-deuteron inelastic scattering.

FUTURE EXPERIMENTS

The above situation can be remedied by using a method proposed by Afnan *et al.* [20], which maximally exploits the mirror symmetry of $A = 3$ nuclei and extracts the F_2^n/F_2^p ratio from DIS measurements off ^{3}H and ^{3}He. Defining the EMC-type ratios for the F_2 structure functions of ^{3}He and ^{3}H (weighted by corresponding isospin factors) by:

$$R(^3\mathrm{He}) = \frac{F_2^{^3\mathrm{He}}}{2F_2^p + F_2^n}\,, \qquad R(^3\mathrm{H}) = \frac{F_2^{^3\mathrm{H}}}{F_2^p + 2F_2^n}\,, \tag{7}$$

one can write the ratio of these as $\mathscr{R}^* = R(^3\mathrm{He})/R(^3\mathrm{H})$, which directly yields the ratio of the free neutron to proton structure functions:

$$\frac{F_2^n}{F_2^p} = \frac{2\mathscr{R}^* - F_2^{^3\mathrm{He}}/F_2^{^3\mathrm{H}}}{2F_2^{^3\mathrm{He}}/F_2^{^3\mathrm{H}} - \mathscr{R}^*}\,. \tag{8}$$

The F_2^n/F_2^p ratio extracted via Equation 8 does not depend on the size of the EMC effect in ^{3}He or ^{3}H, but rather on the *ratio* of the EMC effects in ^{3}He and ^{3}H. Since the neutron and proton distributions in the $A = 3$ nuclei are not dramatically different, this ratio can be calculated reliably and it has been shown that $\mathscr{R}^* \approx 1$ [20, 21, 22, 23].

DIS scattering with the available 6 GeV and the proposed 11 GeV JLab electron beams can provide measurements for the ^{3}H and ^{3}He F_2 structure functions in the x range from 0.10 to 0.82. Using ^{3}H and ^{3}He targets with luminosities of $\sim 5\times10^{37}$ cm^{-2} s^{-1}, and existing JLab spectrometer facilities, the time required for these measurements is less than two weeks. The quality of the projected data for the F_2^n/F_2^p and d/u ratios is shown in Figure 2. The error bars assume a $\pm1\%$ overall systematic experimental error in the measurement of the $\sigma(^3H)/\sigma(^3He)$ inelastic cross section ratio and a theoretical uncertainty in the calculation of $\mathscr{R}^*$ that increases linearly from 0% at $x = 0$ to $\pm1\%$ at $x = 0.82$. The shaded areas in Figure 2 indicate the present uncertainty due to possible nuclear corrections in the extraction of F_2^n/F_2^p and d/u from iclusive DIS proton and deuterium data. It is evident that the proposed measurement will be able to unquestionably distinguish between the present competing extractions of the F_2^n/F_2^p and d/u ratios from proton and deuterium DIS measurements, and determine their values with an unprecedented precision in an almost model-independent way. More experimental details can be found in References [21] and [24].

Another way to measure the neutron structure function is by measuring the cross section for semi-inclusive deep inelastic scattering off deuterium. This is accomplished by measuring backward spectator protons in coincidence with the scattered electrons from the $e+n \to e+p_s+X$ inelastic reaction. The cross section for this process is factorized in terms of the deuteron spectral function S and an effective neutron F_2 structure function:

$$\frac{d\sigma}{d^3p} \sim S(y\cdot p^2)(F_2^n)_{eff}\left(\frac{x}{y}, p^2\cdot Q^2\right), \tag{9}$$

with:

$$y = \frac{M_d - E_s + p_s^z}{M_d}, \qquad p = -\frac{p_t^2}{1-y} - \frac{y}{1-y}\left[M^2 - M_d^2(1-y)\right], \tag{10}$$

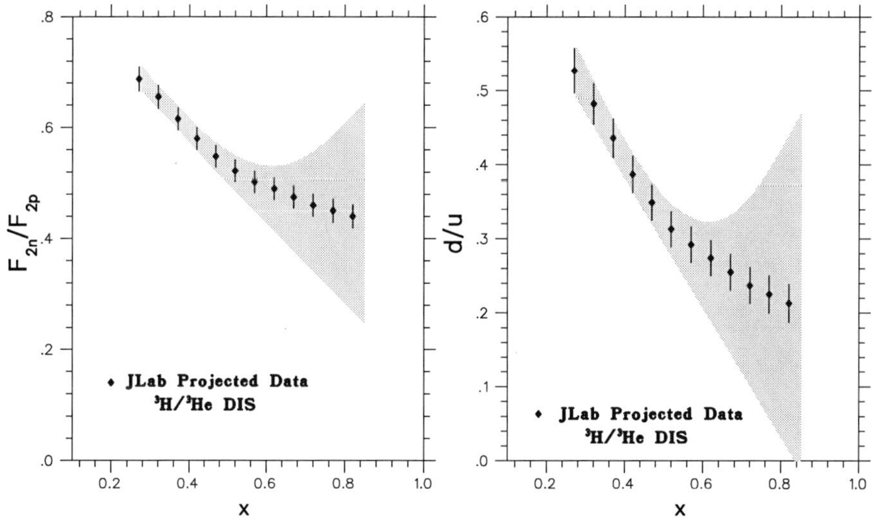

FIGURE 2. Projected data for the F_2^n/F_2^p structure function (left) and d/u quark (right) ratios from the proposed ^{3}H/^{3}He JLab DIS experiment [21, 24]. The error bars include experimental and theoretical systematic uncertainties added in quadrature. The shaded band indicates the present uncertainty due to possible binding effects in deuteron.

where p and p_s are the struck neutron and spectator proton four-momenta (with z and t denoting longitudinal and transverse components), E_s is the proton energy and M_d is the deuteron mass. This method will be used in the approved JLab experiment E03-012, which is expected to data data in the near future [25]. The same measurement was part of the proposed HERA electron-deuteron scattering experimental program with the H1 detector [26]. This experimental approach is based on the isolation of the modifications in the structure of the bound nucleon within the impulse approximation, by choosing kinematics to minimize effects from the deuteron wave function and final-state interactions. It relies on the selection of backward low-energy proton kinematics to minimize: i) production of low-momentum protons from quark fragmentation, and ii) final-state interactions between the spectator proton and the neutron remnant. In addition, off-shell effects appear to be minimal for $p_s < 100$ MeV/c, resulting in $(F_2^n)_{eff} \simeq (F_2^n)_{free}$. An extensive recent theory overview of this method is given in Reference [27].

This JLab experiment will detect scattered electrons in the CLAS JLab Hall B detector. The spectator protons will be detected in a new recoil detector (radial time projection chamber) in the backward direction. Quasi-elastic $d(e,e'p)n$ and elastic e-n scattering with spectator proton detection at low x will be also measured and used to estimate the normalization of the high-x measurements. A reliable control of the high-x results would require measurement of the proton structure function F_2^p by detecting spectator neutrons recoiling in the backward direction, admittedly not an easy task. The projected data from this experiment are similar to the ^{3}He/^{3}H ones shown in Figure 2.

Another promising way of measuring the d/u ratio is by measuring the yields of π^+ and π^- mesons in semi-inclusive electron-proton scattering. At a large energy ratio

$z = E_\pi/\nu$ (with E_π being the π energy), the $u(d)$ quark fragments primarily into a $\pi^+(\pi^-)$. The number of charged pions produced is given in the quark-parton model by :

$$N^{\pi^+}(x,z) \sim 4u(x)[D(z)]_u^{\pi^+} + d(x)[D(z)]_d^{\pi^+}, \tag{11}$$

$$N^{\pi^-}(x,z) \sim 4u(x)[D(z)]_u^{\pi^-} + d(x)[D(z)]_d^{\pi^-}, \tag{12}$$

where the fragmentation functions $D(z)$ (probabilities that a particular type of quark fragments into a pion) obey isospin symmetry:

$$D \equiv [D(z)]_u^{\pi^+} = [D(z)]_d^{\pi^-}, \qquad \bar{D} \equiv [D(z)]_d^{\pi^+} = [D(z)]_u^{\pi^-}. \tag{13}$$

The ratio of pion yields leads to a d/u determination free of experimental systematics, assuming knowledge of the fragmentation functions:

$$R^\pi(x,z) = \frac{N^{\pi^-}}{N^{\pi^+}} = \frac{4\frac{\bar{D}}{D} + \frac{d}{u}}{4 + \frac{d}{u}\frac{\bar{D}}{D}}. \tag{14}$$

Dominance of the leading fragmentation ($D >> \bar{D}$) in the limit $z \to 1$ results in the d/u extraction: $R^\pi(x,z) = d/(4u)$. In a realistic experiment, the data must be integrated over a large z range using previously measured fragmentation functions. A serious high-statistics experimental program is in progress at JLab (experiment E04-114) with the present 6 GeV beam [28], expected to be extended with the proposed 12 GeV beam energy upgrade. A comprehensive recent review of this method is provided in Ref. [29].

A measurement of the d/u ratio is also planned through a high-statistics inelastic neutrino-proton Fermilab Minerνa experiment [30] using the new NuMI beam and a fine-grained detector upstream of the MINOS detector. It is based on the exploration of the different "tastes" of neutrinos and antineutrinos: neutrinos interact primarily with d quarks and antineutrinos with u quarks. The associated ν-p and $\bar{\nu}$-p cross sections are given by:

$$\left[\frac{d\sigma}{dxdy}\right]_{\nu p} \simeq \frac{2MEG^2}{\pi}\left[xd(x) + x(1-y)^2\bar{u}(x)\right], \tag{15}$$

$$\left[\frac{d\sigma}{dxdy}\right]_{\bar{\nu} p} \simeq \frac{2MEG^2}{\pi}\left[x\bar{d}(x) + x(1-y)^2 u(x)\right], \tag{16}$$

where $y = (1 + \cos\theta_{cm})/2]$, with E here being the ν and $\bar{\nu}$ energies, G the Fermi constant, and θ_{cm} the center-of-mass scattering angle. The d/u quantity is determined directly from the ratio of the two measured cross sections and it is obviously free of nuclear binding effects.

Lastly, another way to measure reliably the d/u ratio is by exploring the parity-violating "left"(L)-"right"(R) cross section asymmetry present in deep inelastic scattering of longitudinally polarized electrons off unpolarized protons:

$$A(x,y) = \frac{\sigma(L) - \sigma(R)}{\sigma(L) + \sigma(R)}, \quad y = (E - E')/E. \tag{17}$$

The asymmetry A is expressed in terms of electroweak quark-$Z_\circ$ coupling constants C_{iq} and quark distributions $q_i(x)$:

$$A(x,y) = \frac{GQ^2}{2\sqrt{2}\pi\alpha}\left[a(x) + \frac{1-(1-y)^2}{1+(1-y)^2}b(x)\right], \tag{18}$$

$$a(x) = \frac{3}{4}\left[\frac{2C_{1u}u(x) - C_{1d}d(x)}{u(x)+d(x)/4}\right], \quad b(x) = \frac{3}{4}\left[\frac{2C_{2u}u(x) - C_{2d}d(x)}{u(x)+d(x)/4}\right]. \tag{19}$$

The drawback of this method is the small value of the asymmetry and of the cross section, necessitating large luminosity and a large acceptance spectrometer for a reliable high-statistics measurement. More details on this approach and planned measurements are given in Professor Souder's paper in these Conference Proceedings.

The author would like to thank the HiX-2004 Conference organizers for their kind invitation and to acknowledge the support of the U.S. National Science Foundation (grants PHY-0072384 and PHY-0355181) for this work.

REFERENCES

1. R. E. Taylor, *Rev. Mod. Phys.* **63**, 573 (1991); H. W. Kendall, *Rev. Mod. Phys.* **63**, 597 (1991); J. I. Friedman, *Rev. Mod. Phys.* **63**, 615 (1991).
2. J. D. Bjorken, *Phys. Rev.* **179**, 1547 (1969).
3. R. P. Feynman, *Phys. Rev. Lett.* **23**, 1415 (1969).
4. A. Bodek *et al.*, *Phys. Rev. Lett* **30**, 1087 (1973); E. M. Riordan *et al.*, *Phys. Rev. Lett.* **33**, 561 (1974); J. S. Poucher *et al.*, *Phys. Rev. Lett.* **32**, 118 (1974).
5. A. Bodek *et al.*, *Phys. Rev.* **D20**, 1471 (1979).
6. F. E. Close, *Phys. Lett.* **B43**, 422 (1973).
7. R. Carlitz, *Phys. Lett.* **B58**, 345 (1975).
8. F. E. Close and A. W. Thomas, *Phys. Lett.* **B212**, 227 (1988).
9. H. L. Lai *et al.* (CTEQ Collaboration), *Eur. Phys. J.* **C12**, 375 (2000), and references therein
10. N. Isgur, G. Karl and R. Koniuk, *Phys. Rev. Lett.* **41**, 1269 (1978); N. Isgur, G. Karl and D. W. L. Sprung, *Phys. Rev.* **D23**, 163 (1981).
11. N. Isgur, *Phys. Rev.* **D59**, 034013 (1999).
12. G. R. Farrar and D. R. Jackson, *Phys. Rev. Lett.* **35**, 1416 (1975).
13. S. J. Brodsky, M. Burkardt and I. Schmidt, *Nucl. Phys.* **B441**, 197 (1995).
14. G. B. West, *Phys. Lett.* **B37**, 509 (1971); W. B. Atwood and G. B. West, *Phys. Rev.* **D7**, 773 (1973).
15. W. Melnitchouk and A. W. Thomas, *Phys. Lett.* **B377**, 11 (1996).
16. U. K. Yang and A. Bodek, *Phys. Rev. Lett.* **82**, 2467 (1999).
17. L. W. Whitlow *et al.*, *Phys. Lett.* **B282**, 475 (1992).
18. J. Gomez *et al.*, *Phys. Rev.* **D49**, 4348 (1994); and references therein.
19. L. L. Frankfurt and M. I. Strikman, *Phys. Rep.* **160**, 235 (1988).
20. I. R. Afnan *et al.*, *Phys. Lett.* **B493**, 36 (2000).
21. I. R. Afnan *et al.*, *Phys. Rev.* **C68**, 035201 (2003).
22. E. Pace, G. Salmè and S. Scopetta, *Phys. Rev.* **C64**, 055203 (2001).
23. M. M. Sargsian *et al.*, *Phys. Rev*. **C66**, 024001 (2002).
24. G. G. Petratos *et al.*, *arcXiv:nucl-ex/0010011* (2000).
25. C. Keppel, S. Kuhn and W. Melnitchouk *et al.*, *JLab Proposal 03-012* (2003).
26. T. Alexopoulos *et al.*, *DESY-03-194* (2003).
27. W. Melnitchouk, M. Sargsian and M. Strikman, *Z. Phys.* **A359**, 99 (1997); and references therein.
28. J.-P. Chen, J.-C. Peng, L. Zhu, X. Jiang *et al.*, *JLab Proposal 04-114* (2004).
29. W. Melnitchouk, J. Speth and A. W. Thomas, *Phys. Lett.* **B435**, 420 (1998); and references therein.
30. D. Drakoulakos *et al.*, *Fermilab Minerva Collaboration Proposal* (2004).

Prospects for and Status of the Measurement of High x Parton Distributions at HERA

Zhiqing Zhang

Laboratoire de l'Accélérateur Linéaire, IN2P3-CNRS et Université de Paris-Sud, Bât. 200, BP 34, 91898 Orsay Cedex, France (zhangzq@lal.in2p3.fr)

Abstract. Inclusive neutral and charged current cross sections and structure functions measured at HERA-I are briefly reviewed. The impact of these measurements on parton density functions is discussed. The emphasis is put on the high x region. A first result obtained using the HERA-II data is also shown followed by future prospects for HERA.

INTRODUCTION

The HERA electron- or positron-proton collider is a unique facility for studying neutral and charged current (NC, CC) processes in deep inelastic scattering (DIS) interactions in an extended kinematic region compared to that covered previously by fixed target experiments. HERA-I operation started in 1992 and ended in 2000. Earlier measurements from H1 and ZEUS revealed the strong rise of the structure function (SF) F_2 towards small x. The more precise measurement of NC and CC cross sections is based on three recent data samples of e^+p taken in years from 1994 to 1997 at a center-of-mass (CM) energy of 301 GeV, e^-p in 1998-1999 and e^+p in 1999-2000 at 319 GeV. The corresponding integrated luminosities per experiment of the three data samples are around 40 pb^{-1}, 15 pb^{-1} and 65 pb^{-1}, respectively. The HERA-I data have driven the determination of parton density functions (PDFs) in various global fits for a decade. The universal nature of the PDFs makes precise determination essential in order to provide reliable predictions for Standard Model (SM) processes and for searches for new physics at future hadron-hadron colliders such as the LHC. Since 2002, HERA has entered its second phase of operation (HERA-II) and is being run in higher luminosity mode providing in addition a longitudinally polarized positron beam for H1 and ZEUS. The improved luminosity is due to a finer beam focusing around the interaction points and to larger beam currents.

In the following, I shall first briefly review the precise cross section and SF data measured at HERA-I and their impact on the determination of PDFs and then show a first measurement made using the HERA-II data before discussing future prospects. I will not discuss here other interesting and important physics topics at HERA such as heavy flavor physics, diffraction and searches for new physics.

CP747, *Structure of the Nucleon at Large Bjorken x,* edited by C. Bourrely, Z.-E. Meziani, and J. Soffer

HERA-I ACHIEVEMENTS

Both NC and CC processes can be produced at HERA. The NC interaction is dominated by γ exchange in the t channel at low Q^2 and receives additional contributions from Z exchange and γZ interference when Q^2 values are comparable with the Z mass squared. The CC interaction takes place by W exchange only. The NC cross section may be expressed in terms of three SFs as

$$\frac{d^2\sigma_{NC}(e^{\pm}p)}{dxdQ^2}=\frac{2\pi\alpha^2}{xQ^4}\left[Y_+\tilde{F}_2-y^2\tilde{F}_L\mp Y_-x\tilde{F}_3\right] \quad (1)$$

where $Y_{\pm}=1\pm(1-y)^2$. The CC cross section may be expressed in a similar way.

The dominant SF $\tilde{F}_2$ has been determined from the measured NC cross sections over more than 4 orders of magnitude both in x and Q^2 with an accuracy down to 2% (Fig.1) [1,2]. The contribution of the SF $\tilde{F}_L$ is sizable only at high y and can only be determined indirectly for the moment [3]. The SF $x\tilde{F}_3$ is non-negligible at high Q^2 and is determined [1,2] using both e^+p and e^-p data samples with so far limited precision.

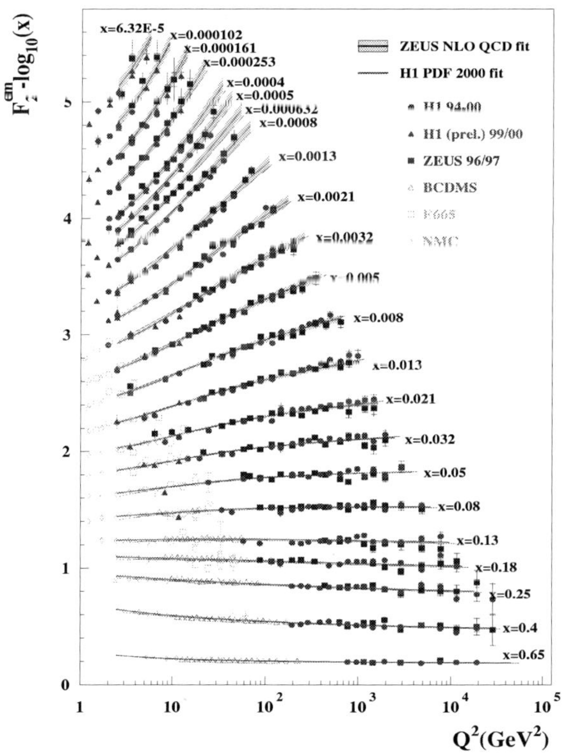

FIGURE 1. The pure electromagnetic (γ exchange) structure function F_2^{em} in $\tilde{F}_2$ measured by HERA experiments H1 and ZEUS and by fixed target experiments BCDMS, E665 and NMC. The curves show

the corresponding expectations from ZEUS next-leading-order (NLO) QCD fit and H1 PDF 2000 fit (see text below).

While the NC SF $\tilde{F}_2$ is essentially flavor blind, the CC cross sections are flavor dependent. In particular the measured CC e^+p cross section data [1,2] can be used to uniquely constrain the down type quark density (mainly d quark) free from any nuclear corrections.

HERA data have been extensively used by the MRST and CTEQ groups [4,5] in their global fits to determine the PDFs in the last decade with steadily increasing precision. The impact of the HERA-I data is best seen in QCD analyses using these data only. Indeed, the fit performed by H1 [1], based on all e^+p and e^-p NC and CC cross section data at low and high Q^2, was used to determine 5 PDF sets xP: xg, xU=x(u+c), xD=x(d+s), $x\overline{U}=x(\overline{u}+\overline{c})$ and $x\overline{D}=x(\overline{d}+\overline{s})$. The PDFs were parameterized with the functional form $xP(x)=A_P x^{B_P}(1-x)^{C_P}[1+D_P x+E_P x^2+F_P x^3+G_P x^4]$ at Q_0^2=4GeV^2. The results of the fit are shown in Fig.2 compared to PDFs determined in a similar analysis from ZEUS [6] using, in addition to cross section data, jet data in order to improve the constraint on the gluon. The agreement between H1 and ZEUS is fairly good and the visible difference in e.g. the gluon distribution may be understood due to different functional forms, heavy flavor treatments and consistency of the data sets.

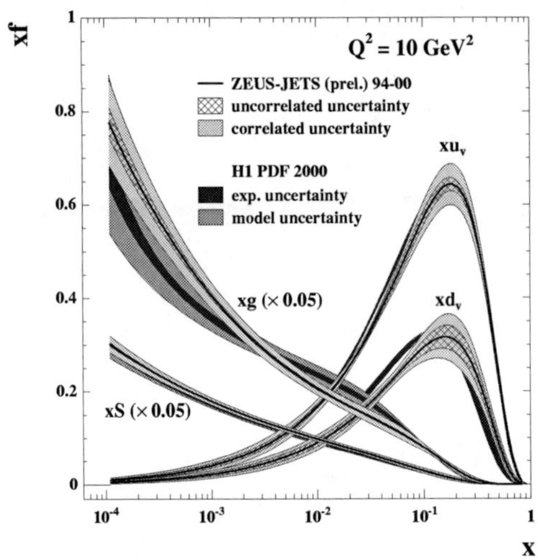

FIGURE 2. A comparison of PDFs determined by H1 (H1 PDF 2000) based on inclusive NC and CC cross section data and by ZEUS (ZEUS-JETS) using jet data in addition to the cross section data. The gluon (xg) and sea (xS) densities are scaled by a factor 20.

Another way [1] to see the constraining power of the HERA cross section data on the PDFs is to extract locally those parton distributions which contribute more than 70% in the relevant cross sections as:

$$xq(x,Q^2)=\sigma(x,Q^2)\left(\frac{xq(x,Q^2)}{\sigma(x,Q^2)}\right)_{fit}. \qquad (2)$$

The resulting xu and xd densities are shown in Fig.3. The determination is shown as a function of x for two values of Q^2 at 3000 and 8000GeV^2 on the upper figures and as a function of Q^2 on the lower figures for x values above 0.08, the high x region where the method applies. Such a determination also compares well with the H1 PDF 2000 fit shown as the shaded bands and with MRST and CTEQ as curves.

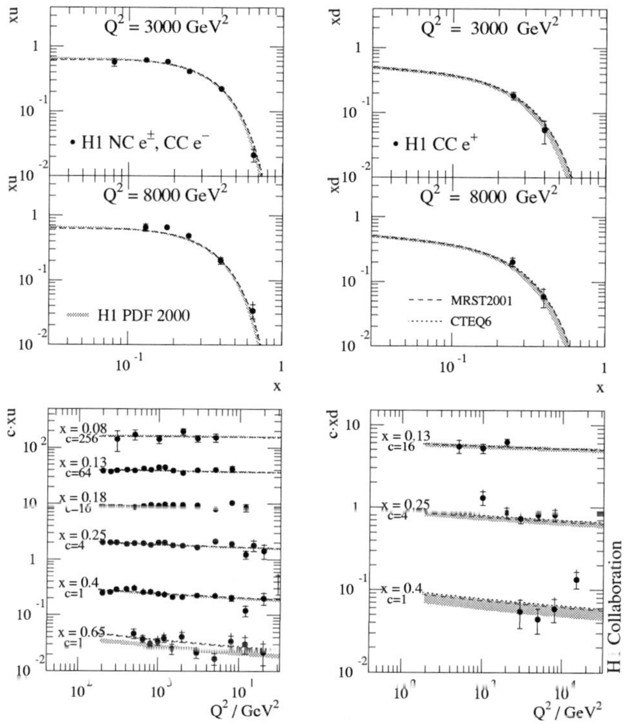

FIGURE 3. The parton densities xu and xd determined in the local extraction method (data point) in comparison with those obtained from the H1 PDF 2000 fit (shaded band) and MRST and CTEQ6 (curve).

HERA-II STATUS AND FUTURE PROSPECTS

In order to increase the luminosity delivered, the HERA accelerator was upgraded in 2001. During the first two years running after the upgrade however, the experiments suffered large backgrounds. The problem has now been solved and an increase in peak luminosity around a factor 3 was achieved in the running of 2004. In addition, two pairs of spin rotators have been installed near the H1 and ZEUS interaction points providing for the first time longitudinally polarized positron beams to these experiments.

HERA-II has already delivered two independent data samples of ~20 pb^{-1} each with a mean polarization value of –44% and 33%, respectively. They have been used by both experiments to measure CC total cross sections. Together with the CC cross section measured [1,2] previously at HERA-I with unpolarized data, the dependence of the CC cross section on polarization is now available for the first time (Fig.4). In the SM, the cross section has a linear dependence as a function of polarization as represented by the full curve based on the expectation from MRST. Due to the absence of right-handed W bosons in the SM, the CC cross section at P_e=−1 is predicted to vanish. Indeed, a straight line fit to the measured cross sections resulted in an extrapolated cross section value at P_e=−1 of (−0.2±1.8[stat]±1.6[syst]) pb, which is consistent with a null contribution. The result of the fit is shown as a dashed line in Fig.4. Future measurements with improved precision will provide a stringent constraint on the contribution of the right-handed W boson and its mass, independent of that from the Tevatron.

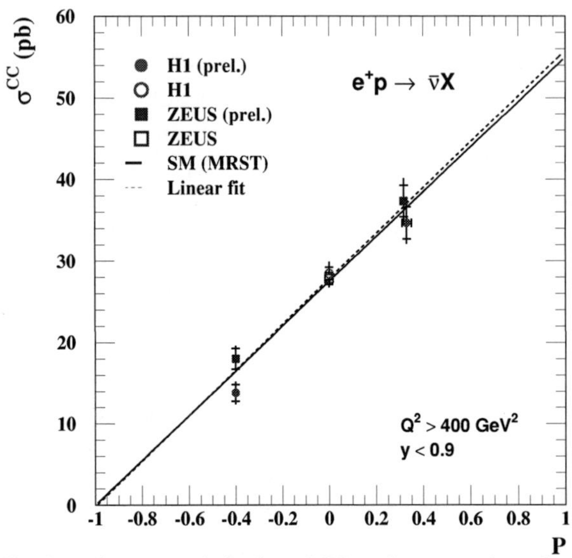

FIGURE 4. The dependence on polarization of CC total cross sections shown for Q^2>400GeV^2 and y<0.9. The full curve corresponds to the expectation from MRST. The dashed line is a straight line fit to the measured cross sections.

The HERA-II operation will end in 2007 according to current planning. In the present envisaged running scenario HERA will operates with an electron beam starting after the summer shutdown 2004 till the end of 2005 and will switch back to positron-proton collisions for the rest of HERA-II. In this scenario, an integrated luminosity of 0.7 fb^{-1} per experiment is expected. This will be a significant gain with respect to the total data taken in HERA-I. Such a large data sample will improve even further our knowledge of the PDFs, in particular at high x and Q^2, the precision of which is currently statistically limited.

Indeed, precision NC cross sections at high Q^2 will improve the determination of $x\tilde{F}_3$ which is sensitive to $2u_v$+d_v with u_v and d_v being the valence quark densities of u and d quarks. Similarly, precision e^+p CC cross sections at high Q^2 will allow the d quark density to be constrained better at high x. Furthermore, from a polarization asymmetry in NC cross sections, one will determine the SF $F_2^{\gamma Z}$ which can be expressed in terms of the d/u ratio. All these data will thus hopefully clarify the controversial value of d/u at x=1 [7].

Apart from the high luminosity running at largest possible beam energies, a direct measurement of F_L will need data at different CM energies. Such a measurement is important to verify the gluon density that is currently constrained mainly indirectly from the scaling violation of the SF data (Fig.1). A lower proton energy running would also help to access high x region at low Q^2.

Beyond HERA-II, there is strong interest [8] in the community to continue the HERA physics programme and to fully exploit its unique potential. In particular it is expected that by replacing the proton beam with a heavy nuclear beam, one can study a new QCD regime in which the gluon parton density becomes so high that saturation would take place. This however needs funding and solutions for a new injector replacing PETRA, the actual injector, which will be rebuilt to be used for XFEL (X-ray Free Electron Laser) from 2007 on.

ACKNOWLEDGMENTS

The author wishes to thank J. Soffer for organizing such a stimulating workshop and colleagues in H1 and ZEUS for the results shown in this contribution.

REFERENCES

1. Adloff, C. et al., [H1 Collaboration], *Eur. Phys. J.* **C30** (2003) 1; *Eur. Phys. J.* **C19** (2001) 269.
2. Chekanov, S. et al., [ZEUS Collaboration], *Eur. Phys. J.* **C32** (2003) 1; accepted by *Phys. Rev. D*; *Phys. Lett.* **B539** (2002) 197; *Eur. Phys. J.* **C28** (2003) 175.
3. Adloff, C. et al., [H1 Collaboration], *Eur. Phys. J.* **C21** (2001) 33.
4. Martin, A.D., Roberts, R.G., Stirling, W.J. and Thorne, R.S., *Eur. Phys. J.* **C28** (2003) 155 and earlier references therein.
5. Pumplin, J. et al., *JHEP* **0207** (2002) 012 and earlier references therein.
6. Cooper-Sarkar, A.M, [ZEUS Collaboration], hep-ph/0407309, Proceedings of the 12th international workshop on deep inelastic scattering, 14-18 April, Strbske Pleso, Slovakia.
7. Zhang, Z., *J. Phys. G.: Nucl. Part. Phys.* **28** (2002) 767.
8. See e.g. http://wwwhera-b.mppmu.mpg.de/hera-3/hera3/ .

The BoNuS Experiment At Jefferson Lab

Stephen Bültmann[1]

Old Dominion University, Physics Department, Norfolk, VA 23529, USA

Abstract. The BoNuS experiment [1] at Jefferson Lab's Hall B is going to measure the structure of the quasi-free neutron via electron scattering off a deuteron target and detection of the slowly recoiling spectator proton at very backward angles. To detect the low momentum backward recoiling protons, the CLAS detector will be augmented with a novel radial time projection chamber featuring gas electron multipliers. Data taking is anticipated for 2005 or 2006.

THE NEUTRON STRUCTURE FUNCTION F_2^N

The structure of the proton has been measured over the last decades to very high precision. In contrast, the structure of the neutron is not directly accessible, because of the instability of the free neutron, and has hence to be inferred from comparative measurements on nuclear targets, like deuterons or ^{3}He nuclei, and protons. When extracting the structure of the free neutron, binding effects inside the nucleus need to be accounted for as well as final state interactions of the scattered neutron and recoiling proton in the nuclear medium.

In Figure 1 (left) the ratio of the structure functions F_2 of the neutron and proton as measured at SLAC in electron scattering of deuterium and hydrogen targets is shown. At values for the Bjørken scaling variable x above 0.6 the same data show a strong dependence on the model corrections being included in the extraction process. These models introduce additional uncertainties in the neutron structure themselves. A measurement using a deuteron target and selecting only neutrons that are bound very weakly will minimize the uncertainties introduced by theoretical models of the deuteron. This is necessary to be able to discriminate among the different predictions for the F_2 ratios shown in Figure 1 (left) and, hence, for the ratio of d- and u-quarks.

The cross-sections for inclusive electro-production in the nucleon resonance region are shown in Figure 1 (right) for hydrogen and deuterium targets at $Q^2 = 1.5$ GeV^2 as measured at Jefferson Lab's Hall C [2]. The well-resolved resonances seen for the hydrogen target are largely smeared out for the deuterium target, with the exception of the $\Delta(1232)$ resonance. However, this resonance is also indistinguishable from the non-resonant background at higher $Q^2 = 2$ GeV^2. The smeared resonant structure is caused by the nuclear binding effects, which have to be modeled when extracting the neutron spectrum from the deuterium measurement. This uncertainty will be largely removed by tagging the spectator proton and, hence, constraining the neutron kinematics.

[1] Representing the BoNuS Collaboration
E-Mail: Stephen.Bueltmann@jlab.org

CP747, *Structure of the Nucleon at Large Bjorken x,* edited by C. Bourrely, Z.-E. Meziani, and J. Soffer

THE BONUS EXPERIMENT

The BoNuS experiment is going to measure the structure of the quasi-free neutron by scattering high-energy electrons off a deuterium gas target and detecting the scattered electron in coincidence with the spectator proton. For an invariant mass of $W > 2$ GeV the upper bound for x will be 0.625, while by loosening this constraint and including parts of the region of the nucleon resonances above $W > 1.5$ GeV will result in an increased upper limit of $x < 0.75$.

Electrons of 4 and 6 GeV beam energy of Jefferson Lab's accelerator will be scattered of the deuterium gas target and the scattered electrons detected by the CLAS detector in Jefferson Lab's Hall B. The most important spectator protons from the scattering process of the electrons off the neutrons inside the deuteron are leaving the target region under very backward angles with respect to the incoming electrons and with recoil momenta of about 100 MeV/c or below. The lowest recoil proton momentum covered by the CLAS detector acceptance is at about 280 MeV/c, so the BoNuS experiment will use a novel recoil detector, which will be installed closely around the target. To prevent absorption of the very low momentum protons, the target will be deuterium gas at 7.5 atmospheres, contained by 50 μm thick Kapton walls, 10 cm long and 1 cm in diameter.

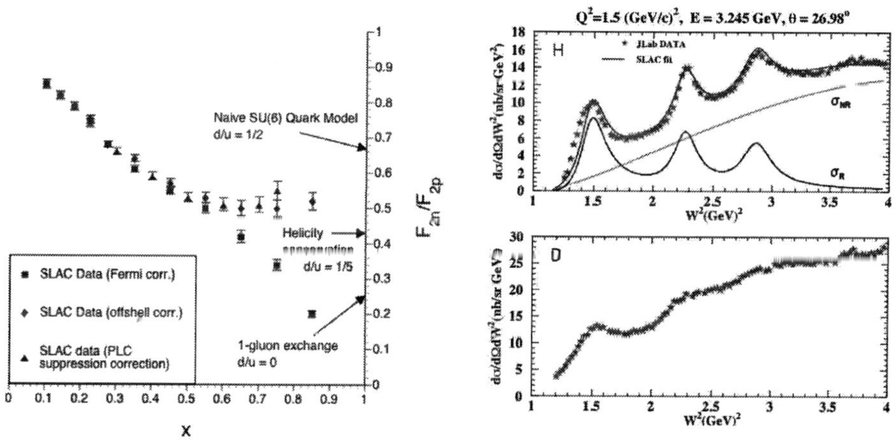

FIGURE 1. Left: Data for the F_2^n/F_2^p ratio from SLAC using different models for the nuclear corrections. Right: The inclusive resonance electro-production cross-section as measured by [2] for hydrogen and deuterium targets.

The recoil detector will be a time projection chamber (TPC) with a 3 cm radial drift region. A radial geometry is made possible by using gas electron multiplier (GEM) foils [3] for amplification, which can be bent into a cylindrical shape around the target axis, previously not possible with the standard wire amplification used in drift chambers. GEM foils are thin, 50 μm thick, non-conductive polyimide foils, perforated by 60 μm diameter holes, and coated on both sides with a thin copper layer. By applying a voltage of about 200 to 300 V across the foil, high electric fields are created across the perforation holes. Electrons from the initial ionization by the particle passing the gas-filled active area of the detector are drifting towards the GEM foil and are creating an avalanche of secondary electrons on passing through the high electric field. The typical single layer GEM gain is about 100, while multiple layers of GEM foils increase the amplification to values above 10^3. Additionally, the multi-layer GEM foils suppress the signal from the slower drifting positive ions. This makes the usage of a gating grid plane in front of the readout plane unnecessary.

The radial TPC (RTPC) [4] will be a 20 cm long double-wall cylinder with inner diameter of 6 cm, a 3 cm radial drift field, and a 1 cm amplification region with three planes of GEM foils. The charge will be collected by about 4,000 pads of 20 mm^2, each, which will provide two-dimensional information for the track reconstruction. The third dimension is determined by the drift time, about two to three microseconds through the 3 cm drift volume, with charge being collected and read out in selectable intervals of probably 50 to 100 nsec.

The RTPC will be placed inside a 30 cm long superconducting solenoid magnet, presently under construction to upgrade the CLAS detector, which will enable the measurement of the proton recoil momentum and also prevent background electrons from Møller scattering to enter the RTPC or the CLAS detector. The Møller electrons will be forced on spiral trajectories around the electron beam directions. Spectator protons will be separated from low momentum pions by the track curvature inside the magnetic field. In addition, the energy deposited by the particle in the active drift region of the RTPC will also be used for particle identification.

At an anticipated beam current of 200 nA, resulting in a luminosity of about $4{\cdot}10^{33}$ cm^{-2} sec^{-1}, the electron trigger rate is expected to be about 1 kHz. Multiple tracks from the background, like pions with large to moderate momentum, will mostly occupy different regions of the RTPC and can be identified. TPC's are very well suited for high particle rates with multi-track events.

The charge of the 4,000 pads will be read out by front end electronics designed for the TPC of the ALICE heavy ion experiment at CERN's Large Hadron Collider [5].

THE BONUS PROTOTYPE RTPC

Based on commercially available GEM foils of $10{\times}10$ cm^2 size (3M corporation), a prototype RTPC was constructed with the correct radii, but of 10 cm length and a radial section of 105°, only. A 16-channel prototype readout board of the ALICE front-end electronics was used to acquire data via a personal computer. The prototype RPTC was tested with γ-rays from a ^{55}Fe radioactive source and with cosmic rays. For a more realistic test with slow protons, it was installed in a secondary proton beam line of the Triangle Universities Nuclear Laboratory (TUNL) providing slow protons of about 100 MeV/*c*.

In Figure 2 an event of a proton passing under a small angle through the prototype RTPC is shown. Besides the two-dimensional position information of the readout pads, the pad charge sampling in 100 nsec intervals provides a third dimension, as electrons created close to the entrance of the prototype RTPC need longer to drift to the readout pads compared to drift electrons created close to the pads.

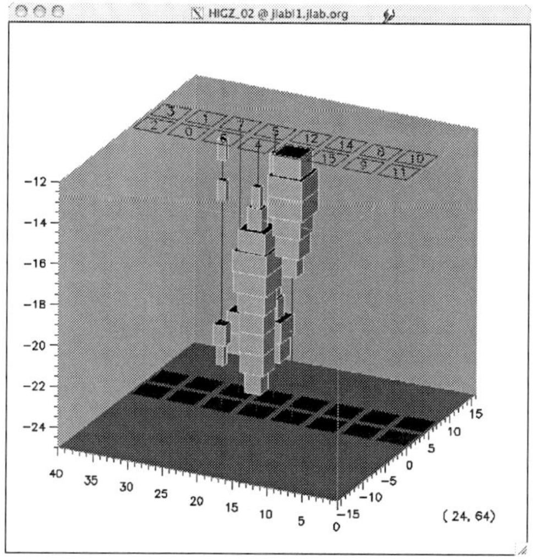

FIGURE 2. Charge deposit of passing proton track at TUNL. The horizontal axes are the 2×8 pad positions and the vertical axis is time in 100 nsec intervals.

OUTLOOK

A measurement of the neutron structure to even higher values of x will be possible with the planned upgrade of Jefferson Lab's accelerator to 11 GeV electron beam energy. The BoNuS collaboration is proposing to use the RTPC and the upgraded CLAS++ detector to measure the neutron structure up to $x < 0.8$ under the same conditions as described above and additionally at higher momentum of the spectator proton (250 – 700 MeV/c) with a liquid deuterium target and an increased luminosity of about 10^{35} cm^{-2} sec^{-1}.

ACKNOWLEDGMENTS

This work has been supported in part by the US Department of Energy under contract numbers DE-FG02-96ER40960 (ODU) and DE-AC05-84ER40150 (JLab).

REFERENCES

1. Dodge, G., *et al.*, *The Structure of the Free Neutron via Spectator Tagging*, BoNuS proposal E03-012 to Jefferson Lab (PAC 23), 2003.
2. Niculescu, I., *et al.*, Phys. Rev. Lett. 85, 2000, p. 1182 and p. 1186.
3. Sauli, F., Nucl. Instr. and Meth. A386, 1997, p. 531.
4. Fenker, H., in *Proc. of the 42nd Int. Winter Meeting on Nuclear Physics*, Bormio, Italy, 2004.
5. Musa, L., *et al.*, in *Proc. of the IEEE Nuclear Science Symposium*, Portland, Oregon, USA, 2003.

Non–Singlet QCD Analysis of F_2 in 3-Loops

J. Blümlein*, H. Böttcher* and A. Guffanti*

*DESY, Platanenallee 6, D–15738 Zeuthen, Germany

Abstract. Results of a non–singlet QCD analysis of the structure function $F_2(x,Q^2)$ in 3–loop order based on the non–singlet world data are presented. Correlated errors are determined and their propagation through the evolution equations is performed analytically. The value for $\alpha_s(M_Z)$ is determined to be $0.1135+0.0023-0.0026$ compatible with results from other QCD analyses.

INTRODUCTION

A consistent 3-loop QCD analysis of the unpolarized structure function $F_2(x,Q^2)$ can be carried out after having the recently completed next-to-next-to-leading order (NNLO) anomalous dimensions available [1], in addition to the 2-loop Wilson coefficients [2]. Two NNLO QCD analyses have been performed previously [3], [4]. In those analyses singlet– and non–singlet evolution was dealt with in parallel. The present analysis concentrates on the non–singlet evolution only in order to firstly obtain an accurate as possible picture for the valence quark distributions and the value of $\alpha_s(M_Z^2)$ with correlated errors in 3–loop accuracy. Analyzing only the non–singlet data has the advantage that large gluon and sea quark effects remain decoupled. First results of this analysis have been published already in [5].

FORMALISM

In Mellin–N space the non–singlet (NS) parts of a structure function $F_i(N,Q^2)$ are given by

$$F_i^{\pm,\mathrm{v}}(N,Q^2) = \left[1+C_{1i}(N)a+C_{2i}(N)a^2\right] f^{\pm,\mathrm{v}}(N,Q^2)\,, \tag{1}$$

where the $C_{ki}(N)$ are the corresponding Wilson coefficients and $f^{\pm,\mathrm{v}}(N,Q^2)$ are the non–singlet quark combinations. The symbol a denotes the strong coupling constant normalized to $a(Q^2)=\alpha_s(Q^2)/4\pi$. The quark combinations to be considered are

$$\Delta^{\pm} = (u\pm\bar{u})-(d\pm\bar{d}) \quad \text{and} \quad \mathrm{v}=(u-\bar{u})+(d-\bar{d})\,. \tag{2}$$

The non–singlet parts of F_2 are proportional to

$$F_2^{NS} \propto \frac{1}{3}\Delta^+, \quad F_2^{p,\mathrm{v}} \propto \tfrac{5}{18}\mathrm{v}+\tfrac{1}{6}\Delta^-, \quad \text{and} \quad F_2^{d,\mathrm{v}} \propto \tfrac{5}{18}\mathrm{v}\,. \tag{3}$$

The structure function F_2^{NS} is the non–singlet part of F_2, while the other combinations represent the valence approximations for F_2^p and F_2^d, respectively. All these combinations evolve as $+$-combinations in Q^2. The relevant parton densities to be determined in a non–singlet QCD analysis are $u_v(x,Q^2)$, $d_v(x,Q^2)$, and $(\bar{d}-\bar{u})(x,Q^2)$.

CP747, *Structure of the Nucleon at Large Bjorken x*, edited by C. Bourrely, Z.-E. Meziani, and J. Soffer

PARAMETERIZATION

The parameterizations of the above mentioned parton densities at the input scale of $Q_0^2 = 4.0$ GeV^2 are determined as follows:

$$xu_v(x,Q_0^2) = A_{u_v} x^{a_u} (1-x)^{b_u} (1 - 1.108x^{\frac{1}{2}} + 26.283x)\,, \tag{4}$$

$$xd_v(x,Q_0^2) = A_{d_v} x^{a_d} (1-x)^{b_d} (1 + 0.895x^{\frac{1}{2}} + 18.179x)\,. \tag{5}$$

Here the values for the coefficients in the polynomial terms given as numbers are obtained by a fit and are then kept fixed since their respective errors are still large. Furthermore, the analysis requires different $\bar{u}$ and $\bar{d}$ quark densities. We adopted the choice of Ref. [6] which gives a good description of the $(\bar{d} - \bar{u})$ data from E866 [7]:

$$x(\bar{d} - \bar{u})(x,Q_0^2) = 1.195x^{1.24}(1-x)^{9.10}(1 + 14.05x - 45.52x^2)\,. \tag{6}$$

The normalization constants A_{u_v} and A_{d_v} are fixed by the conservation of the number of valence quarks: $\int_0^1 u_v(x)dx = 2$, $\int_0^1 d_v(x)dx = 1$. The remaining four parameters, the low–x and high–x parameters a and b for u_v and d_v, respectively, are to be determined in the fit along with the value for Λ_{QCD}.

DATA AND ERROR CALCULATION

The data used in the present analysis come from BCDMS [8], SLAC [9], NMC [10], H1 [11], and ZEUS [12]. The BCDMS data were recalculated replacing R_{QCD} by R_{1998} [13]. All deuteron data were corrected for Fermi motion and off–shell effects [14]. The non–singlet structure function F_2^{NS} was constructed according to the relation $F_2^{NS} = 2(F_2^p - F_2^d)$ from proton and deuteron data measured at the same x and Q^2 values. This distribution was used in the region $x < 0.3$, while the measured $F_2^{p,d}$ was used in the region $x > 0.3$ which is expected to be dominated by valence quark contributions. In addition, a Q^2 region of $Q^2 > 4.0$ GeV^2 was chosen and a cut in the hadronic mass $W^2 > 12.5$ GeV^2 was applied in order to diminish higher twist effects. After all these cuts the results presented here are based on 762 data points, 322 for F_2^p, 232 for F_2^d, and 208 for F_2^{NS}. We allowed for a relative normalization shift between the data sets within the overall normalization uncertainties quoted by the experiments or assumed accordingly. These normalization shifts were fitted once and then kept fixed.

Correlated error calculation: The systematic errors known to be partly correlated are not treated separately here. For all data sets we used the simplest procedure by adding the statistical and the total systematic errors in quadrature being aware of the fact that this will eventually overestimate the errors. The fully correlated 1σ experimental error bands are determined via Gaussian error propagation demanding a positive definite covariance matrix in the fit. The derivatives of the parton distributions with respect to the variable parameters needed for the error propagation can be calculated analytically at the input scale Q_0^2. Their values at Q^2 are given by evolution.

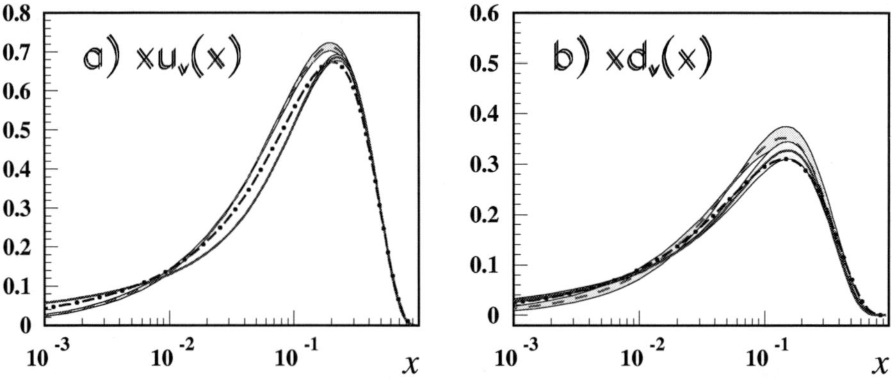

FIGURE 1. The parton densities a) xu_v and b) xd_v at the input scale $Q_0^2 = 4.0$ GeV^2 (solid line) compared to results obtained by MRST (dashed–dotted line) [3] and Alekhin (dashed line) [4]. The shaded areas represent the fully correlated 1σ statistical error bands.

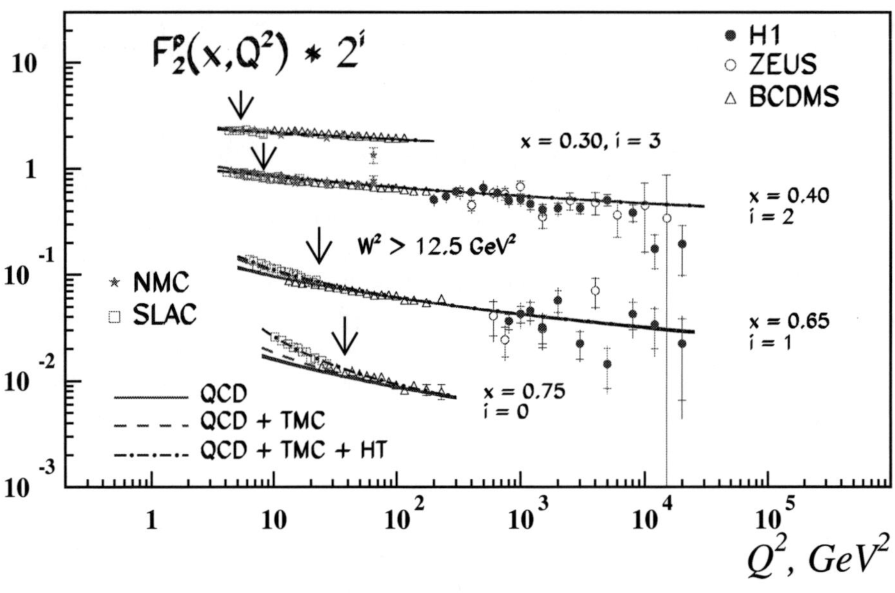

FIGURE 2. The structure function $F_2^p(x,Q^2)$ as function of Q^2 and x. The curves are the results of this analysis (QCD fit: solid line, TMC added: dashed line, HT added: dashed-dotted line). The shaded areas represent the fully correlated 1σ statistical error bands of the QCD curves.The arrows indicate the point $W^2 = 12.5$ GeV^2 above which the QCD fit was performed.

RESULTS

The QCD fit parameter values obtained are as follows: for u_v: $a = 0.314 \pm 0.007$, $b = 4.199 \pm 0.032$, for d_v: $a = 0.413 \pm 0.047$, $b = 6.196 \pm 0.332$, and $\Lambda_{QCD}^{(4)} = 227 \pm 30$ MeV with $\chi^2/ndf = 652/757 = 0.86$. The value of $\Lambda_{QCD}^{(4)}$ is quite stable against a variation of the Q^2–cut to the data between 4.0 GeV^2 and 10.0 GeV^2. From $\Lambda_{QCD}^{(4)}$ the following value of α_s is extracted:

$$\alpha_s(M_Z^2) = 0.1135 \begin{array}{l} +0.0023 \\ -0.0026 \end{array} \text{(expt)}.$$

This value is within the errors in good agreement with results from other NLO/NNLO QCD analyses, see Ref. [5], and with a recent value for the world average of 0.1182 ± 0.0027 [15]. The resulting parton densities $xu_v(x)$ and $xd_v(x)$ at the input scale of $Q_0^2 = 4.0$ GeV^2 are presented in Figure 1. Comparison with results from global analyses shows satisfactory agreement. In describing the proton and deuteron F_2 data best target mass corrections (TMC) [16] and higher twist (HT) contributions have to be added. The HT contributions are treated in a phenomenological way as outlined in Ref. [17] and have been determined in the region 4.0 GeV$^2 < W^2 <$ 12.5 GeV^2 by keeping the QCD fit parameters fixed. In Figure 2 is shown $F_2^p(x, Q^2)$ as an example.

CONCLUSIONS

A non–singlet QCD analysis of the structure function $F_2(x, Q^2)$ based on the non–singlet world data has been performed at 3–loop order. The value determined for $\alpha_s(M_Z^2)$ is compatible within the errors with results from other QCD analyses and with the world average. New parameterizations of the parton densities u_v and d_v including their fully correlated 1σ errors are derived. An exellent description of the F_2 data is achieved.

Acknowledgment. This work was supported in part by DFG Sonderforschungsbereich Transregio 9, Computergestützte Theoretische Physik and EU grant HPRN–CT–2000–00149.

REFERENCES

1. S. Moch, J. A. M. Vermaseren, and A. Vogt, Nucl. Phys. **B688** (2004) 101.
2. W. L. van Neerven and A. Vogt, Nucl. Phys. **B568** (2000) 263 and references therein.
3. A. D. Martin et al., hep-ph/0307262.
4. S. Alekhin, Phys. Rev. **D68** (2003) 014002.
5. J. Blümlein, H. Böttcher, and A. Guffanti, hep-ph/0407089.
6. A. D. Martin et al., Eur. Phys. J.**C23** (2002) 73.
7. R. S. Towell et al., Phys. Rev. **D64** (2001) 052002.
8. A. C. Benvenuti et al., Phys. Lett. **B223** (1989) 485; Phys. Lett. **B237** (1990) 592.
9. L. W. Whitlow et al., Phys. Lett. **B282** (1992) 475.
10. M. Arneodo et al., Nucl. Phys. **B483** (1997) 3.
11. C. Adloff et al., Eur. Phys. **C21** (2001) 33; Eur. Phys. **C30** (2003) 1.
12. J. Breitweg et al., F. Phys. **C7** (1999) 609; S. Chekanov et al., Eur. Phys. **C21** (2001) 443.
13. K. Abe et al., Phys. Lett. **B452** (1999) 194.
14. W. Melnitchouk and A. W. Thomas, Phys. Lett. **B377** (1996) 11; Phys. Rev. **C52** (1995) 3373.
15. S. Bethke, hep-ex/0407021; J. Phys. **G26** (2000) R27.
16. H Georgi and H. D. Politzer, Phys. Rev. **D14** (1976) 1829.
17. M. Virchaux and A. Milsztajn, Phys. Lett. **B274** (1992) 221.

POLARIZED STRUCTURE FUNCTIONS—I

Statistical Polarized Parton Distributions at Large-x

Claude Bourrely

Centre de Physique Théorique (UMR 6207)[1]*,*
CNRS-Luminy case 907,
13288 Marseille cedex 9, France

Abstract. We present a new set of polarized parton distributions constructed in a statistical physical picture of the nucleon. The chiral peoperties of QCD lead to precise relations between quarks and antiquarks distributions and the importance of the Pauli exclusion principle is also emphasized. The model involves a small number of free parameters which are determined by a global next-to-leading order QCD analysis of unpolarized and polarized deep-inelastic scattering data. We discuss the implications of experimental data on the structure of polarized distributions.

INTRODUCTION

An experimental study of deep-inelastic scattering (DIS) of leptons on hadrons during the last twenty years has produced measurements of cross sections and structure functions over a large kinematic domain, and as a consequence a better knowledge of the partons distributions functions (PDF) in the unpolarized and polarized cases. Although, we have no theory to compute the PDF, a large variety of phenomelogical functions has been proposed in the literature [1], and their success in describing experimental data has provided a better explanation of the hadrons structure and a test of pQCD.

The above *classic* PDF are generally parametrized as a polynomial in the Bjorken variable x at an input scale Q_0, in order to introduce more physical aspects in the description of PDF, we propose a new statistical approach where the nucleon is viewed as a gas of massless partons (quarks, antiquarks, gluons) in equilibrium at a given temperature in a finite size volume. Moreover, in contrast with other parametrizations, we consider from the beginning that the fundamental entities are polarized parton distributions defined by their helicities. In the process of building our PDF, we take into account experimental informations well established by now. For instance, data exhibit a clear evidence for a flavor-asymmetric light sea, *i.e.*, $\bar{d} > \bar{u}$, which can be understood in terms of the Pauli exclusion principle, based on the fact that the proton contains two u quarks and only one d quark. The signs of the polarized light quarks distributions are essentially well established, $\Delta u > 0$ and $\Delta d < 0$, while we just begin to uncover a flavor asymmetry, for the polarized light sea, namely $\Delta\bar{u} \neq \Delta\bar{d}$.

---

[1] UMR 6207 is Unité Mixte de Recherche du CNRS and of Universités Aix-Marseille I and Aix-Marseille II and of Université du Sud Toulon-Var, laboratoire affilié à la FRUMAM.

CP747, *Structure of the Nucleon at Large Bjorken x,* edited by C. Bourrely, Z.-E. Meziani, and J. Soffer

CONSTRUCTION OF THE STATISTICAL PDF

Our motivation is to use the statistical approach to build up : $q_i, \Delta q_i, \bar{q}_i, \Delta\bar{q}_i, G$ and ΔG, by means of a very small number of free parameters. We propose a simple description of the parton distributions $p(x)$, at an input energy scale Q_0^2, proportional to $1/\left[\exp\left[(x-X_{0p})/\bar{x}\right] \pm 1\right]$, the *plus* sign for quarks and antiquarks, corresponds to a Fermi-Dirac distribution and the *minus* sign for gluons, corresponds to a Bose-Einstein distribution. Here, X_{0p} is a constant which plays the role of the *thermodynamical potential* of the parton p and $\bar{x}$ is the *universal temperature*, which is supposed to be the same for all partons. Since quarks carry a spin-1/2, it is natural to consider that the basic distributions are $q_i^{\pm}(x)$, corresponding to a quark of flavor i and helicity parallel or antiparallel to the nucleon helicity. Clearly one has : $q_i = q_i^+ + q_i^-$ and $\Delta q_i = q_i^+ - q_i^-$ and similarly for antiquarks and gluons.

From the chiral structure of QCD, we have two important properties which allow to relate quark and antiquark distributions and to restrict the gluon distribution :

- The potential of a quark q_i^h of helicity h is opposite to the potential of the corresponding antiquark $\bar{q}_i^{-h}$ of helicity *-h*,

$$X_{0q}^h = -X_{0\bar{q}}^{-h}. \tag{1}$$

- The potential of the gluon G is zero, $X_{0G} = 0$.

From well established features of the u and d quark distributions extracted from DIS data, we anticipate some simple relations between the potentials:

- $u(x)$ dominates over $d(x)$, therefore one can expect: $X_{0u}^+ + X_{0u}^- > X_{0d}^+ + X_{0d}^-$
- $\Delta u(x) > 0$, therefore $X_{0u}^+ > X_{0u}^-$
- $\Delta d(x) < 0$, therefore $X_{0d}^- > X_{0d}^+$.

So we expect X_{0u}^+ to be the largest thermodynamical potential and X_{0d}^+ the smallest one. In fact, we have the following ordering $X_{0u}^+ > X_{0d}^- \sim X_{0u}^- > X_{0d}^+$, these inequalities are consistent with the previous determinations of the potentials [2]. By using Eq. (1), this ordering leads immediately to some important consequences for antiquarks, namely, $\bar{d}(x) > \bar{u}(x)$, the flavor symmetry breaking which also follows from the Pauli exclusion principle, as recalled above. This was already confirmed by the violation of the Gottfried sum rule [3, 4]. Notice that since $\bar{u}^+(x) \sim \bar{d}^+(x)$, we have $\Delta\bar{u}(x) - \Delta\bar{d}(x) \sim \bar{d}(x) - \bar{u}(x)$, so the flavor symmetry breaking is almost the same for unpolarized and polarized distributions.

Let us now complete the description of our parametrization. The small x region is characterized by a rapid rise of the distributions as $x \to 0$, so they should be dominated by a universal diffractive term, flavor and helicity independent, coming from the Pomeron universality. Therefore we must add a term of the form $\tilde{A}x^{\tilde{b}}/[\exp(x/\bar{x}) + 1]$, where $\tilde{b} < 0$ and $\tilde{A}$ is a normalization constant. In summary, for the light quarks q = *u, d* of helicity

$h = \pm$, at the input energy scale $Q_0^2 = 4\text{GeV}^2$, we take

$$xq^h(x, Q_0^2) = \frac{AX_{0q}^h x^b}{\exp[(x - X_{0q}^h)/\bar{x}] + 1} + \frac{\tilde{A}x^{\tilde{b}}}{\exp(x/\bar{x}) + 1}, \quad (2)$$

and similarly for the light antiquarks

$$x\bar{q}^h(x, Q_0^2) = \frac{\bar{A}(X_{0q}^{-h})^{-1} x^{2b}}{\exp[(x + X_{0q}^{-h})/\bar{x}] + 1} + \frac{\tilde{A}x^{\tilde{b}}}{\exp(x/\bar{x}) + 1}. \quad (3)$$

For the strange quarks and antiquarks, we take the particular choice

$$xs(x, Q_0^2) = x\bar{s}(x, Q_0^2) = \frac{1}{4}[x\bar{u}(x, Q_0^2) + x\bar{d}(x, Q_0^2)], \quad (4)$$

$$x\Delta s(x, Q_0^2) = x\Delta\bar{s}(x, Q_0^2) = \frac{1}{3}[x\Delta\bar{d}(x, Q_0^2) - x\Delta\bar{u}(x, Q_0^2)]. \quad (5)$$

This choice gives rise to a large negative $\Delta s(x, Q_0^2)$. The charm quarks c , both unpolarized and polarized, are set to zero at $Q_0^2 = 4\text{GeV}^2$. Finally concerning the gluon distribution, as indicated above, we use a Bose-Einstein expression given by

$$xG(x, Q_0^2) = \frac{A_G x^{\tilde{b}+1}}{\exp(x/\bar{x}) - 1}, \quad (6)$$

with a vanishing potential and the universal temperature $\bar{x}$. For the sake of completeness, we also need to specify the polarized gluon distribution where we take the particular choice, $x\Delta G(x, Q_0^2) = 0$.

To summarize our parametrization involves a total of *eight* free parameters

$$\bar{x}\ , X_{0u}^{+}\ , X_{0u}^{-}\ , X_{0d}^{-}\ , X_{0d}^{+}\ , b\ , \tilde{b} \text{ and } \tilde{A}\ .$$

These parameters, are determined at NLO by a fitting procedure described in Ref. [5, 6].

We have defined the PDF in terms of Fermi functions at the input scale Q_0^2, does this property still persist at any given Q^2. We have proven in Ref. [7] that the statistical PDF can be approximated by a linear combination of Fermi functions which depend explicitely on x and Q^2. In the next section we will concentrate on the large x results obtained at NLO with the statistical PDF.

EXPERIMENTAL TESTS FOR (UN)-POLARIZED DIS

We first consider μp and ep DIS for which several experiments have yielded a large number of data points on the structure function $F_2^p(x, Q^2)$. In Fig. 1 (left) we compare our calculations with measurements made by NMC, BCDMS, ZEUS and H1 collaborations, for fixed Q^2 and large x values. We have also a quite good description of the neutron

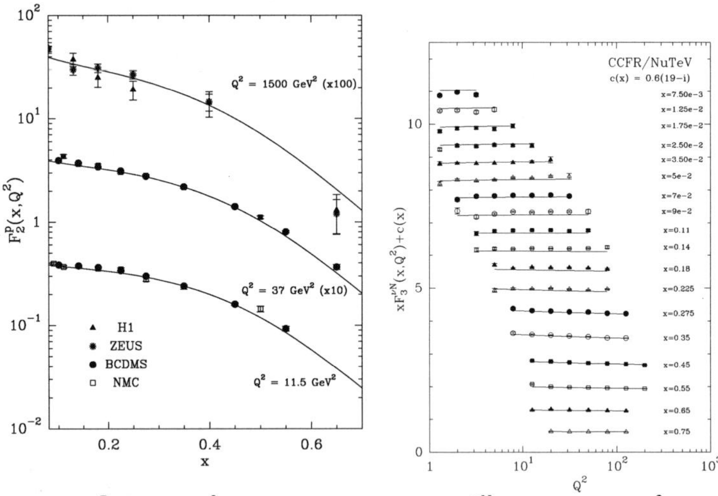

FIGURE 1. Left, F_2^P for fixed Q^2 as a function of x. Right, $xF_3^{\nu N}$ as a function of Q^2 for fixed x bin .

structure function $F_2^n(x,Q^2)$ data [5]. The high statistic νN DIS data from CCFR/NuTeV allows to extract the $xF_3^{\nu N}(x,Q^2)$ structure function, which is successfully compared to our results on Fig. 1 (right), we have also a very good description of the ZEUS and H1 data in the range $1500 \leq Q^2 \leq 310^5$ [8, 9].

Since our approach is based on the direct construction of quarks and antiquarks PDF of a given helicity, we immediately obtained Δq_i and $\Delta\bar{q}_i$ for each flavor. We display in Fig. 2 (left) the polarized structure functions $g_1^{p,n,d}$ at large x, from different experiments on proton, deuterium and helium targets, evolved at a fixed value $Q^2 = 5\text{GeV}^2$. The x dependence is in fair agreement with our results. We notice, in particular, for g_1^n, the perfect agreement with the 3 very precise data points recently measured at Jefferson Lab (Jlab-E99).

In Fig. 2 (right) the asymmetries $A_1^{p,n}$ are displayed with the world data at $Q^2 = 4\text{GeV}^2$ and restricted to $x > 0.1$. An interesting feature is the behavior of the asymmetries when $x \to 1$, in our model we find that $A_1^{p,n} < 1$ due to the fact that $\Delta u/u = 0.77$ and $\Delta d/d = -0.46$. For the A_1^n asymmetry measured by JLab-E99 at $x = 0.33, 0.47, 0.6$, our model agrees with their data at the two extreme points, but at $x = 0.47$ its value is above by 1σ. We have also computed the $g_2^{p,n}$ asymmetries using the Wandzura-Wilczek relation, a comparison with data from SLAC and Jefferson Lab. show a reasonable agreement owing to the experimental errors.

Recently, HERMES collaboration has obtained a set of polarized PDF at $Q^2 = 2.5\text{GeV}^2$, (see L. De Nardo, these proceedings) a comparison with our results is displayed in Fig. 3 (left), the agreement is satisfactory within the experimental errors.

Let us comme back to the important question of the flavor asymmetry of the light antiquarks. Our determination of $\bar{u}(x,Q^2)$ and $\bar{d}(x,Q^2)$ is perfectly consistent with the violation of the Gottfired sum rule, for which we found $I_G = 0.2493$ for $Q^2 = 4\text{GeV}^2$. Nervertheless there remains an open problem with the x distribution of the ratio $\bar{d}/\bar{u}$

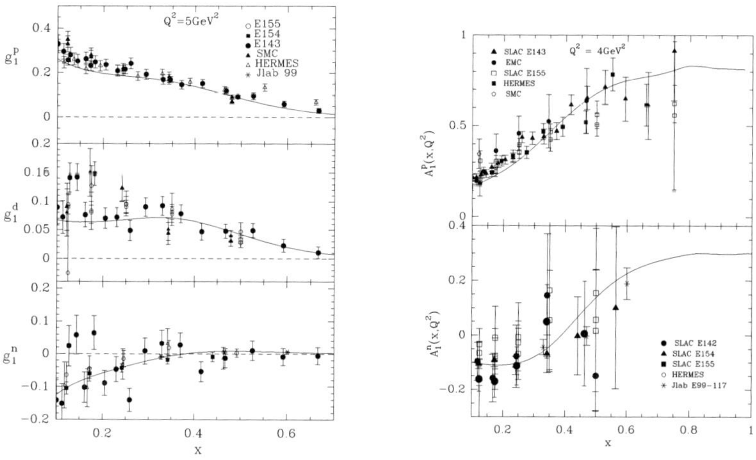

FIGURE 2. $g_1^{p,n,d}$ and $A_1^{p,n}$ for fixed Q^2 as a function of x.

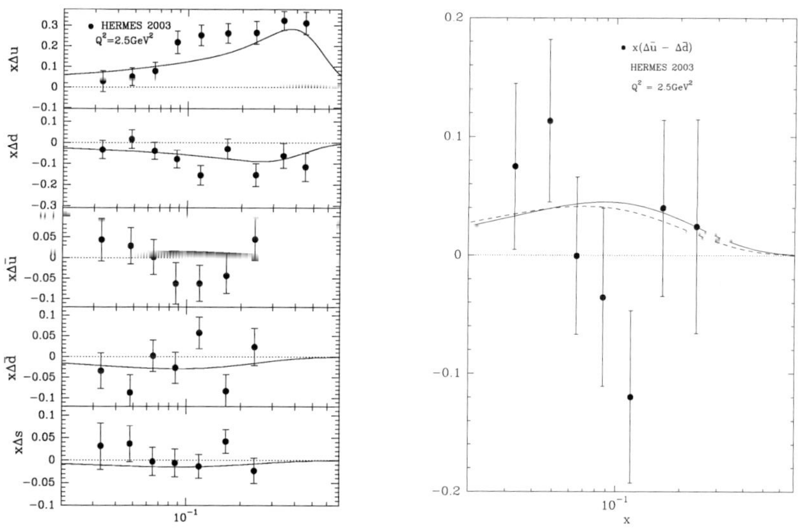

FIGURE 3. Left, comparison between polarized PDF and Hermes data, Right, $x(\Delta\bar{u} - \Delta\bar{d})$ compared with Hermes data at $Q^2 = 2.5\text{GeV}^2$ solid line and $Q^2 = 54\text{GeV}^2$ dashed line. .

for $x > 0.2$, in connection with the E866/NuSea collaboration. They have found at $Q^2 = 54\text{GeV}^2$ that $\bar{d}/\bar{u}$ is decreasing above $x > 0.2$ while in our model this ratio is, according to the Pauli principle, above 1 for any value of x. One possibility to clarify the situation is to measure the ratio of the unpolarized cross sections for the production of $W^{\pm}$ in pp collisions, which will directly probe the behavior of $\bar{d}/\bar{u}$ ratio. As an illustration, we have computed this ratio using MRST2002 PDF, where $\bar{d}/\bar{u} < 1$ for $x > 0.2$, and compare with our result, a significant gap at $y = 0$ exists. An other test is to consider the polarized difference $x[\Delta\bar{u}(x) - \Delta\bar{d}(x)]$, taking into account the properties of the polarized PDF in our model, we can show that if $\bar{d}/\bar{u} > 1$ for $x > 0.2$, the previous difference is positive, in the opposite case it becomes negative. The HERMES collaboration has released some values of this difference shown in Fig. 3 (right), however, no definite conclusion can be drawn due the large errors, nevertheless, above $x > 0.1$, it seems that the trend of data favor positive values. In the figure, the solid curve corresponds to $Q^2 = 2.5\text{GeV}^2$, the dashed curve to $Q^2 = 54\text{GeV}^2$, showing a small variation. An extended measurement is scheduled by the E906 collaboration (see A. Towell, these proceedings).

Let us mention an other way to obtain a good flavor separation by measuring the parity-violating asymmetry in $W^{\pm}$ production at RHIC-BNL. Calculations in the statistical model at $\sqrt{s} = 500\text{GeV}$ near a rapidity $y = +1$ show that $A_L^{PV}(W^+) \sim \Delta u/u$ and that $A_L^{PV}(W^-) \sim \Delta d/d$ (evaluated at $x = 0.435$). Similarly, near $y = -1$, $A_L^{PV}(W^+) \sim -\Delta\bar{d}/\bar{d}$ and $A_L^{PV}(W^-) \sim -\Delta\bar{u}/\bar{u}$ (see Ref. [5] for more details).

We have presented a few samples of the predictions obtained with the statistical model, a full set of results is available in Ref. [9].

REFERENCES

1. A detailed description of PDF models : CETQ, GRV, MRS and others, can be found in `http://durpdg.dur.ac.uk/hepdata/pdf.html`.
2. C. Bourrely, F. Buccella, G. Miele, G. Migliore, J. Soffer and V. Tibullo, *Z. Phys.* C **62**, 431 (1994); C. Bourrely and J. Soffer, *Phys. Rev.* D **51**, 2108 (1995); C. Bourrely and J. Soffer, *Nucl. Phys.* B **445**, 341 (1995).
3. K. Gottfried, *Phys. Rev. Lett.* **18**, 1154 (1967).
4. New Muon Collaboration, M. Arneodo *et al.*, *Phys. Rev.* D **50**, R1 (1994) and references therein.
5. C. Bourrely, F. Buccella and J. Soffer, *Eur. Phys. J.* C **23**, 487 (2002).
6. C. Bourrely, F. Buccella and J. Soffer, *Mod. Phys. Lett.* **18**, 771 (2003)
7. C. Bourrely, *Numerical approximation of statistical parton densities functions.* `http://www.cpt.univ-mrs.fr/~bourrely/research/bbs-dir/pdflib/partondf.pdf` (internal report).
8. C. Bourrely, F. Buccella and J. Soffer, in preparation
9. C. Bourrely, F. Buccella and J. Soffer, *A statistical approach for unpolarized and polarized parton distributions and fragmentations : a comparison with experiments.* `http://www.cpt.univ-mrs.fr/~bourrely/research/bbs-dir/pdflib/gallery.pdf` (internal report).

Status of the RHIC Spin Program

L.C. Bland

Brookhaven National Laboratory, Upton, NY USA

Abstract. The Relativistic Heavy Ion Collider (RHIC) at Brookhaven National Laboratory has been developing the capability of accelerating, storing and colliding high-energy polarized proton beams over the past several years. During this development phase, important first measurements of cross sections and spin asymmetries for neutral pions produced in polarized proton collisions at $\sqrt{s}$=200 GeV have been completed by STAR and PHENIX, the two large collider experiments at RHIC. This contribution reports on progress of the RHIC spin program and provides an outlook for the future.

INTRODUCTION

To date, our knowledge of how the spin of the proton is distributed amongst its quark and gluon constituents comes from studies of deep-inelastic scattering (DIS) of polarized electrons and muons from polarized proton, deuteron and ^{3}He targets. Global analyses of this data (*e.g.*, Ref. [1] and references to the data therein) have provided a map of the Bjorken-x dependence of the quark (and anti-quark) polarization. These studies have left us with many puzzles, the most significant of which is how the proton gets its spin from its constituents. The spin physics program at the Relativistic Heavy Ion Collider (RHIC) at Brookhaven National Laboratory [2] is an ambitious effort to address the puzzles by measuring spin asymmetries of particles produced in polarized proton collisions at center of mass energy in the range $200 \leq \sqrt{s} \leq 500$ GeV. The object is to utilize polarized quarks as a probe of the polarization of the constituents of the proton in scattering processes that are described by perturbative QCD. To achieve this objective, it is necessary to establish that pQCD quantitatively describes experimental cross sections and particle correlations at these collision energies.

One of the long term objectives of the RHIC spin program is to determine the Bjorken x dependence of the helicity asymmetry distribution of gluons within a longitudinally polarized proton, $\Delta G(x)$. Completion of double helicity asymmetry (A_{LL}) measurements for particles produced in different pseudorapidity (η) ranges and as a function of transverse momentum (p_T) can result in a detailed map of gluon polarization over a broad range of Bjorken x. Inclusive meson and jet production probes $\Delta G(x)$ through multiple partonic subprocesses with contributions from a broad range of x_{gluon} as specified by the standard convolution integrals for hard scattering processes. Di-hadron and di-jet production can provide additional constraints on the kinematics, thereby changing the admixture of partonic subprocesses and providing tighter constraints on the Bjorken x values of the partons that participate in the hard scattering. Prompt photon production is expected to be dominated by the QCD Compton process ($qg \rightarrow q\gamma$), thereby providing a more selective probe of $\Delta G(x)$. Further information can be provided by the measure-

CP747, *Structure of the Nucleon at Large Bjorken x*, edited by C. Bourrely, Z.-E. Meziani, and J. Soffer

ment of A_{LL} for particle correlations, such as γ+jet coincidences [3]. At $\sqrt{s}$=500 GeV, measurements of the parity violating single spin asymmetry A_L for $W^{\pm}$ production will result in the flavor separation of quark polarization at large Bjorken x and anti-quark polarization at moderately small x, particularly when the daughter leptons from $W^{\pm}$ decay are detected at large rapidity. The unknown transversity structure function can be probed via measurement of the transverse spin correlation coefficient (A_{TT}) for inclusive jet production.

HOW RHIC SPIN WORKS

High-energy electrons develop transverse polarization in a storage ring because of a small spin flip probability in the sychrotron radiation they emit. Muons are produced through the decays of pion beams, and so are naturally polarized because the weak interaction violates parity. To provide significant polarization for intense proton beams at high energy, it is necessary to use atomic methods at an ion source and then preserve the polarization through the acceleration sequence. The development of Siberian snakes [4] has enabled the preservation of polarization while accelerating protons to high energies through numerous depolarizing resonances in synchrotrons. To date, the highest energy polarized proton beam produced in a synchrotron is at RHIC and is 100 GeV. Previously, the highest energy polarized proton collisions were studied using 200 GeV proton beams in fixed target experiments. The beams were produced by the parity-violating decays of hyperons [5] and the total center-of-mass energy for those experiments was $\sqrt{s}$=20 GeV. At RHIC, the polarized proton beams collide, enabling spin physics experiments at collision energies $200 \leq \sqrt{s} \leq 500$ GeV. There are plans to accelerate polarized proton beams with energies greater than 200 GeV in the upcoming RHIC run as part of the development of the capability for studying spin asymmetries for $W^{\pm}$ production. Exquisite care must be taken throughout the entire acceleration chain to ensure that the polarization is preserved. Significant progress has been made in the development of these capabilities in the first years of RHIC operations.

Also differing from polarized DIS experiments is the need to develop methods to measure the polarization of high energy polarized protons. Precision measurements of spin observables requires accurate determination of the beam polarization. For polarized lepton beams used in DIS, quantum electrodynamics can be used to accurately predict cross sections and spin observables for scattering processes. Spin dependent scattering processes involving polarized protons must be calibrated by experiment since robust theoretical calculations of spin observables at small momentum transfer, where interaction rates are large, cannot be reliably made from first principles.

The RHIC polarimeters are based on measuring proton elastic scattering from carbon at very small four momentum transfer, in the range $0.006 \leq -t \leq 0.03$ (GeV/c)^2 [6]. An analyzing power (A_N) is expected because of the interference of the Coulomb spin-flip amplitude and the spin-indepedent strong interaction amplitude; hence, the polarimetry process is referred to as Coulomb-Nuclear Interference (CNI). The observed $-t$ dependence of the measured spin-dependent asymmetry, $P_{beam} \times A_N$, suggests substantial contributions from the hadronic spin flip amplitude, meaning that independent measurements are necessary to determine the beam polarization (P_{beam}). Precision measurements were made for the first time at 100 GeV in RHIC run 4 (April through May, 2004) by exploiting identical particle symmetries in polarized proton elastic scattering from a polarized gas jet target [7, 8]. The contribution of the spin dependendent strong interaction amplitude to small $|t|$ elastic scattering is of interest in its own right, and can be extracted from spin observables measured at RHIC in either fixed target experiments with the polarized gas jet target or with carbon ribbon targets, or by studying elastic scattering of the polarized colliding beams [9], which is the object of the pp2pp experiment at RHIC.

The first polarized proton collisions at $\sqrt{s}$=200 GeV were achieved during RHIC run 2 (January through May, 2002). Typical collision luminosities were $\sim 0.5 \times 10^{30}$ $cm^{-2}s^{-1}$ and the average polarization was ~16%, inferred from assumptions about the beam energy dependence of the CNI analyzing power [10]. Significant polarization loss in the RHIC injector resulted because the rate of acceleration in the AGS was two times smaller than expected because of equipment failure. An integrated luminosity of 0.3 pb^{-1} was observed at the STAR interaction point. Only vertical beam polarization was available.

Polarized proton collisions at $\sqrt{s}$=200 GeV were also achieved during RHIC run 3 (March through May, 2003). Typical collision luminosities observed with all subsystems operational at STAR were 3×10^{30} $cm^{-2}s^{-1}$ and the beam polarization averaged ~25%, based again on assumptions about the beam energy dependence of the effective analyzing power of the polarimeter. In this run, helical dipole magnets located either side of PHENIX and STAR were used to produce longitudinal polarization at the interaction points. Local polarimeters were developed by both experiments to establish the absence of transverse polarization components for the two beams at the interaction point concurrent with measurement of spin-dependent asymmetries by the CNI polarimeters that are located in parts of the ring where the beam polarization is vertical. The combined results from the local polarimeters and the CNI polarimeters established that the colliding beams at STAR and PHENIX had longitudinal polarization. Integrated luminosities of 0.5 pb^{-1} with vertical polarization and 0.4 pb^{-1} with longitudinal polarization were observed at the STAR interaction point.

Time was devoted during run 4 to accelerator development interleaved with the polarization calibration experiment. A new betatron tune for RHIC was developed to reduce beam loss induced by the influence of one beam on the other at the interaction points. With unpolarized beam, collision luminosities in excess of $10^{31}cm^{-2}s^{-1}$ were achieved during tests. It was also established that beam polarization could be preserved with the new tune of the accelerator. Commissioning of a new warm-bore partial helical snake in the AGS was successful. At the end of the run there was a successful demonstration of repeated stores in RHIC with an average collision luminosity of 4×10^{30} $cm^{-2}s^{-1}$ with average polarization in both beams of ~45%.

For the immediate future, a long polarized proton run is planned for the RHIC run starting in November, 2004. Performance comparable to that demonstrated during run 4 is expected to result in proton collisions with an average polarization of 45% and an average integrated luminosity of 1 pb^{-1} per week at a collision energy of $\sqrt{s}$=200 GeV. As discussed below, this should result in good measurements of A_{LL} for inclusive π^0 and jet production, thereby providing sensitivity to possible gluon polarization. A new superconducting partial helical snake magnet for the AGS is planned to be commissioned during this run. It is anticipated that this device will ultimately permit full transmission of the polarization from the ion source into RHIC at the design intensity of 2×10^{11} protons per bunch. Plans have been formulated to resolve the vacuum breakdown issue within RHIC which presently limits luminosity. When these developments are complete, it is expected that design luminosity and polarization for $\vec{p}+\vec{p}$ collisions will be realized.

WHAT HAS BEEN LEARNED TO DATE?

The first polarized proton collisions at RHIC at $\sqrt{s}$=200 GeV resulted in measurements of the cross section for neutral pion production at both PHENIX and STAR (Fig. 1). The results were for quite different π^0 rapidities, with PHENIX reporting the p_T dependence of the cross section near $\eta \approx 0$ and STAR reporting the Feynman x dependence of the cross section at $<\eta>$=3.3 and $<\eta>$=3.8. These data have been compared with state of the art perturbative QCD calculations [11] computed at next-to-leading order. The calculations use the CTEQ6M [12] parton distribution functions and the KKP [13] and Kretzer [14] fragmentation functions.

The theoretical calculations are found to quantitatively represent the measured cross sections at $\sqrt{s}$=200 GeV over a broad range of transverse momentum that extends down to $p_T \approx 1$ GeV/c. Agreement between calculation and data for transverse momenta between $1 \leq p_T \leq 10$ GeV/c had also been observed for inclusive particle production at the Tevatron at much higher $\sqrt{s}$ [13]. The results from RHIC form part of a picture of the pseudorapidity and collision energy dependence of π^0 production. Comparison of NLO pQCD calculations to data obtained at lower $\sqrt{s}$ [17] shows a systematic tendency for more quantitative agreement as $\sqrt{s}$ increases. It is also the case that the agreement extends over an increasingly broader range of η as $\sqrt{s}$ increases. As shown in Fig. 1, at RHIC collision energies, agreement between first principles calculations and measured cross sections extends over a broad rapidity interval. This agreement suggests that rapidity dependence of particle production can be exploited to significantly alter the admixture of gg, qg and qq' subprocess contributions and to emphasize partonic contributions from different regions of Bjorken x.

Understanding π^0 production with the electromagnetic calorimeters in PHENIX and from STAR is also an important prerequisite for using these devices for prompt photon detection. We can anticipate results for cross sections [18] and spin observables for prompt photons in the near future.

In addition to inclusive cross sections, the first RHIC runs have produced results for particle correlations. At mid-rapidity, measurements of azimuthal correlations of pairs of charged hadrons produced at midrapidity have been reported for p+p, d+Au and Au+Au collisions at $\sqrt{s_{NN}}$=200 GeV [19]. For p+p collisions the correlations have the pattern expected for a partonic scattering origin to high-p_T particle production; namely, a relative narrow peak when the hadrons have a small azimuthal angle difference ('near-side') ($\Delta\phi$) and a broader, but well developed, peak when they are separated by $\Delta\phi = 180°$ ('away-side'). These studies have been extended to hadron pairs separated by large

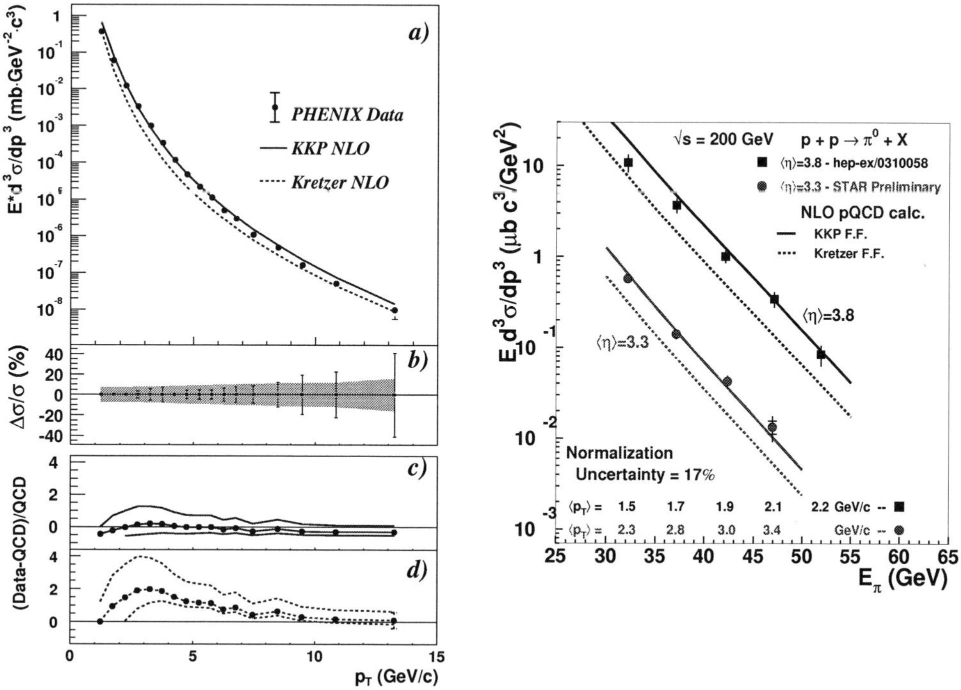

FIGURE 1. Results for *(left)* the invariant cross section for neutral pions produced at midrapidity by the PHENIX collaboration [15] and *(right)* the invariant cross section for neutral pions produced at large rapidity by the STAR collaboration [16] in p+p collisions at $\sqrt{s}$=200 GeV in comparison with next-to-leading order perturbative QCD calculations.

rapidity intervals [20]. Strong azimuthal correlations are observed for $\pi^0 - h^{\pm}$ hadron pairs separated by four units of rapidity. To date, comparisons of azimuthal correlations with theoretical predictions have been limited. One analysis [21] has established the need for multiple-soft gluon emissions to understand the width and shape of the away-side di-hadron $\Delta\phi$ distribution. It is fully anticipated that di-hadron correlations will provide a wealth of information when confronted against NLO pQCD calculations. Furthermore, future measurements of transverse and longitudinal spin observables for di-hadron, di-jet and γ+jet events will lead to many interesting results.

The first measurements of spin observables in π^0 production have been made as RHIC develops the polarization and intensity of the colliding proton beams. For π^0 produced at large rapidity, a large analyzing power has been observed in transversely polarized proton collisions at $\sqrt{s}$=200 GeV [16] (left side of Fig. 2). The pQCD picture that describes the unpolarized cross section in these kinematics identifies the dominant partonic process as large Bjorken-x quarks (from the polarized proton) interacting with soft gluons from the other proton. At large x, global analyses have established the quarks are highly polarized within a longitudinally polarized proton. If transversity behaves in a similar fashion, then the large A_N could result from highly polarized quarks that undergo

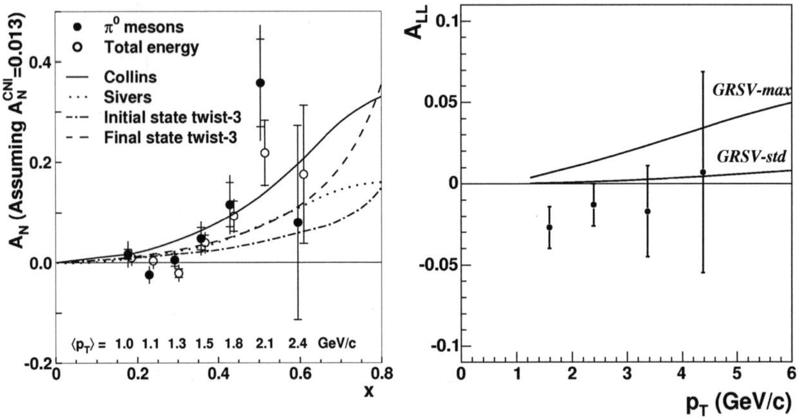

FIGURE 2. The first spin asymmetries from RHIC from $\vec{p}+\vec{p}$ collisions at $\sqrt{s}$=200 GeV. *(Left)* The Feynman x dependence of the analyzing power for large rapidity π^0 production in comparison to QCD model expectations. *(Right)* The p_T dependence of the longitudinal spin asymmetry, A_{LL} for π^0 production at midrapidity.

a spin- and transverse-momentum (k_T) dependent fragmentation into π^0 (Collins effect) [22, 23]. It could also be that the quarks involved in the particle production are from a spin- and k_T-dependent distribution function associated with the polarized proton (Sivers effect) [24, 25]. There are also twist-3 mechanisms that can give rise to the observed A_N [26]. Further experiments are required to disentangle these mechanisms. Possibilities include the measurement of the spin dependence of azimuthally correlated hadron pairs (*e.g.*, π^0-$h^{\pm}$). Extension of these measurements to negative x_F could isolate contributions from the Sivers function associated with gluons, since gluons cannot contribute to transversity within the proton, thereby eliminating the Collins effect. It would also be useful to separately study the x_F and p_T dependence of A_N. We can also expect results from the BRAHMS experiment for the x_F dependence of A_N for charged pions [27].

The A_N results for $\vec{p}+\vec{p}$ collisions at $\sqrt{s}$=200 GeV have an x_F dependence quite similar to that observed at $\sqrt{s}$=20 GeV [28] and at lower energies [29]. One of the challenges is to obtain a robust theoretical understanding of the dynamics responsible for this phenomena at energies where NLO pQCD provides a good framework for understanding the cross section and at lower energies where the theory fails to explain data [17].

The right side of Fig. 2 shows the first measurment of A_{LL} for π^0 production at RHIC, recently reported by the PHENIX collaboration [30]. Data from midrapidity π^0 and jet production promise to provide insight into gluon polarization. First measurements of A_{LL} for midrapidity π^0 production in $\vec{p}+\vec{p}$ interactions at $\sqrt{s}$=20 GeV were made by the

E704 collaboration [31]. The first A_{LL} measurement at RHIC at $\sqrt{s}$=200 GeV is based on an integrated luminosity of only 0.22 pb^{-1} with average beam polarization of 27%. Even with these limited statistics the π^0 A_{LL} measurement has accuracy comparable to the uncertainty on gluon polarization deduced from inclusive DIS measurements, as reflected by the two curves in the figure. The GRSV-max curve is computed [32] assuming 100% gluon polarization at the input scale, Q_0^2=0.6 GeV^2. The GRSV-std curve represents the best global fit [33] to the inclusive polarized DIS data. Given the improved performance of polarized proton operations at RHIC, there is promise for at least an order of magnitude improvement in the statistical accuracy of A_{LL} in the upcoming RHIC run.

What does this mean for gluon polarization? Inclusive observables are computed as a convolution of distribution functions, hard scattering cross section and fragmentation functions. Consequently, $\Delta G(x)$ must be extracted from a global theoretical analysis, in a manner similar to that used for unpolarized parton distribution functions. The midrapidity data probes Bjorken x values that are on average equal to $x_T = 2p_T/\sqrt{s}$. Tests of the resulting $\Delta G(x)$ can be made by measurements of the rapidity dependence of A_{LL} for π^0 production; midrapidity inclusive jet production, that would eliminate the fragmentation function from the convolution integral; and di-hadron correlations, that can narrow the range of contributions from the distribution functions depending on the kinematics of the hadron pair. Measurements of A_{LL} for prompt photon production are expected in the future, when the luminosity and polarization of the RHIC beams are further developed.

SUMMARY AND OUTLOOK

RHIC is the world's first polarized proton collider. The first polarized proton collisions have resulted in cross section data that show fixed-order perturbative QCD is a robust framework for understanding inclusive particle production at $\sqrt{s}$=200 GeV. The first spin asymmetry measurements have begun for inclusive π^0 production. The analyzing power at large x_F is found to be large. Additional measurements are needed to disentangle contributions from different mechanisms. The first measurements of the double helicity asymmetry (A_{LL}) for midrapidity π^0 production have been reported. The expectations are that significantly more precise measurements can be completed in the upcoming RHIC run. The new data will address whether gluon polarization makes significant contributions to the spin of the proton. Future measurements of A_{LL} for direct photon production will become feasible as the luminosity and polarization at RHIC improves. Future runs at $\sqrt{s}$=500 GeV will measure parity violating spin asymmetries for $W^{\pm}$ production.

ACKNOWLEDGMENTS

The progress in the RHIC spin program has been made by a very large group of people. This report is a summary of their work. Groups involved are the Collider-Accelerator Department at BNL, with primary responsibility for developing the capability to accelerate, store and collide polarized proton beams and the experiment collaborations; STAR and PHENIX, with primary responsibility for building, operating and analyzing the results from the collider detectors; and the polarimetry group that have developed techniques for measuring the beam polarization. The RHIC spin project has received important support from the the Department of Energy, the RIKEN institute and the National Science Foundation.

I would like to thank G. Bunce, A. Ogawa and G. Rakness for their careful reading of this manuscript.

REFERENCES

1. J. Blümlein and H. Böttcher, Nucl. Phys. **B636**,225 (2002).
2. Gerry Bunce, Naohito Saito, Jacques Soffer and Werner Vogelsang, Ann. Rev. Nucl. Part. Sci. **50**, 525 (2000).
3. L. C. Bland, in *Physics with a High Luminosity Polarized Electron Ion Collider*, eds. L. C. Bland, T. Londergan and A. Szczepaniak (World Scientific, Singapore, 2000). Also available at hep-ex/9907058.
4. Ya.S Derbenev and AM. Kondratenko, Sov. Phys. JETP **37**, 968 (1973); Ya.S Derbenev *et al.*, Part. Acc. **8**, 115 (1978).
5. D. P. Grosnick *et al.*, Nucl. Instr. Meth. **A290** (1990) 269.
6. J. Tojo *et al.*, Phys. Rev. Lett. **89** (2002) 052302; O. Jinnouchi *et al.*, AIP Conf. Proc. bf 675 (2003) 817.
7. A. Bravar *et al*, 16th Int. Spin Physics Symposium (SPIN 2004), to be published in the proceedings.
8. H. Okada *et al*, 16th Int. Spin Physics Symposium (SPIN 2004), to be published in the proceedings.
9. I. Alekseev *et all*, 16th Int. Spin Physics Symposium (SPIN 2004), to be published in the proceedings.
10. T. L. Trueman, hep-ph/0203013.
11. F. Aversa *et al.*, Nucl. Phys. **B327** (1989) 105; B. Jager *et al.*, Phys. Rev. D **67** (2003) 054005; D. de Florian, *ibid.* **67** (2003) 054004.
12. J. Pumplin *et al.*, J. High Energy Phys. **07** (2002) 012.
13. B. A. Kniehl *et al.*, Nucl. Phys. **B597** (2001) 337.
14. S. Kretzer, Phys. Rev. D **62** (2000) 054001.
15. S. S. Adler, et al. (PHENIX collaboration), Phys. Rev. Lett. **91** (2003) 241803.
16. J. Adams, et al. (STAR collaboration), Phys. Rev. Lett. **92** (2004) 171801.
17. C. Bourrely and J. Soffer, Eur. Phys. J. C **36** (2004) 371 and hep-ph/0311110.
18. K. Okada *et al.* (PHENIX collaboration), 16th Int. Spin Physics Symposium (SPIN 2004), to be published in the proceedings.
19. C. Adler *et al.* (STAR collaboration), Phys. Rev. Lett. **91** (2003) 172302; J. Adams *et al.* (STAR collaboration), Phys. Rev. Lett. 91 (2003) 072304.
20. A. Ogawa, contribution to the 12th International Workshop on Deep Inelastic Scattering (to be published), nucl-ex/0408004.
21. Daniel Boer and Werner Vogelsang, Phys.Rev. D69 (2004) 094025.
22. J. Collins, Nucl. Phys. **B396** (1993) 161.
23. M. Anselmino, M. Boglione, and F. Murgia, Phys. Rev. D **60**, 054027 (1999).
24. D. Sivers, Phys. Rev. D **41** (1990) 83; **43** (1991) 261.
25. M. Anselmino, M. Boglione, and F. Murgia, Phys. Lett. B **362**, 164 (1995); M. Anselmino and F. Murgia, *ibid.* **442**, 470 (1998).
26. A. Efremov and O. Teryaev, Phys. Lett. **150B** (1985) 383; J. Qiu and G. Sterman, Phys. Rev. D **59** (1998) 014004; Y. Koike, AIP Conf. Proc. **675** (2003) 449.
27. F. Videbaek *et al.* (BRAHMS collaboration), 2004 Fall Meeting of the Division of Nuclear Physics.
28. B.E. Bonner *et al.*, Phys. Rev. Lett. **61** (1988) 1918; D.L. Adams *et al.*, Phys. Lett. B **261** (1991) 201; **264** (1991) 462; Z. Phys. C **56** (1992) 181; A. Bravar *et al*, Phys. Rev. Lett. **77** (1996) 2626.

29. R. D. Klem *et al.*, Phys. Rev. Lett. **36** (1976) 929; W. H. Dragoset *et al.*, Phys. Rev. D **18** (1978) 3939; S. Saroff *et al.*, Phys. Rev. Lett. **64** (1990) 995; B. E. Bonner *et al.*, Phys. Rev. D **41** (1990) 13; K. Krueger *et al.*, Phys. Lett. B **459** (1999) 412; C. E. Allgower *et al.*, Phys. Rev. D **65** (2002) 092008.
30. S. S. Adler, et al. (PHENIX collaboration), Phys. Rev. Lett. **93** (2004) 202002.
31. D. L. Adams *et al.*, Phys. Lett. **B261** (1991) 197.
32. B. Jäger *et al.*, Phys. Rev. D **67** (2003) 054005.
33. M. Glück *et al.*, Phys. Rev. D **63** (2001) 094005.

THE COMPASS EXPERIMENT AT CERN

A. MAGNON

DAPNIA/SPhN, CEA-Saclay, F91191 Gif-Sur-Yvette Cedex

On Behalf of the COMPASS Collaboration

Abstract. The COMPASS experiment at the CERN SPS has a broad physics program focused on the study of the spin structure of the nucleon and on hadron spectroscopy. Key measurements for the spin program are the gluon contribution to the spin of the nucleon, flavor dependent quark spin distribution, and the measurement of the transverse spin structure function. The apparatus consists of a two-stage spectrometer designed for high data rates and equipped with high-resolution tracking, particle identification, electromagnetic and hadronic calorimetry. Data taking has started in 2002. Following the CERN SPS shut down in 2005, the experiment will resume data taking in 2006 and is planned to continue (at least) until 2010. Few hundreds of Terabytes of data are put on tape each year. Out of this large amount of data first important physics results have been obtained.

PACS: <13.85.Hd - 14.40.Lb - 25.30.Mr

INTRODUCTION

The COMPASS Collaboration brings together about 250 physicists from 27 Institutes from Europe, Russia, Japan and India. The experiment [1] is located at the M2 beam line at the CERN SPS. The primary physics goals are twofold:

1- A rich spin program with a polarised muon beam and a polarised target. It is aimed to provide a direct measurement of the gluon polarisation $\Delta G/G$ by measuring the spin dependent asymmetry of the photon gluon fusion process, that can be accessed by detecting open-charm events and high-p_T correlated hadron pairs. The flavour separated spin structure functions of the nucleon in Deep Inelastic Scattering (DIS) are also measured in the longitudinal polarisation mode. With transverse target polarisation, the still unmeasured transversity spin structure function h_1 will be investigated. Not requiring polarisation, diffractive vector meson production can be studied in parallel. Finally the prospect to study the Generalized Parton Distribution functions by the means of Deep Virtual Compton Scattering and also Highly Exclusive Meson Production in the scattering of high energy muons is investigated for the future.

2- A wide program of hadron spectroscopy will be performed by using hadronic beams. COMPASS plans to study π and K polarisabilities (Primakoff reaction), allowing for a test of chiral perturbation theory predictions; extensive meson spectroscopy to investigate the presence of exotics states; collection of large samples of semi-leptonic decays of charmed mesons and baryons to determine form factors and test the predictions of heavy quark effective theory.

The first construction phase ended with a commissioning run in 2001. Data were taken during 2002, 2003 and 2004. Following the SPS shut down during 2005, data taking will resume in 2006 and will continue beyond 2006.

CP747, *Structure of the Nucleon at Large Bjorken x*, edited by C. Bourrely, Z.-E. Meziani, and J. Soffer

QUARK AND GLUON SPIN

The spin of the nucleon is the total angular momentum of its constituents. For a nucleon with a spin projection $+1/2$ along its flight direction, we have: $\frac{1}{2} = \frac{1}{2}\Delta\Sigma + \Delta G + L_z^q + L_z^g$ where $\frac{1}{2}\Delta\Sigma$ is the contribution from the spin of quarks, ΔG is the contribution from the spin of gluons, and $L_z^q + L_z^g$ is the orbital angular momentum of quarks and gluons. EMC, SMC and SLAC have obtained for the flavor singlet axial matrix element a_0 the following value: $a_0 = 0.27 \pm 0.13$ [2]. The naive approach which assumes $a_0 = \Delta\Sigma$ leads to the surprising result that the spin of the quarks accounts for only a small fraction of the nucleon spin. However, it has been shown that a_0 gets contribution from both $\Delta\Sigma$ and ΔG. Therefore, as shown in Fig. 1 the small a_0 is compatible with a sizeable $\Delta\Sigma$ if ΔG is large.

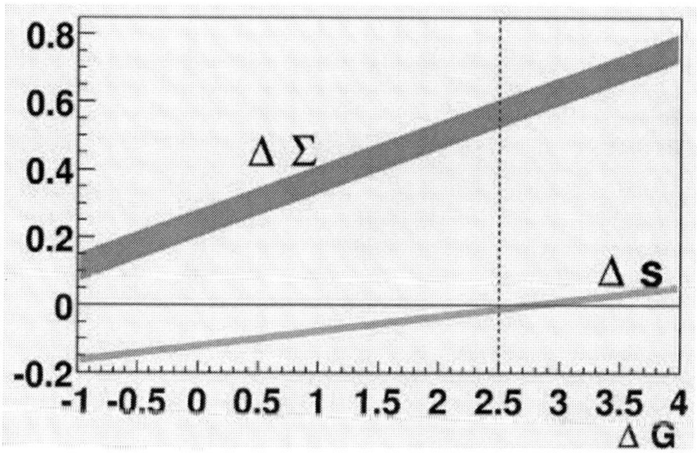

FIGURE 1. Allowed values for $\Delta\Sigma$ and Δs as a function of the hypothesis made for ΔG.

As ΔG is involved in the DGLAP evolution equations [3] of the spin structure function g_1, it can, in principle, be obtained by a QCD fit to the g_1 data. However, g_1 was up to now only measured in fixed target experiments where the small lever arm in Q^2 as well as the present data accuracy prevent to obtain significant constraints on ΔG [4]. It is therefore imperative to perform direct measurements of ΔG. Such measurements of $\frac{\Delta G}{G}$ have been started by four experiments: HERMES and COMPASS using polarised lepton-nucleon scattering on fixed target and STAR and PHENIX at RHIC using polarised p-p scattering.

THE COMPASS SPECTROMETER

The COMPASS spectrometer shown in Fig. 2 is an about 50 m long two-stage set-up [5]. Muons of 160 GeV/c momentum and 80 % average polarisation impinge on a double-cell solid-state polarised target filled with polarised ^6LiD [6]. $2 \cdot 10^8$ muons are delivered during the 4.8 s long spill with a repetition time of 16.8 s. Their momenta are measured individually and their trajectory are determined by scintillating fiber hodoscopes and silicon strips detectors upstream of the target. The first stage, the Large Angle Spectrometer is arranged around a first spectrometer magnet SM1. Particles emerging from the target are tracked by gazeous detectors: GEM and MicroMegas detectors for the innermost region close to the beam and by Drift Chambers and straw trackers for the outer region. The innermost region is again covered by scintillating fiber hodoscopes. The first spectrometer ends with particle identification detectors: the RICH, the hadronic calorimeter HCAL1 and the muon identification wall MW1.

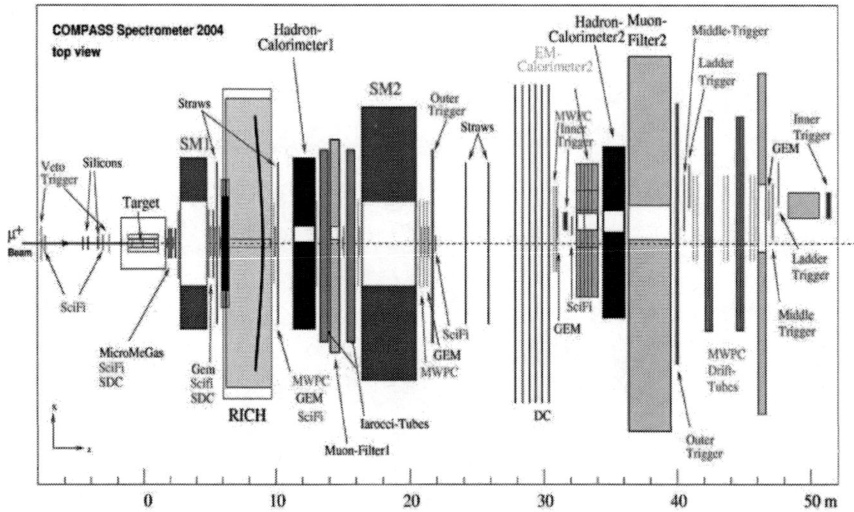

FIGURE 2. The COMPASS spectrometer.

The second stage, the Small Angle Spectrometer, is arranged around the SM2 spectrometer magnet and tracking is performed by multi-wire proportional chambers and GEM detectors. Again, a hadron calorimeter HCAL2 and a muon identification MW2 complete the setup. Trigger hodoscopes and a subsequent logics select the interesting phase space of scattered muon. Calorimeter information is added for a certain class of triggers.

FIRST PHYSICS RESULTS FROM 2002 AND 2003 DATA

Exclusive ρ^0 production

A large number of ρ^0 and also Phi mesons are recorded which allows us, by putting a narrow cut on the missing energy, to select highly exclusive events. In the case of ρ^0 mesons, analysis of the angular distributions of the decay products allows us to perform high precision measurements of spin density matrix elements. An illustration is shown in Fig. 3 a) where the values obtained for the matrix element r^{04}_{00} which describes the production of longitudinally polarised ρ^0 from longitudinally polarised virtual photons are compared to existing results. Assuming s-channel helicity conservation one can express the ratio of Longitudinal to Transverse cross-section for that process $R = \sigma_L/\sigma_T$ as a function of r^{04}_{00}. The results for R are shown in Fig. 3 b) compared with existing results [7] [8]. The COMPASS results extend the existing ones to lower Q^2, by about one order of magnitude. In the region of overlap they have an improved accuracy.

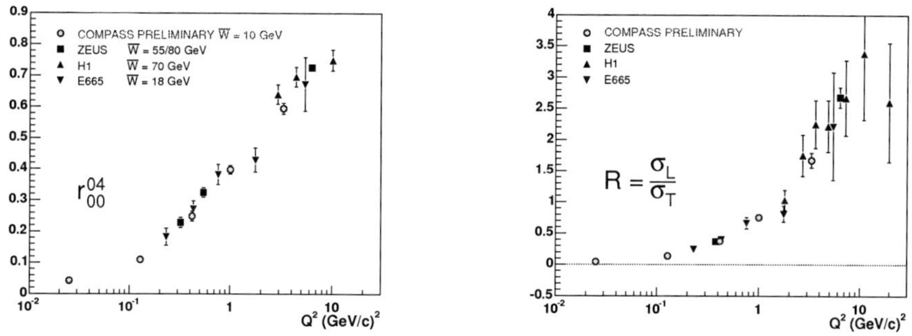

a) b)

FIGURE 3. a) the spin density matrix element r^{04}_{00} and, b) the ratio of Longitudinal to Transverse cross-sections for exclusive ρ^0 meson production.

Transverse spin Collins asymmetry

Semi-inclusive DIS provides the possibility to measure the transverse polarised parton distribution function $\Delta_T q(x)$. This function generates a modulation in the yield of produced hadrons which can be expressed as $(1 + A_{Coll} \cdot sin(\Phi_C))$ where Φ_C and A_{Coll} are the Collins angle and Collins asymmetry respectively [9]. A_{Coll} has been measured on our ^6LiD target polarised in the transverse mode. Fig. 4 shows the preliminary values of A_{Coll} obtained from our 2002 data when a positive or a negative leading hadron is selected, as a function of x_{Bj} and as a function of the fraction of energy z carried by the leading hadron [10]. Given the present accuracy, all measured values for A_{Coll} are compatible with zero.

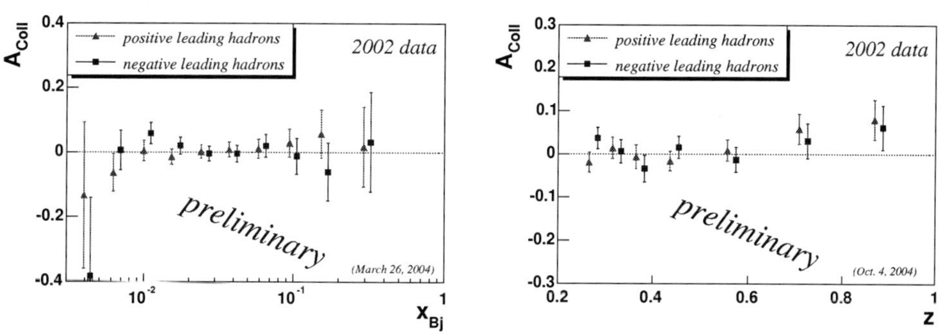

FIGURE 4. The A_{Coll} transverse spin asymmetry for h^+ and h^- leading hadrons.

Double spin longitudinal inclusive asymmetry A_1^d

By requiring only a high energy scattered muon, one can measure the double spin longitudinal inclusive asymmetry on the deuteron A_1^d from which one derives the first spin structure function g_1^d of deuteron. Fig. 5 shows the results, from the 2002 and 2003 data, obtained from the scattering of longitudinally polarised muons on a longitudinally polarised ^6LiD target [11]. The much shorter overall data taking time used by COMPASS, compared to SMC is more than compensated by the higher luminosity of COMPASS and the higher dilution factor of the polarised material. A significant improvement of the errors in the low x_{Bj} region has been achieved.

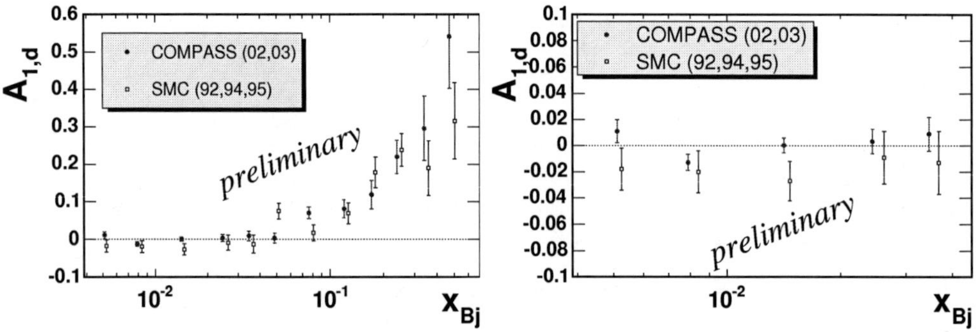

FIGURE 5. Double spin longitudinal inclusive asymmetry A_1^d from 2002 and 2003 data.

$\Delta G/G$ determination from open-charm

The gluon polarisation $\Delta G/G$ is studied using the Photon Gluon Fusion (PGF) process $g\gamma^* \rightarrow q\bar{q}$ shown in Fig.6 a).

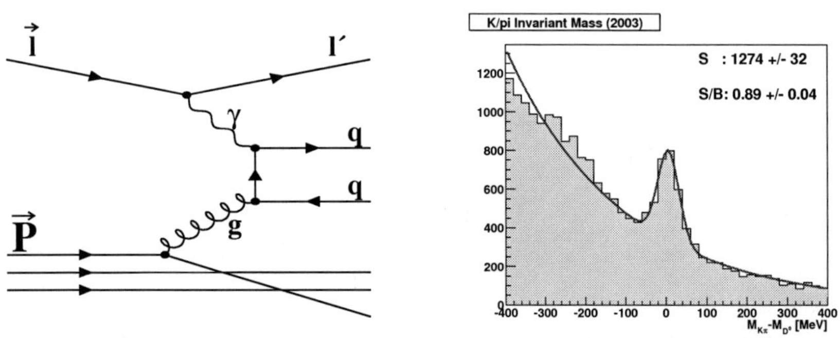

a) Photon-Gluon Fusion process b) D^0 signal in $K\pi$ invariant mass

FIGURE 6. a) the Photon Gluon Fusion process. b) observation of the PGF in the open-charm channel via $D^* \rightarrow D^0\pi \rightarrow K\pi\pi$ production (invariant mass of $K\pi$ pairs).

The production of charmed mesons D^0 provides a clean identification of the PGF process. However, D^0 are difficult to detect because in our long targets, the charm production vertex cannot be reconstructed with sufficient accuracy. D^0 detection is helped by requiring that they originate from a "parent" D^*: $D^* \to D^0\pi \to K\pi\pi$ called D^* tagging. Fig. 6 b) shows for the 2003 data [12], the invariant mass plot for D^0 using D^* tagging. The longitudinal spin asymmetry has not yet been evaluated from this limited sample of open charm events. However, from this sample, we are able to draw a projection (subsection below) for the uncertainty on $\Delta G/G$ expected from the full statistics on open charm events available at the end of 2004.

$\Delta G/G$ determination from high p_T hadrons pairs

The PGF process can also be tagged by the production of a pair of hadrons with high transverse momenta (high-p_T). This process is more abundant but plagued by a background from competing processes. The photon can interact with the nucleon directly or be resolved to a hadronic state (Vector Meson or $q\bar{q}$) where a parton from this hadronic state interacts perturbatively with a parton from the nucleon. Selecting events with $Q^2 > 1$ (GeV/c)^2 strongly reduces the contribution from resolved photons which can then be neglected. Fig. 7 shows the three main direct processes which contribute to the cross-section for production of a pair of hadrons.

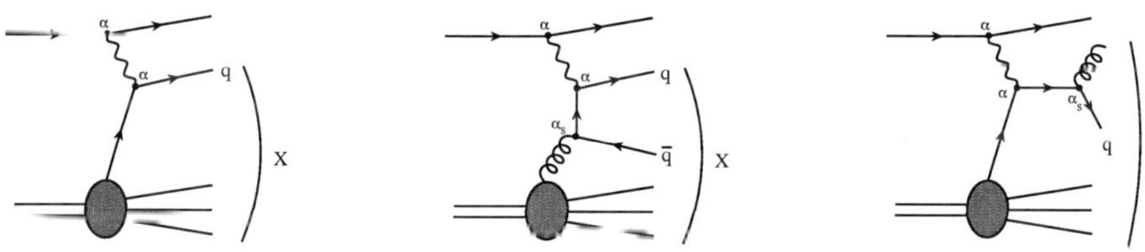

a) Leading order DIS   b) Photon-Gluon Fusion   c) QCD Compton

FIGURE 7. The 3 direct processes contributing at order α_s to the hadrons pair production

The gluon polarisation in the nucleon $\frac{\Delta G}{G}$ is extracted from the cross-section asymmetry of the PGF: $A_{LL}^{\gamma^* d(PGF)} = < \hat{a}_{LL}^{PGF} > \frac{\Delta G}{G}$ where the asymmetry $< \hat{a}_{LL}^{PGF} >$ of the partonic (photon-gluon) reaction can be calculated. Selecting events with a pair of high-p_T hadrons, one removes most of the Leading Order DIS background. However, the QCD Compton and a fraction of LO DIS processes remain and a Monte-Carlo simulation is necessary to estimate their contribution. The following cuts have been applied: $x_F > 1$ and $z > 0.1$ to select current fragmentation; $Q^2 > 1$ (GeV/c)^2, $x_{Bj} < 0.05$, $0.1 < y < 0.9$ on the lepton kinematics; $p_{t1(2)} > 0.7$ GeV/c, $p_{t1}^2 + p_{t2}^2 > 2.5$ (GeV/c)^2, $m(h_1 h_2) > 1.5$ GeV/c^2 on the leading (h_1) and the next-to-leading (h_2) hadrons. With this selection and combining the 2002 and 2003 data, the (virtual-photon) longitudinal double-spin asymmetry for high-p_T hadron pairs at $Q^2 > 1$ (GeV/c)^2 is [12]:

$$A^{\gamma^* d} = -0.015 \pm 0.08(stat.) \pm 0.013(syst.) \tag{1}$$

Using LEPTO Monte-Carlo a determination of the fraction of PGF events leads to $\sigma^{PGF}/\sigma^{tot} = 0.34 \pm 0.07(syst.)$ where the systematic error is estimated by varying the p_t^2 cut and using standard and modified sets of fragmentation functions. Using this estimate, one obtains for the gluon polarisation:

$$\frac{\Delta G}{G} = 0.06 \pm 0.31(stat.) \pm 0.06(syst.) \quad (2)$$

corresponding to an average value for gluon x of $< x_G >= 0.13$

Projections for $\sigma(\Delta G/G)$

Analysis of the additional data from the 2004 run combined with some recent improvements in the hardware and more importantly in the analysis software will enhance the overall figure of merit of the experiment. Concerning the high-p_T events, releasing the $Q^2 > 1$ (GeV/c)^2 cut increases the statistics by a factor of $\simeq 10$. The statistical error quoted below for this case was obtained by taking into account the dilution due to resolved photons, however it does not include their contribution to the asymmetry. The projections for the achievable error on $\Delta G/G$ from the 2002 to 2004 data are as follows:

$$\sigma(\Delta G/G) = 0.05 \;\; (high\ p_T\ all\ Q^2); \;\; \sigma(\Delta G/G) = 0.16 \;\; (high\ p_T\ Q^2 > 1) \quad (3)$$

$$\sigma(\Delta G/G) = 0.25 \;\; (open\ charm) \quad (4)$$

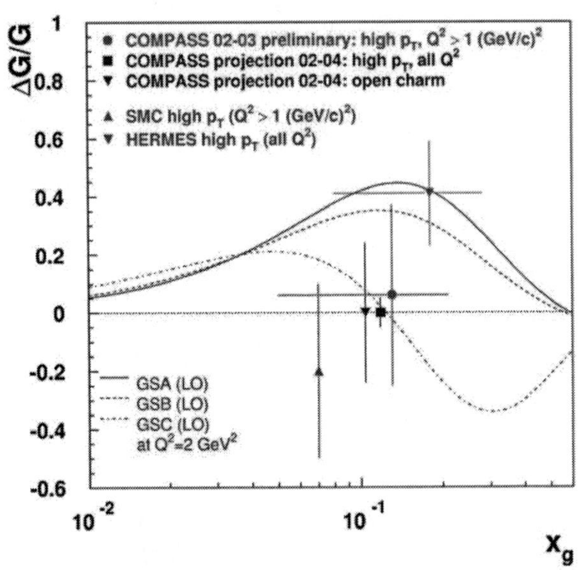

FIGURE 8. First COMPASS result on $\Delta G/G$, compared with existing results and projections

The COMPASS preliminary determination of $\Delta G/G$ from high-p_T hadrons pairs using $Q^2 > 1$ (GeV/c)^2 events is shown in Fig. 8 along with a result derived from a similar analysis of all the SMC data [13] and also a HERMES result [14] obtained without Q^2 cut. Note that the uncertainty quoted by HERMES does not include any systematic error from possible background subtraction. The above projections are also shown.

COMPASS BEYOND 2005

COMPASS is expected to resume data taking in 2006, following the shut down of CERN accelerators in 2005 required for LHC installation. An important meeting, organized by CERN SPSC will examine the future of fixed target experiments at CERN at the end of September [15]. Both the ongoing program with muon beam and the program with hadron beam will be reviewed. In addition, an expression of interest to study GPDs using high energy muon beyond 2010 will also be discussed.

REFERENCES

1. G. Baum *et al.*, COMPASS Proposal, (1996) *CERN-SPSLC-96-14, SPSC/P 297*
2. B. Lampe and E. Reya, Spin physics and polarized structure functions, *Phys. Rep.*, 332 1-163 2000.
3. G. Altarelli *Phys. Rep.* 81, 1 (1982) and references therein
4. J. Blümlein and H. Böttcher *J.Phys.G: Nucl.Part. Phys.* 29, 1959-1961 (2003)
5. G.K. Mallot *Nucl. Instr. Meth* A518, 121 (2004) and references therein
6. F. Gautheron contribution to Spin2004, October 10-16 Trieste, Italy, The COMPASS polarized target
7. D. Neyret contribution to Spin2004, October 10-16 Trieste, Italy, Results on exclusive ρ^0 production from COMPASS at CERN
8. E. Burtin contribution to Baryons04, October 25-29 Ecole Polytechnique, Palaiseau, France, Results on exclusive ρ^0 production from COMPASS at CERN
9. J. Collins *Nucl. Phys.* B396, 161 (1993)
10. P. Pagano contribution to Spin2004, October 10-16 Trieste, Italy, Measurements of Collins and Sivers asymmetries at COMPASS
11. D. Peshekhonov contribution to Spin2004, October 10-16 Trieste, Italy, Inclusive spin dependent asymmetry $A_1(x,Q^2)$
12. C. Schill contribution to Spin2004, October 10-16 Trieste, Italy, Measurement of the gluon polarisation $\Delta G/G$ at the COMPASS experiment.
13. B. Adeva *et al Phys. Rev.* D70, 012002 (2004)
14. A. Airapetian *et al., Phys. Rev. Lett.* 84, 2584 (2000)
15. see: http://committees.web.cern.ch/Committees/SPSC/Villars-info.html

POLARIZED STRUCTURE FUNCTIONS—II

Neutron Spin Structure Measurements in JLab Hall A

Jian-ping Chen, for the JLab Hall A and E94-010 Collaborations

Jefferson Lab, Newport News, Virginia 23606, USA

Abstract. Recent progress from Jefferson Lab has significantly improved our understanding of the nucleon spin structure in the high-x region. Results from two experiments in Hall A are presented. The first experiment is a precision measurement of the neutron spin asymmetry, A_1^n, in the high-x (valence quark) region. The results show for the first time that A_1^n becomes positive at large x, strongly breaking SU(6) (spin-flavor) symmetry. The data trend is in good agreement with SU(6)-breaking valence quark models and with the fits to the previous world data. Combining the A_1^n results with the world A_1^p data, the up and down quark spins distributions in the nucleon were extracted. The results showed that for the proton the valence down quark spin is in the opposite direction from that of the proton, in disagreement with predictions of leading-order perturbative QCD models, which neglect quark orbital angular momentum.

In the second experiment, the Q^2 dependence of the moments and sum rules of the spin structure in the low to intermediate Q^2 region were measured, providing a unique bridge linking the quark-gluon picture of the nucleon and the coherent hadronic picture.

INTRODUCTION

Since the 'spin crisis' [1], substantial efforts, both theoretical and experimental, have been devoted to understanding the nucleon's spin puzzle. A new generation of experiments were carried out in the 1990s at SLAC, CERN and DESY. These experiments concluded that the quarks carry about $20-30\%$ of the nucleon spin. The rest of the nucleon spin should come from the quark orbital angular momentum (OAM) and the gluon total angular momentum. There is almost no direct experimental information available on the quark OAM and the gluon total angular momentum. The Bjorken sum rule[2], a fundamental sum rule of the spin structure function based on QCD, was verified to an accuracy of better than 10%. Attempts have been made to extract the parton spin distributions from global analyses of the polarized deep-inelastic-scattering data. The uncertainties are much larger than those of the unpolarized parton distribution due to the fact that the polarized data coverage is much less extensive than that of the unpolarized data.

Recently, the high polarized luminosity available at Jefferson Lab (JLab) has allowed the study of the nucleon spin structure at an unprecedented precision, enabling us to access the hard-to-reach valence quark (high-x) region and also to accurately map the intermediate to low Q^2 region.

The high-x region is of special interest, because this is where the valence quark contributions are expected to dominate. With sea quarks and explicit gluon contributions expected not to be important, it is a clean region to test our understanding of nucleon

CP747, *Structure of the Nucleon at Large Bjorken x,* edited by C. Bourrely, Z.-E. Meziani, and J. Soffer

structure. Relativistic constituent quark models [4] should be applicable in this region and perturbative QCD [6] is also able to make predictions in the large x ($x \to 1$) limit. JLab experiment E99-117 measured, with high precision, the spin asymmetry A_1^n in the high-x region and extracted polarized quark distributions.

Related to the Bjorken sum rule, at $Q^2 = 0$ there is another sum rule for spin structure, the Gerasimov-Drell-Hearn (GDH) sum rule [7]. A generalized GDH sum rule [8] connects the GDH sum rule with the Bjorken sum rule and provides a clean way to test theories with experiments over the entire Q^2 range. JLab experiments measured the generalized GDH sum in the low to intermediate Q^2 region. These results and the results on the moments of the spin structure functions provide a bridge between the quark-gluon picture at high Q^2 to the coherent hadronic picture at low Q^2. In particular, at the low end of the Q^2 range, the results were compared with Chiral Perturbation Theory calculations and provided tests of the current understanding of the chiral dynamics of QCD.

INCLUSIVE POLARIZED ELECTRON-NUCLEON SCATTERING

For inclusive polarized electron scattering off a polarized nucleon target, the cross section depends on four structure functions, $F_1(Q^2,x)$, $F_2(Q^2,x)$, $g_1(Q^2,x)$ and $g_2(Q^2,x)$, where F_1 and F_2 are the unpolarized structure functions and g_1 and g_2 the polarized structure functions. In the quark-parton model, F_1 or F_2 gives the quark momentum distribution and g_1 gives the quark spin distribution. Another physics quantity of interest is the virtual photon-nucleon asymmetry A_1

$$A_1 = \frac{g_1 - (Q^2/\nu^2)g_2}{F_1} \approx \frac{g_1}{F_1}. \tag{1}$$

SPIN ASYMMETRIES IN THE HIGH-X REGION

To a first approximation, the constituent quarks in the nucleon are described by the SU(6) wavefunctions. The SU(6) symmetry leads to the following predictions:

$$A_1^p = 5/9; \;\; A_1^n = 0; \;\; \Delta u/u = 2/3; \;\; \Delta d/d = -1/3; \;\; \text{and} \;\; F_2^n/F_2^p = 2/3. \tag{2}$$

These predictions are not expected to work in the low-x region because the sea quarks and gluon contributions are not included. Experimental data on F_2^n/F_2^p agree poorly with the SU(6) quark model predictions even in the high-x region ($x > 0.4$), which is a sign that SU(6) symmetry is broken for valence quarks. However, a recent analysis [3] revealed a possible large uncertainty associated with the nuclear corrections to the F_2^n/F_2^p data in the high-x region, which makes the observed SU(6) breaking in F_2^n/F_2^p less significant.

Relativistic Constituent Quark Models (RCQM) with broken SU(6) symmetry, e.g., the hyperfine interaction model [4], lead to a dominance of a 'diquark' configuration [5] with the diquark spin $S = 0$ at high x. This implies that as $x \to 1$:

$$A_1^p \to 1; \;\; A_1^n \to 1; \;\; \Delta u/u \to 1; \;\; \text{and} \;\; \Delta d/d \to -1/3. \tag{3}$$

In the RCQM, the relativistic effect takes into account the quark orbital angular momentum and reduces the valence quark contributions to the nucleon spin from 1 to about $0.7-0.75$.

Another approach is with leading-order pQCD [6], which assumes the quark orbital angular momentum is negligible and leads to hadron helicity conservation. It yields the same limiting values for A_1^p and A_1^n as previously, but different limiting values for $\Delta u/u$ and $\Delta d/d$:

$$\Delta u/u \rightarrow 1; \text{ and } \Delta d/d \rightarrow 1. \tag{4}$$

Not only are the limiting values at $x \rightarrow 1$ important, but also the behavior in the high-x region. How A_1^n and A_1^p approach their limiting values when x approaches 1 is sensitive to the dynamics of the valence quarks.

In 2001, JLab experiment E99-117 [9] was carried out in Hall A to measure A_1^n with high precision in the x region from 0.33 to 0.61 (Q^2 from 2.7 to 4.8 GeV^2). Asymmetries from inclusive scattering of a highly polarized 5.7 GeV electron beam on a high pressure ($>$ 10 atm) (both longitudinal and transversely) polarized ^{3}He target were measured. Beam polarization was measured with a Møller polarimeter and a Compton polarimeter. The average beam polarization was $78\% \times (1 \pm 0.03)$. The ^{3}He target was polarized by spin exchange with optically pumped Rubidium. The average in-beam polarization was $40\% \times (1 \pm 0.04)$. The scattered electrons were detected with two high-precision spectrometers with their standard detector packages (scintillators for trigger, vertical drift chambers for tracking, gas Cherenkov counters and shower counters for particle identification).

Parallel and perpendicular asymmetries were extracted for ^{3}He. After taking into account the beam and target polarization and the dilution factor, they were combined to form $A_1^{^3He}$. Using the most recent model [10], nuclear corrections were applied to extract A_1^n. Final results on A_1^n are shown in the left panel of Fig. 1. For clarity, not all theoretical predictions are shown. A more complete list is given in Ref. [9]

The experiment greatly improved the precision of data in the high-x region. This is the first evidence that A_1^n becomes positive at large x, showing clear SU(6) symmetry breaking. The results are in good agreement with the LSS 2001 pQCD fit to previous world data [11] (solid curve) and the statistical model [12] (long-dashed curve). The trend of the data is consistent with the RCQM predictions (the shaded band). The data disagree with the predictions from the leading-order pQCD models (short-dashed and dash-dotted curves).

Assuming the strange sea quark contributions are negligible in the region $x > 0.3$, the polarized quark distribution functions $\Delta u/u$ and $\Delta d/d$ were extracted from our neutron data combined with the world proton data. The results are shown in the right panel of Fig. 1, along with predictions from the RCQM (dot-dashed curves), leading-order pQCD (short-dashed curves), the LSS 2001 fits (solid curves) and the statistical model (long-dashed curves). The results agree well with RCQM predictions as well as the LSS 2001 fits and statistical models but are in significant disagreement with the predictions from the leading-order pQCD models assuming hadron helicity conservation. This suggests that effects beyond leading-order pQCD, such as the quark orbital angular momentum, may play an important role in this kinematic region.

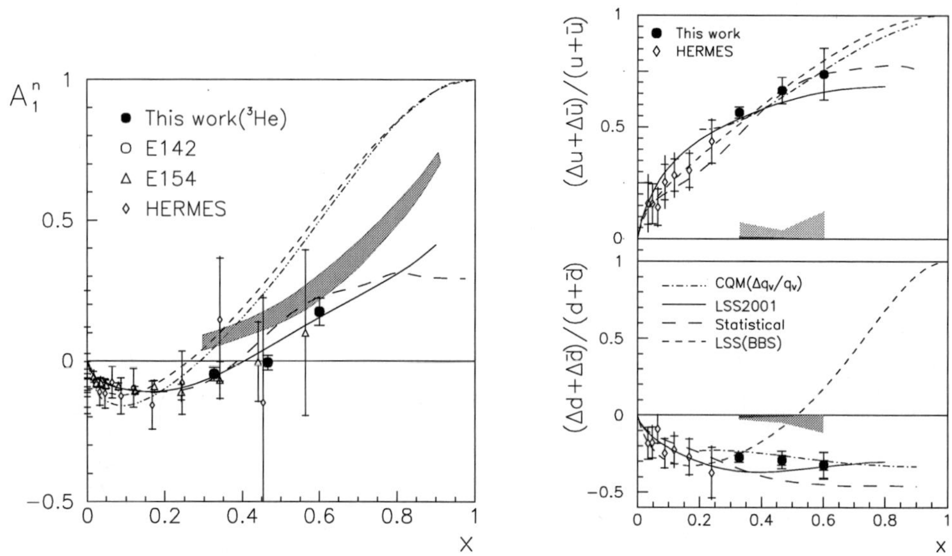

Fig. 1: A_1^n, $\Delta u/u$ and $\Delta d/d$ results compared with the world data and theoretical predictions.

GENERALIZED GDH SUM AND MOMENTS OF THE SPIN STRUCTURE FUNCTIONS

JLab E94-010 [13] measured the generalized GDH sum and the moments of the neutron spin structure functions Γ_1 and Γ_2 in the low to intermediate Q^2 range using a polarized electron beam on a polarized ^{3}He target. The measurement of doubly-polarized inclusive cross sections was performed at five beam energies from 0.86 to 5.1 GeV at a scattering angle of 15.5°. Parallel and perpendicular cross-section differences were obtained, from which g_1, g_2 and σ_{TT}, σ_{LT} for ^{3}He were extracted. Interpolation to constant Q^2 values was performed and the GDH integrals were formed from pion threshold to $W^2 = 4$ GeV^2. Finally, nuclear corrections [14] were applied, to extract the GDH integrals for the neutron. The results are shown in the left-top panel of Fig. 2. The higher energy contributions (for W^2 from 4 to 1000 GeV^2) were estimated using the parameterization of Thomas and Bianchi [15].

These data show a smooth but dramatic change in the value of the generalized GDH sum from what was observed at high Q^2. While not unexpected from phenomenological models, these data illustrate the sensitivity to the transition from partonic to hadronic behavior. The measured values of the first moment of g_1^n are shown in the left-middle panel of Fig. 2, along with the world data from SLAC and HERMES. Also shown are Chiral Perturbation Theory calculations and several model predictions. These data provide a precision data base for twist expansion analysis at the higher end of the Q^2 range, a check for Chiral Perturbation Theory (ChPT) calculations [16] at the low end of the Q^2 range, and establish an important benchmark against which one can compare future calculations (such as Lattice Gauge Theory calculations). The measured values of the first moment of g_2^n are shown in left-bottom panel of Fig. 2. These results indicate the first validation of the Burkhardt-Cottingham sum rule [17], $\Gamma_2 = 0$.

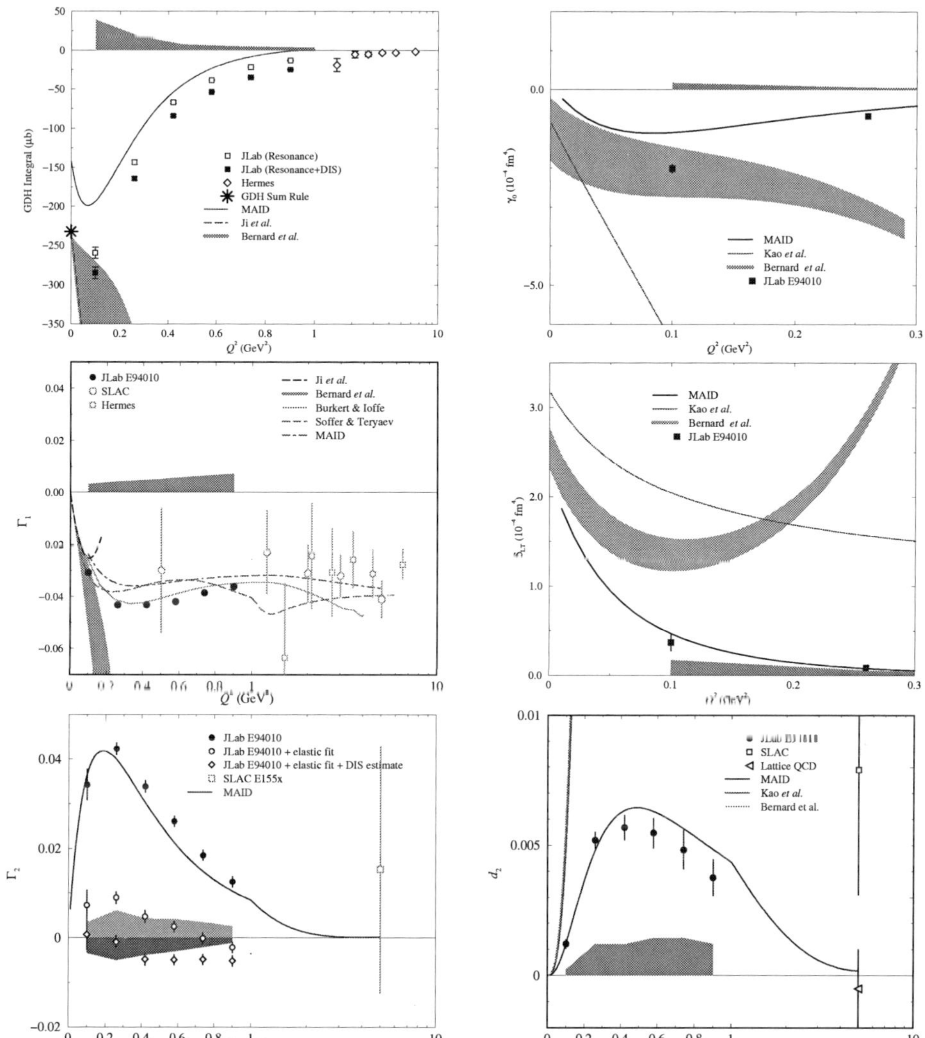

Fig. 2: Comparisons of E94-010 results with world data, ChPT calculations and model calculations.

Higher (x^2 weighted) moments of the spin structure functions are related to generalized forward spin polarizabilities γ_0, δ_{LT} and the color polarizability d_2 [13]. The right panels of Fig. 2 show the E94-010 results on γ_0, δ_{LT} and d_2, and the ChPT calculations at low Q^2, MAID model [18] predictions at low to intermediate Q^2 and the Lattice QCD prediction at high Q^2. The relativistic baryon ChPT with resonance shows good agreement with the data for γ_0 at $Q^2 = 0.1$ GeV^2. However, ChPT calculations deviate significantly from the data for δ_{LT}, which was expected to be an excellent candidate to check Chiral dynamics of QCD since it was not sensitive to the dominating resonance (Δ) contributions. This disagreement presents a real challenge to theorists.

A new experiment [19] will extend the generalized GDH sum measurements to very low Q^2 (down to $Q^2 = 0.02$ GeV^2), below the turn-around point predicted by calculations (at $Q^2 \approx 0.1$ GeV^2). ChPT calculations will be extensively tested at low Q^2 where they are expected to be applicable. Extrapolation to the real photon point provides an alternative way to test the original GDH sum rule. Data taking was completed in the summer of 2003. Analysis is underway.

In summary, with high polarized luminosity, JLab has provided a set of high-precision data to study the nucleon spin structure in a wide kinematic range, which sheds light on the valence quark structure and helped to understand the transition region between perturbative and non-perturbative regions of QCD.

ACKNOWLEDGMENTS

The work presented was supported in part by the U. S. Department of Energy (DOE) contract DE-AC05-84ER40150 Modification NO. M175, under which the Southeastern Universities Research Association operates the Thomas Jefferson National Accelerator Facility.

REFERENCES

1. B. W. Filippone and X. Ji, Adv. Nucl. Phys. **26**, 1 (2001)
2. J. D. Bjorken, Phys. Rev. **148**, 1467 (1966); Phys. Rev. D**1**, 465 (1970)
3. W. Melnitchouk and A. W. Thomas, Phys. Lett. **B 377**, 11 (1996); G. Petratos, these proceedings
4. N. Isgur, Phys. Rev. D**59**, 034013 (1999)
5. B. -Q. Ma, Phys. Lett. **B 375**, 320 (1996)
6. S. Brodsky, M Burkhardt and I. Schmidt, Nucl. Phys. B**441**, 197 (1995)
7. S. B. Gerasimov, Yad. Fiz. **2**, 839 (1965).
   S. D. Drell and A. C. Hearn, Phys. Rev. Lett. **162**, 1520 (1966)
8. X. Ji and J. Osborne, J. Phys. G**27**, 127 (2001)
9. X. Zheng, *et al.*, Phys. Rev. Lett. **92**, 012004 (2004); X. Zheng, *et al.*, nucl-ex/0405006, to be published in Phys. Rev. C
10. F. Bissey, *et al.*, Phys. Rev. C **65**, 064317 (2002)
11. E. Leader, A. V. Sidorov and D. B. Stamenov, Eur. Phys. J. **C 23**, 479 (2002)
12. C. Bourrely, J. Soffer and F. Buccella, Eur. Phys. J. **C 23**, 487 (2002)
13. M. Amarian, *et al.*, Phys. Rev. Lett. **89**, 242301 (2002); M. Amarian, *et al.*, Phys. Rev. Lett. **92**, 022301 (2004); M. Amarian, *et al.*, nucl-ex/0406005, to be published in Phys. Rev. Lett.
14. C. Ciofi degli Atti and S. Scopetta, Nucl. Phys. B**404**, 223 (1997)
15. E. Thomas and N. Bianchi, Nucl. Phys. B**82** (Proc. Suppl.), 256 (2000)
16. X. Ji, C. Kao, and J. Osborne, Phys. Lett. B **472**, 1 (2000); C. W. Kao, T. Spitzenberg and M. Vanderhaeghen, Phys. Rev. D **67**, 016001 (2003); V. Bernard, T. Hemmert and Ulf-G. Meissner, Phys. Rev. D **67**, 076008
17. H. Burkhardt and W. N. Cottingham, Ann. Phys. **56**, 453 (1970)
18. D. Drechsel, S. Kamalov and L. Tiator, Phys. Rev. D **63**, 114010 (2001)
19. JLab E97-110, J. P. Chen, A. Deur and F. Garibaldi, *et al.*

Spin structure of the proton and deuteron with CLAS at Jefferson Lab

P.E. Bosted for the CLAS Collaboration

Jefferson Lab, Newport News VA 23606 USA

Abstract. The EG1 experiment at Jefferson Lab used the CLAS detector to measure inclusive, semi-inclusive, and exclusive reactions with longitudinally polarized electrons scattering from longitudinally polarized protons and deuterons. Preliminary results for g_1^p and g_1^d are presented for a subset of the inclusive data. The data span a large kinematic range, $0.02 < Q^2 < 5$ GeV^2 and $0.05 < x < 1$, covering the region where nucleon resonances dominate as well as the traditional DIS region.

The spin-dependent structure function $g_1(x,Q^2)$ for inelastic lepton-nucleon scattering is of fundamental importance in understanding the quark and gluon spin structure of the proton and neutron in the deep inelastic region of high Q^2, as well as the resonant structure at low Q^2 and excitation energy. The goal of the CLAS EG1 experiment was to make precise measurements over the entire (x,Q^2) range currently accessible at Jefferson Lab in a single experiment, using both polarized proton and deuteron targets. Results from an initial data run in 1998 (EG1a) have recently been published [1, 2]. We report here on some of the data from the much higher statistics run in 2000.

Beam currents ranged from 0.3 nA to 6 nA. The helicity was frequently reversed to minimize instrumental asymmetries. The 1-cm-long polarized target cells contained pre irradiated granules of $^{15}NH_3$ or $^{15}ND_3$ immersed in liquid He at 1 K in a uniform magnetic field of 5 T. Microwaves near 140 GHz were used to drive the hyperfine transition which aligns (or anti-aligns) the nucleon spins with the magnetic field, producing proton polarizations of typically 70% (30%) for NH_3 (ND_3). The sub-millimeter diameter electron beam spot was rastered over the 1.8 cm^2 front surface of the target to uniformly distribute beam heating and radiation damage. The target polarization direction was periodically reversed.

The product of target polarization P_t and beam polarization P_B was determined using the well-known double-spin asymmetry in ep elastic (quasielastic) scattering for NH3 (ND3). Results from single arm (e only) and double arm (both e and p detected) measurements were found to be consistent within statistical errors, and the average for the experiment was $P_bP_t \approx 0.5$ for protons and $P_bP_t \approx 0.2$ for deuterons.

Scattered electrons were tracked in the CLAS [3] toroidal field using drift chambers, and separated from hadrons by requiring a substantial signal in the gas Cherenkov counters and most of the particle energy to be deposited in the electromagnetic calorimeter. Data were taken with approximately equal running times with positive and negative torus polarity (corresponding to electrons bending inward or outward) with beam energies near 1.7, 2.4, 4.2, and 5.7 GeV. Approximately 2.3×10^{10} triggers were registered,

CP747, *Structure of the Nucleon at Large Bjorken x*, edited by C. Bourrely, Z.-E. Meziani, and J. Soffer

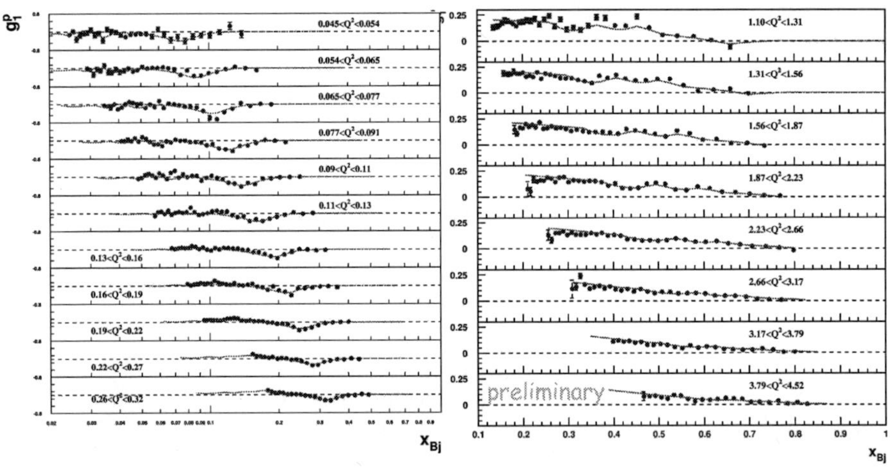

FIGURE 1. Preliminary results for g_1^p from the CLAS EG1 experiment using an incident electron energy of 1.6 GeV (left panel) and 5.7 GeV (right panel). The solid curves represent the EG1 fit to the data. The dashed curves indicate zero.

yielding about 10^{10} cleanly identified electrons. The present results are based on approximately half this total, as the 2.4 and 4.2 GeV data are still being processed.

The experimental asymmetries $A_{\parallel}$ were determined from

$$A_{\parallel} = \left(\frac{N_{-} - N_{+}}{N_{-} + N_{+}}\right) \frac{1}{f P_b P_t f_{RC}} + A_{RC}, \tag{1}$$

where the target polarization is parallel to the beam direction, N_{-} (N_{+}) is the number of scattered electrons per incident charge for negative (positive) beam helicity, f is the dilution factor representing the fraction of measured events originating from polarizeable hydrogen within the target, and f_{RC} and A_{RC} take into account radiative corrections. Small corrections for for the polarized nitrogen nuclei were also made.

The dilution factor f was determined by careful comparison of counting rates from the NH_3 and ND_3 targets with a target in which the polarized granules were replaced with a known amount of C, and another target with only liquid helium and the target windows. For NH_3 (ND_3), f varied with x and Q^2 between 0.12 (0.2) and 0.16 (0.24),

A correction to the asymmetries was made for pions misidentified as electrons (typically 2% of electron candidates, but up to 10% in the lowest x bins). The correction

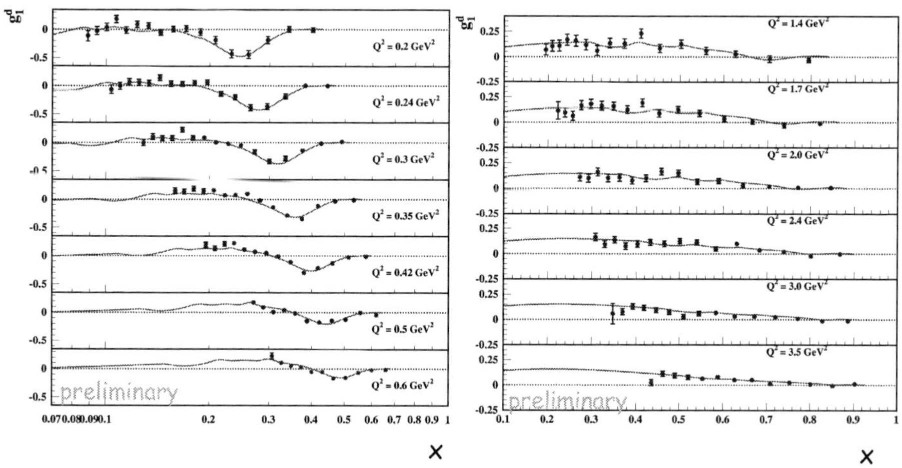

FIGURE 2. Same as Fig. 1, except for g_1^d .

used the asymmetry measured for a large sample of inclusive pions, which was found to be close to zero at all kinematics. An additional correction was made for electrons from pair-symmetric processes (such as e^+/e^- pair production from photons) measured using identified positrons. The measured pair-symmetric $A_{||}$ was consistent with zero at all kinematics, so the correction is equivalent to a dilution factor correction of typically 10% at the lowest E' used, decreasing rapidly to a negligible correction at higher E'. No corrections have been applied for kinematic resolution.

The internal radiative corrections for $A_{||}$ were evaluated using the formulae of Kuchto and Shumeiko [4]. The cross sections entering the asymmetry were 'externally radiated' according to Tsai [5]. Comparison of Born and fully radiated asymmetries allowed us to extract the asymmetry corrections f_{RC} and A_{RC}. By splitting the radiative correction into these two parts, we can propagate consistently the experimental error to the extracted Born asymmetries for the corresponding kinematic bins, in the presence of 'dilution' from elastic and inelastic radiative tails.

The ratio of polarized to unpolarized structure functions was determined from $A_{||}$ using

$$g_1/F_1 = A_{||}/d + (g_2/F_1)[(2Mx)/(2E-\nu)], \qquad (2)$$

where the virtual photon de-polarization factor $d = [(1-\varepsilon)(2-y)]/\{y[1+\varepsilon R(x,Q^2)]\}$, $y = \nu/E$, and $\nu = E - E'$, where E is the incident and E' is the scattered electron energy

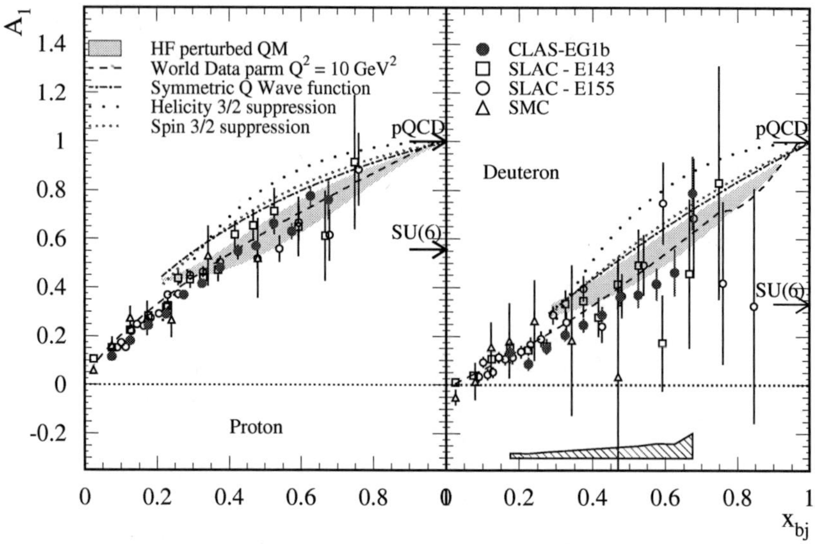

FIGURE 3. Preliminary EG1 results for A_1^p (left) and A_1^d (right) at high x for $Q^2 > 1.4$ GeV^2 and $W > 2$ GeV, compared to several quark models (with all have the limit $A_1 \to 1$ as $x \to 1$, [7] and the SU(6) spin-flavor symmetry limit [8]. Data from SLAC [9, 10] and SMC [11] are also shown.

in the lab frame, $\varepsilon^{-1} = 1+2[1+\gamma^{-2}]\tan^2(\theta/2)$, $\gamma^2 = Q^2/\nu^2$, θ is the electron scattering angle, M is the nucleon mass, and $R(x,Q^2) = [F_2(x,Q^2)(1+\gamma^2)]/[2xF_1(x,Q^2)] - 1$ is typically 0.2 for the kinematics of this experiment. For the contribution of the transverse spin structure function g_2 we used the twist-two result of Wandzura and Wilczek (g_2^{WW}) [6]

$$g_2^{WW}(x,Q^2) = -g_1(x,Q^2) + \int_x^1 g_1(\xi,Q^2)d\xi/\xi, \tag{3}$$

(evaluated using an empirical fit to g_1/F_1) at high W, and a unitary isobar model (MAID) in the resonance region. Values of g_1 where obtained from g_1/F_1 using a fit to F_1.

Preliminary results for g_1^p and g_1^d are shown in Figs. 1 and 2 as a function of x at several values of Q^2. For the present experiment, most systematic errors (beam polarization, target polarization, fraction of polarizeable nucleons in the target) for a given target are common to all data and correspond to an overall relative normalization error of about 5%. The remaining systematic errors (model dependence of radiative corrections, model uncertainties for $R(x,Q^2)$, model dependence of f) vary smoothly with x in a locally correlated fashion, ranging from a few percent for mid-range x bins, up to 15% of the $E = 5.7$ GeV proton data for the highest and lowest bins.

At high x, g_1 is seen to decrease with increasing Q^2 for both proton and deuteron targets. Although space limitations do not allow for a figure in this proceedings, a plot of the ratios g_1/F_1 shows almost no dependence on Q^2 at fixed x for $Q^2 > 1.5$ GeV^2, indicating that the QCD evolution of the two structure functions is very similar. At lower Q^2, the data fall below the high Q^2 limit, corresponding to the excitation of the $\Delta(1232)$

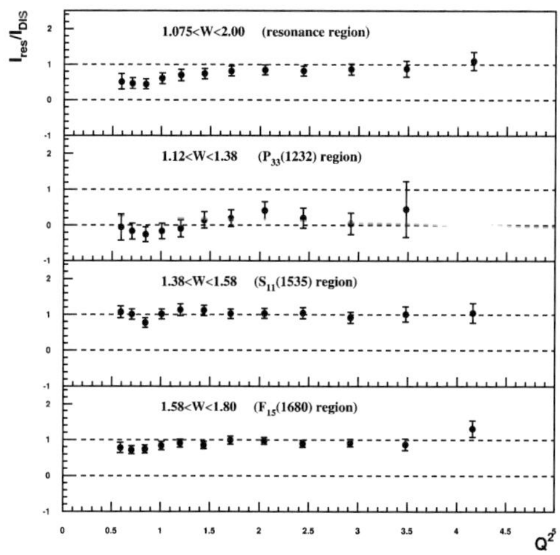

FIGURE 4. Very preliminary CLAS results for the ratio the integral of g_1^p over the entire resonance region (top) and three individual resonance regions (below),to the corresponding DIS integral in the appropriate region of Nachmann ξ, as a function Q^2 in GeV^2.

resonance. The small and sometimes negative values of g_1 in the $\Delta(1232)$ region are expected since this is predominantly a spin-flip transition. On the low x side of the $\Delta(1232)$ dip, positive enhancements can be seen corresponding to the predominately non-spin-flip transition to the $S_{11}(1520)$ resonance.

The CLAS data provide a considerable improvement in precision in the high x region, which is of particular interest in distinguishing between various models of the valence spin structure of the nucleon (see Fig. 3). The preliminary results for $A_1 = g_1/F_1 - \gamma^2 g_2/F_1$ clearly show a preference for the pQCD limit at $x = 1$ [7], compared to the SU(6) spin-flavor symmetry limit [8].

A subject that has seen re-awakened interest lately is quark-hadron duality, in which structure functions averaged over suitable ranges in excitation energy become independent of Q^2 when an appropriate scaling variable is chosen. Although quark-hadron duality appears to be fairly well validated in the case of spin-averaged response function F_1 and F_2 in the region of the $\Delta(1232)$ resonance [12], it is manifestly broken in the case of g_1, which is negative at the peak of the Δ at low Q^2, but positive at high Q^2, as shown in Fig. 4 for the proton. In this figure, the integral of g_1 over several regions in W is compared with high Q^2 parametrizations in the DIS region, taking target mass effects into account using the Nachmann scaling variable ξ instead of x. Even at $Q^2 = 5$ GeV^2, the $\Delta(1232)$ integral lies below the DIS curve. This can be partially restored by pairing the $\Delta(1232)$ (lowest lying spin-3/2 state) with the *ep* elastic peak (lowest lying spin-1/2 state), but in this case spin-duality is approached from above unity, rather than from below. On the other hand, spin-duality appears to work remarkably well for higher mass resonance regions for $Q^2 > 1$ GeV^2, as well as for the entire resonance region for $Q^2 > 1.5$ GeV^2.

Combined with other data, the new CLAS data have been used to study the extended GDH Sum Rule [13] and the transition to Ellis-Jaffe [14] and Bjorken Sum [15] Rules at higher Q^2. Using reasonable Regge models to extrapolate to $x = 0$, both $\Gamma_1^p(Q^2) = \int_0^{1-} g_1^p(x, Q^2)dx$ and $\Gamma_1^d(Q^2) = \int_0^{1-} g_1^d(x, Q^2)dx$ show a clear change of sign from being negative at low Q^2 as predicted by the GDH Sum Rule, to being positive at high Q^2, with the cross-over point occurring at about $Q^2 = 0.3$ GeV^2 for the proton and 0.5 GeV^2 for the deuteron.

In summary, the new high statistics measurements of g_1^p and g_1^d show great promise to shed light on several of the central issues in nucleon spin structure: the Q^2 evolution of the structure functions, the high-x behavior, the Q^2 dependence of the first moments of g_1, and studies of duality. Semi-inclusive scattering results hold promise for flavor decomposition of the PDF's, and studies of fragmentation and single-spin asymmetries. More information on specific nucleon resonances will be obtained, particularly with the study of specific exclusive final states. For more information on all of these topics, the interested reader is invited to consult the many talks given at the recent GDH 2004 conference [16]

REFERENCES

1. The CLAS Collaboration, R. Fatemi *et al.*, nucl-ex/0306019 (2003).
2. The CLAS Collaboration, J. Yun *et al.*, Phys. Rev. **C 67**, 055204 (2003).
3. B. Mecking *et al.*, Nucl. Instrum. Meth. **A503**, 513 (2003).
4. T. V. Kuchto and N. M. Shumeiko, Nucl. Phys. **B219**, 412 (1983); I. V. Akusevich and N. M. Shumeiko, J. Phys. **G20**, 513 (1994).
5. Y. S. Tsai, Report No. SLAC–PUB–848, 1971; Y. S. Tsai, Rev. Mod. Phys. 46, 815 (1974).
6. S. Wandzura and F. Wilczek, Phys. Lett. **B72**,195 (1977).
7. G. Farrar and D. Jackson, Phys. Rev. Lett. **35**, 1416 (1975); N. Isgur, Phys. Rev. D **59** 034013 (1999); F.E. Close and W. Melnitchouk, Phys. Rev. C **68**, 035210 (2003).
8. F.E. Close , Phys. Lett. **43**, 42 (1973).
9. SLAC E143, K. Abe *et al.*, Phys. Lett. **B364**, 61 (1995); Phys. Rev. D **58**, 112003 (1998).
10. SLAC E155, P.L. Anthony *et al.*, Phys. Lett. **B493**, 19 (2000).
11. SMC, D. Adeva *et al.*, Phys. Rev. D58, 112001 (1998).
12. I. Niculescu *et al.*, Phys. Rev. Lett. **85**, 1186 (2000).
13. S. D. Drell and A. C. Hearn, Phys. Rev. Lett 16, 908 (1966); S. B. Gerasimov, Yad. Fiz. 2, 598 (1966).
14. J. Ellis and R. Jaffe, Phys. Rev. D9, 1444 (1974); D10, 1669 (1974) (E).
15. J. D. Bjorken, Phys. Rev. 148, 1467 (1966) ; Phys. Rev. D1, 1376 (1970).
16. Proceedings of the The 3rd International Symposium on the Gerasimov-Drell-Hearn Sum Rule and its extensions, Old Dominion University, Norfolk, Virginia, June 2-5, 2004.

Polarized Structure Functions: Proton/Deuteron Measurements in Hall C

Oscar A. Rondon

University of Virginia, Charlottesville, VA 22901

Abstract. The study of the nucleon polarized structure functions has matured beyond the inclusive measurements of the past to the investigation of all eight quark distribution functions in the nucleon. Jefferson Lab's Hall C program of polarized structure functions studies started with a measurement of the proton and deuteron spin structure in the resonances at $Q^2 \sim 1.3$ [GeV/c]^2. This work will be extended for the proton to more than 5 [GeV/c]^2 for both DIS and the resonances in the upcoming *SANE* experiment. *SANE* will use a novel non-magnetic very large solid angle detector, *BETA*. Semi-inclusive asymmetries will be measured to determine the flavor composition of the nucleon spin in the recently approved *Semi – SANE* experiment. The 11 GeV energy upgrade will open new opportunities to study other functions, such as the transversity, Collins and Sievers functions, using vertical polarized targets.

Nucleon structure can be described in terms of eight quark distribution functions, three of which are independent of the quark's transverse momentum, and are of leading twist. These three functions are the well measured unpolarized F_1 and polarized g_1 structure functions (SF's), and the yet to be determined transversity h_1, or δ. The five transverse momentum dependent SF's include the polarized transverse SF $g_T = g_1 + g_2$, the Collins $h_T^\perp$ fragmentation function and Sievers $f_{1T}^\perp$ distribution function. Only three of these distribution functions can be explored in inclusive inelastic lepton-nucleon scattering (F_1, g_1 and g_2). The others require the measurement of semi-inclusive polarized asymmetries, with one hadron detected in coincidence with the lepton.

Jefferson Lab's Hall C program of polarized nucleon structure measurements started in the year 2002 with an inclusive measurement (E 01-006) of g_1 and g_2 in the resonances for protons and deuterons at intermediate four-momentum transfer $Q^2 \sim 1.3$ [GeV/c]^2. The *Spin Asymmetries on the Nucleon Experiment*(*SANE*) will extend these inclusive measurements for the proton to $Q^2 \sim 6$ [GeV/c]^2, in the Bjorken x range of 0.3 to 0.8. Semi-inclusive polarized asymmetries to study the quark flavor composition of the nucleon spin will be measured in another experiment which will share with *SANE* the *Big Electron Telescope Array* (*BETA*) to detect electrons.

TJNAF E-01-006 - RESONANCES SPIN STRUCTURE (*RSS*)

The leading twist spin structure functions (SSF's) $g_1$and g_2, and their related spin asymmetries A_1 and A_2 require the measurement of two asymmetries for different relative orientations of the target spin and the lepton beam helicity. The usual approach is to measure the asymmetries with the nucleon spin parallel ($A_{||}$) and perpendicular ($A_\perp$)

CP747, *Structure of the Nucleon at Large Bjorken x*, edited by C. Bourrely, Z.-E. Meziani, and J. Soffer

to the beam helicity. There is a significant scarcity of $A_\perp$ measurements for much of the Q^2-x kinematic plane for $x >\sim 0.5$, which is the focus of this workshop, in particular for Q^2 below ~ 6 [GeV/c]^2.

Much of the high x region for Q^2 up to ~ 10 [GeV/c]^2 corresponds to final state invariant mass $W \leq 2$ GeV, which is dominated by the nucleon resonances. The first Hall C experiment on the SSF's, *RSS* [1], concentrated on a precision measurement of proton and deuteron A_1 and A_2 in the resonances. The main goals of this experiment are to study the W dependence of the SSF's, look for the onset of polarized local duality, and explore the contribution of twist-3 processes to g_2.

RSS measured $A_\parallel$ and $A_\perp$ for protons and deuterons by scattering 5.76 GeV polarized electrons accelerated with CEBAF on polarized frozen ammonia (NH_3) and deuterated ammonia (ND_3) targets. The scattered electrons were detected in the High Momentum Spectrometer-HMS parked at 13.15°. Two settings of the HMS central momentum were used, 4.73 and 4.09 GeV/c. The kinematic range covered with these parameters extends from 0.6 GeV to 2 GeV in W at $Q^2 \approx 1.3 \pm 0.3$.

The measured asymmetries are obtained from the ratio of the difference over the sum of charge normalized event counts for opposite beam helicities, N_L, N_R. This ratio is normalized by the beam polarization P_b, target polarization P_t and dilution f of the polarized nuclei by unpolarized nucleons in the target (nitrogen, liquid He, etc.)

$$A_{\parallel,\perp} = \frac{1}{C_N f_{RC}} \left(\frac{N_L - N_R}{N_L + N_R} \frac{1}{f P_b P_t} - C_D \right) + A_{RC}. \tag{1}$$

The asymmetry is also corrected for the contributions of radiative effects, f_{RC}, A_{RC} and the small nitrogen polarization C_N, C_D.

The usable beam current was limited by the cooling power of the polarized target to 200 nA or less. In addition, the beam needed to be rastered over a 2 cm diameter circle to spread the damage induced by the beam radiation and reduce the resulting rapid local polarization loss. The beam polarization was measured with the Hall's Møller spectrometer. For the perpendicular asymmetry part of the experiment the polarization was 66%±0.5% and 71%±0.5% for the parallel data.

The out of plane offset due to the beam rastering requires a special treatment in the event reconstruction. A special set of spectrometer optics reconstruction matrix elements was used to correct for the out of plane offset in a consistent way in the data analyzer and the Montecarlo simulation.

The polarized target operated at 5 T and 1 K. Dynamic nuclear polarization of the ammonia by microwave irradiation produced in-beam proton polarizations of 70% and deuteron polarizations of 20%. The polarization was measured continuously during the experiment by the NMR technique. The intense magnetic field of the target produces large deflections in the trajectories of the scattered electrons, especially in the perpendicular configuration. These deflections need to be known very accurately to obtain the

---

[1] *RSS* Collaboration (JLab E-01-006): U. Basel, Florida International U., Hampton U., U. Massachusetts, U. Maryland, Mississippi S. U., North Carolina A&T U., U. of N. C. at Wilmington, Norfolk State U., Old Dominion U., S. U. New Orleans, U. of Tel-Aviv, TJNAF, U. of Virginia, Virginia P. I. & S. U., Yerevan Physics I. Spokespersons: O. A. Rondon (Virginia) and M. Jones (TJNAF).

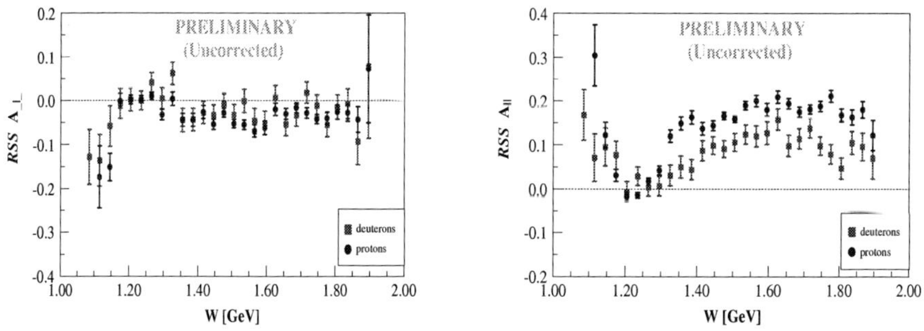

FIGURE 1. Preliminary measured asymmetry for the perpendicular confi guration (left) and the parallel one (right) for protons (circles) and deuterons (squares), without corrections (see text for details.

correct kinematics for the reconstructed events. *RSS* is unique in that it is the only experiment to date to have measured high resolution perpendicular asymmetries on solid polarized targets in the $e-p$ elastic and nucleon resonances regions.

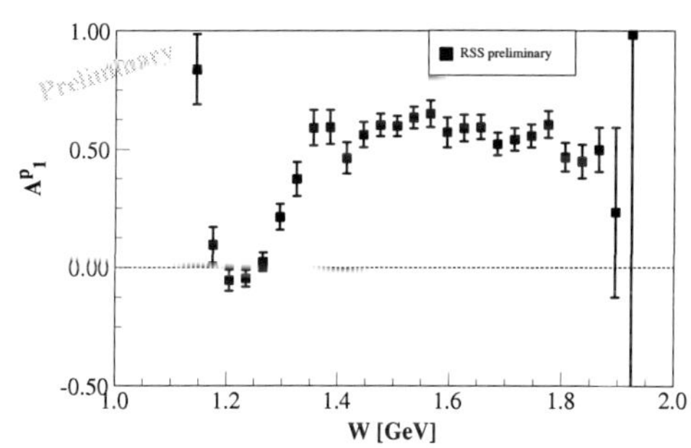

FIGURE 2. Preliminary proton A_1.

The dilution of the asymmetries by the unpolarized nucleons is calculated using a Montecarlo simulation of the yields for each material. Model Born cross sections from the *QFS* [1] software package were combined with parameterizations [2] of single-arm proton and deuteron electro-nuclear cross sections measured in Hall C. The fraction by volume of ammonia in the target cells is specific for every cell load, ranging from about 50% to about 60%. A nominal 50% value was used for the preliminary dilution factors.

The uncorrected preliminary $A_\perp$ and $A_\parallel$ are shown in Figure 1. These asymmetries are based on the first pass version of the data analysis and preliminary dilution factors. Corrections for individual cells' packing fractions, nitrogen asymmetry and radiative effects have not been applied. Since A_1 is mainly dependent on $A_{\|}$, which is not very sensitive to the details of the target field (the angle between the field and the scattered electrons is only 13°), we have used the preliminary $A_{\|}$ and $A_{\perp}$ to obtain preliminary, uncorrected results for the proton A_1, which are shown in Figure 2.

TJNAF E-03-009 - SPIN ASYMMETRIES ON THE NUCLEON EXPERIMENT – *SANE*

SANE [2] will extend the measurements of the preceding experiment into the Q^2 range of 2.5[GeV/c]^2 to 6.5 [Gev/c]^2 for proton targets, in the DIS and resonances, for $0.3 \leq x \leq 0.8$. The goals of *SANE* are to measure both g_2 and A_1 over a broad range of x and Q^2 to study their x dependence at constant Q^2, and Q^2 dependence at fixed x, in order to compute the moments of the SSF's. The third moments are related to the d_2 twist-3 quark matrix element $d_2 = \int_0^1 x^2 (2g_1 + 3g_2) dx$, which can be calculated in Lattice QCD. Polarized or spin local duality will be tested down to the second resonance, too.

Polarized electron beams of 6 and 4.8 GeV energy will be scattered on polarized solid ammonia targets. The proton spins will be aligned parallel to the beam and at an 80° angle to measure parallel and near-perpendicular asymmetries, for a model independent separation of g_1 and g_2 .

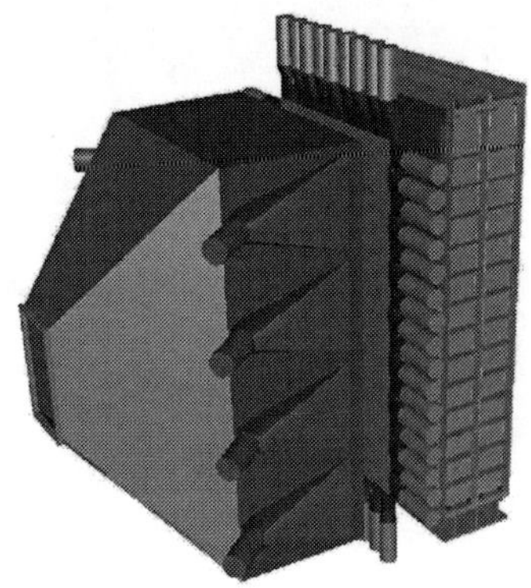

FIGURE 3. Artist view of *BETA*. The sideways pyramid shaped object at left is the gas Cherenkov, the stack of rectangular blocks on the right is BigCal. The tracking hodoscope is sandwiched in between.

To compensate for the decrease in cross section with increasing Q^2 and x, *SANE* is building a new type of non-magnetic large solid angle detector for electrons, the *Big Electron Telescope Array* (*BETA*). This detector will be aligned with its axis at 40° to the left of the beam. The HMS will be aligned at 40° to the right of the beam to measure the small asymmetry for the the positrons produced in pair-symmetric processes, which

---

[2] TJNAF E03-109: U. Basel, Florida International U., Hampton U., Norfolk State U., IHEP Protvino, U. of Regina, Rensselaer P. I., Seoul U., Temple U, TJNAF, U. of Virginia, C. of William & Mary, Yerevan Physics I. Spokespersons: O. Rondon (Virginia), Z-E. Meziani (Temple), S. Choi (Seoul).

contaminate the electro-nuclear events, especially at low momenta. *BETA* consists of three main subsystems:

BigCal is a 1700-element lead glass calorimeter being built by the BigCal collaboration for the high Q^2 measurement of the proton elastic form factor ratio G_E^p/G_M^p [3]. It is the main component of the detector package, covering a 194 msr solid angle (after cuts). Its energy resolution is $dE/E \sim 5\%/\sqrt{E}$. The uniformity of the response of all the elements will be monitored by a Lucite plate excited by laser light [4].

Gas Cherenkov will be the main particle identification device, with expected 1000:1 pion rejection capability in combination with BigCal. The Cherenkov matches BigCal's solid angle. It is being built by the Temple University group [5].

Tracking Hodoscope based on Cherenkov light emission in transparent solids (Lucite or quartz - choice of material under study). The hodoscope will add an extra level of background rejection by tracking events in the calorimeter back to the target.

Figure 3 is an artist's view of *BETA*. The final design and element configuration is being actively studied at the time of this report.

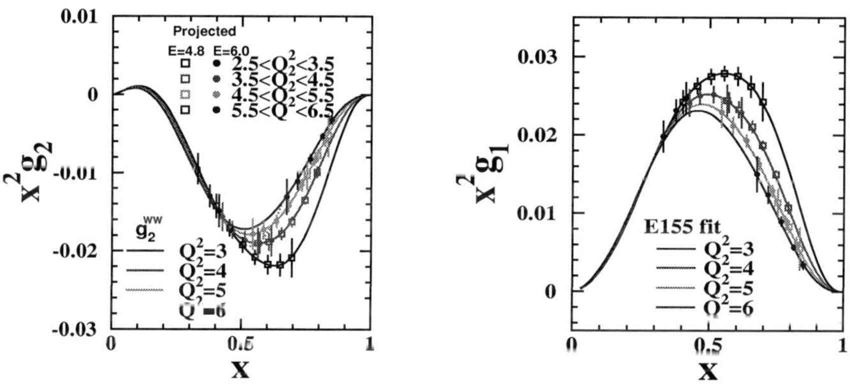

FIGURE 4. Expected results for the integrands of the third moments of g_2 (left) and g_1(right). Both the x dependence at constant Q^2 and the Q^2 dependence at fixed x will be tested.

Examples of the expected *SANE* results for g_2 and g_1 are shown in Figure 4. With *SANE*'s results, the current statistical error on d_2 will be cut in half. More examples for g_2, A_1 and tests of polarized local duality, not shown here due to space limitations, can be found at *SANE*'s Web site `http://www.jlab.org/~rondon/sane/`.

TJNAF E-04-013 - FLAVOR DECOMPOSITION OF NUCLEON SPIN STRUCTURE

The contributions of each quark flavor to the nucleon spin can only be measured in electro-nuclear scattering by tagging recoil hadrons (pions, kaons), that are detected in coincidence with the lepton. Experiment 04-113[3] will measure proton and deuteron semi-inclusive spin asymmetries in polarized DIS reactions $\vec{p}(\vec{e},e'h)$ and $\vec{d}(\vec{e},e'h)$, where h symbolizes $\pi^{\pm}, K^{\pm}$. The measurement will extend over the ranges $1.2[\mathrm{GeV/c}]^2 \leq Q^2 \leq 3.2[\mathrm{GeV/c}]^2$ and $0.12 \leq x \leq 0.43$, for hadrons carrying a fraction $0.5 \leq z(= E_h/\nu) \leq 0.7$ of the energy transfer ν. Polarized electrons of 6 GeV beam energy will be scattered from longitudinally polarized protons in NH_3 and deuterons in LiD. The scattered electrons will be detected in *BETA*, parked at 30°, in coincidence with hadrons detected in the HMS aligned at 10.8°.

The valence quark helicities will be obtained using the LO and NLO Leader-Christova method [6], in which the difference of asymmetries for both signs of the hadron charges $A_{1p,d}^{h^+-h^-}$ is combined with inclusive results on the SSF's to get the polarized light sea flavor asymmetry. In addition, by forming the semi-inclusive asymmetries at fixed z in terms of fragmentation function (FF) ratios, the five quark flavor helicity distributions can be obtained from a global fit to all fixed z data, using known parton distribution functions (PDF). The combined asymmetry sums $A_{1p,d}^{h^++h^-}$ and the inclusive $A_1^{p,d}$ data can be compared to test factorization of the asymmetries into PDF's times FF's.

The experiment is expected to improve the current errors on flavor helicities by factors of 3 or better, and to clearly establish whether the polarized light sea is asymmetric. The wealth of results expected from this experiment cannot be properly presented in the limited space available. For extensive detailed information the reader is referred to the proposal document [7].

ACKNOWLEDGMENTS

This work was supported by Department of Energy contract DE-FG02-96ER40950 and by the Institute of Nuclear and Particle Physics of the University of Virginia.

REFERENCES

1. Lightbody, J., and O'Connell, J., *Computers in Physics, May/June 1988*, pp. 57–64 (1988).
2. Keppel, C., private communication (2003).
3. Brash, E., Jones, M., Perdrisat, C., and Punjabi, V., Measurement of G_E^p/G_M^p to Q^2=9GeV^2 via Recoil Polarization, Tech. rep., TJNAF Proposal E-04-108 (2004).
4. Frlez, E. (2004), URL `http://www.jlab.org/~choi/Sane/Frlez.pdf`.
5. Meziani, Z.-E. (2004), URL `http://www.jlab.org/~rondon/sane/mtg1/cherenkov.pdf`.
6. Christova, E., and Leader, E., *Nucl. Phys.*, **B607**, 369–390 (2001).
7. Jiang, X., Flavor Decomposition of Nucleon Spin Structure, Tech. rep., TJNAF Proposal E-04-113 (2004), URL `http://www.jlab.org/exp_prog/generated/apphallc.html`.

---

[3] TJNAF E03-109: Argonne National Lab, Duke U., Florida International U., Hampton U., U. Kentucky, U. Maryland, U. Massachusetts, Norfolk State U., Old Dominion U., IHEP Protvino, U. of Regina, Rensselaer P. I., Rutgers U., Seoul U., Temple U, TJNAF, U. of Virginia, C. of William & Mary, Yerevan Physics I. Spokespersons: X. Jiang (Rutgers), P. Bosted (TJNAF), D. Day (Virginia), M. Jones (TJNAF).

How High is High x?

Jerrold Franklin

Department of Physics, Temple University, PA 19122-6082

Abstract. A general three quark bound state satisfying the Pauli principle, and conserving angular momentum and isospin, is used to investigate the spin structure of nucleons at high x. It is shown that, if any one quark in a proton dominates as $x \to 1$, it must be a u quark with the same spin direction as the proton. Then both A_{1p} and A_{1n}, as well as the quark spin distribution $\Delta u/u$, will approach 1 as $x \to 1$. The ratio F_{1n}/F_{1p} will approach $\frac{1}{4}$ (not $\frac{3}{7}$). The spin distribution $\Delta d/d$ does not approach 1, but is bound by the limits $-\frac{1}{3} \leq \Delta d/d \leq 0$ for all x.

The region of high Bjorken x in deep inelastic scattering (DIS) has become of great interest recently because of the possibility of testing theoretical predictions for the asymmetry parameters $A_{1p}(x)$ and $A_{1n}(x)$ as $x \to 1$. A recent experiment has measured A_{1n} up to $x = 0.6$[1], and planned experiments may be able to measure the asymmetries at higher x. Also, there is a renewed recognition that the absence of sea and explicit gluon effects for x above 0.3 makes the high x region a good testing ground for valence quark theories of DIS. In this talk, we examine the general properties of nucleon quarks (u and d) obeying the Pauli principle in a general three quark bound state, and show that there are strong model independent constraints relating the behavior of unpolarized and polarized DIS at high x.

We start by constructing the most general spin $\frac{1}{2}$, isospin $\frac{1}{2}$ bound state of three nucleon quarks. For a spin up proton, this can be written as[2][1]

$$\psi(\text{uud}) = N\text{uud}\{\uparrow\uparrow\downarrow [\phi(123)+\phi(213)] - \uparrow\downarrow\uparrow \phi(123) - \downarrow\uparrow\uparrow \phi(213)\}. \tag{1}$$

This general proton wave function depends on a single three-body spatial function $\phi(123)$, which satisfies the constraint

$$\phi(123)+\phi(213) = \phi(132)+\phi(231). \tag{2}$$

The three quark coordinates could be the position, the momentum or, to apply directly to DIS, the Bjorken x of each quark. The normalization N is model dependent because of interference between $\phi(123)$ and $\phi(213)$. The state in Eq. (1) is to be considered as the uud part of a completely symmetrized wave function of the form

$$\Psi = \frac{1}{\sqrt{3}}[\psi(\text{uud}) + \psi(\text{udu}) + \psi(\text{duu})]. \tag{3}$$

[1] This state does not include total orbital angular momentum, which we will consider later.

CP747, *Structure of the Nucleon at Large Bjorken x,* edited by C. Bourrely, Z.-E. Meziani, and J. Soffer

In this symmetrization, the state $\psi(\text{udu})$ is obtained from $\psi(\text{uud})$ by interchanging the second and third spin and space coordinates as well as the indicated quark type, with a similar interchange of the first and third coordinates for $\psi(\text{duu})$. For most purposes, including calculating structure functions for DIS, it is sufficient to use only the uud order[3].[2] Using all three quark orders would just involve doing the same calculation three times.

We arrived at the proton wave function in Eq. (1) by the following steps:

1. We assume that the quarks are antisymmetric in the color degree of freedom, so that the Pauli principle for the two u quarks requires the spatial and spin functions to appear in the combination $[\uparrow\downarrow\uparrow\,\phi(123)+\downarrow\uparrow\uparrow\,\phi(213)]$ in the last two terms of Eq. (1).
2. For the three quarks to be in a $J=\frac{1}{2}$ spin state, application of the step-up operator J_+ on the state $\psi(\text{uud})$ should give zero. This requires the combination $\uparrow\uparrow\downarrow\,[\psi(123)+\psi(213)]$ for the first term in Eq. (1).
3. The quark-quark forces are charge independent, and the uud state in Eq. (1) represents a state of isospin $\frac{1}{2}$. This means that applying the type raising operator d$\rightarrow$u should give zero. Doing so gives

$$\psi(\text{uuu}) = N\text{uuu}\{\uparrow\uparrow\downarrow\,[\phi(123)+\phi(213)] - \uparrow\downarrow\uparrow\,\phi(123) - \downarrow\uparrow\uparrow\,\phi(213)\}. \quad (4)$$

We use the fact that the wave function is actually completely symmetrized to rewrite Eq. (4) as

$$\psi(\text{uuu}) = N\text{uuu}\,\uparrow\uparrow\downarrow\,[\phi(123)+\phi(213)-\phi(132)-\phi(231)]. \quad (5)$$

This will equal zero if the constraint of Eq. (2) holds. This completes the derivation of Eq. (1).

The wave function for a spin down proton is given by the spin lowering operation $\uparrow\rightarrow\downarrow$ on Eq. (1). The neutron wave function is given by the type lowering operation u$\rightarrow$d on Eq. (1). This gives, after rewriting the result in the ddu order,

$$\psi(\text{ddu}) = -N\text{ddu}\{\uparrow\uparrow\downarrow\,[\phi(123)+\phi(213)] - \uparrow\downarrow\uparrow\,\phi(123) - \downarrow\uparrow\uparrow\,\phi(213)\}. \quad (6)$$

As an example of the completely general nature of the wave function in Eq. (1), the mixed symmetry wave function $|N_M>$ in the Isgur-Karl model[5] is given by Eq. (1) with

$$\phi(123) = \Phi^{\lambda}(2,3,1), \quad (7)$$

where $\Phi^{\lambda}(2,3,1)$ is the Isgur-Karl mixed symmetry spatial function that is symmetric in its first two coordinates. If the function $\phi(123)$ in Eq. (1) were completely symmetric in all three coordinates, then Eq. (1) would represent the simplest form of the wave function for the SU(6) symmetric quark model. Although SU(6) wave functions have often been

---

[2] The uud representation of the proton wave function has been called the "uds basis" in its application to the Isgur-Karl quark model. See, for instance, Ref. [4] .

given with various linear combinations of the u and d quarks and spin states, they *all* can be rewritten in the form of Eq. (1)[3, 6, 7].

In order to use the general three body wave function of Eq. (1) for DIS, we integrate the spatial functions over the coordinates of the unstruck spectator quarks. For the general case, this leads to the quark spin distributions:

$$u\uparrow(x,Q^2) = 2N^2[2\phi(\mathbf{1}23)^2+\phi(2\mathbf{1}3)^2+2\phi(\mathbf{1}23)\phi(2\mathbf{1}3)] \quad (8)$$
$$u\downarrow(x,Q^2) = 2N^2\phi(1\mathbf{2}3)^2 \quad (9)$$
$$d\uparrow(x,Q^2) = 2N^2\phi(12\mathbf{3})^2 \quad (10)$$
$$d\downarrow(x,Q^2) = 2N^2[\phi(12\mathbf{3})^2+\phi(12\mathbf{3})\phi(21\mathbf{3})]. \quad (11)$$

In each case, the coordinate in boldface represents the struck quark, and the other two coordinates are integrated over. The x variable in the quark distribution is that of the struck quark. The dependence of the quark distributions on Q^2, the four-momentum transfer to the nucleon squared, follows from the mechanics of deep inelastic scattering. We will drop the Q^2 variable from the notation, but its presence should be kept in mind. Our general analysis applies for any Q^2 large enough to eliminate interference between two different struck quarks. Because we are only interested here in the high x region, we do not include sea quarks or explicit gluonic effects which are expected to be small for $x > 0.3$.

Inspection of the constraint equation (2) shows that if any quark dominates at high x, then it must be the first variable in $\phi(123)$ that is dominant. If the second or third variable were dominant, then the left and right hand sides of Eq. (2) could not agree. Only the spin up u quark distribution (for a spin up proton) of Eq. (8) depends on the first variable in $\phi(123)$, so only that quark can dominate as $x\to 1$. That is, if any one quark in a proton dominates as $x\to 1$, it must be the u quark with the same spin as the proton. This means that as $x\to 1$, the ratio F_{1n}/F_{1p} will approach $\frac{1}{4}$, and the asymmetry functions A_{1p} and A_{1n}, as well as the quark spin distribution $\Delta u/u$, will approach 1.

The ratio $\Delta d/d$. cannot approach 1 for any x. From Eqs. (10) and (11) we see that

$$\frac{\Delta d(x)}{d(x)} = \frac{-\phi(12\mathbf{3})\phi(21\mathbf{3})}{[2\phi(12\mathbf{3})^2+\phi(12\mathbf{3})\phi(21\mathbf{3})]}. \quad (12)$$

This ratio depends on the interference term in the numerator. It will remain negative if SU(6) breaking is not so radical that the numerator changes sign, which is unlikely since that would require a node in the three quark wave function. Thus, the prediction is that $-\frac{1}{3}\leq\Delta d/d\leq 0$ for all x, for any three quark model. Measurements of $\Delta d/d$ up to x=0.6 are within these limits[1].

There have been suggestions from perturbative QCD that a spin up quark dominates in a spin up proton as $x\to 1$[8, 9]. These references assume that the spin up quark can be either a d or a u quark. This leads to predictions that F_{1n}/F_{1p} will approach $\frac{3}{7}$ as $x\to 1$, and $\Delta d/d$ will approach 1. But, as we have shown above, the only way for the wave function to be dominated by a spin up quark is if it is the $u\uparrow$ quark and not the $d\uparrow$ quark. Thus F_{1n}/F_{1p} must approach $\frac{1}{4}$, and $\Delta d/d$ cannot become positive for any x.

We have not explicitly included orbital angular momentum, but we do not expect it to affect our analysis as $x\to 1$. This is because the ratio of perpendicular to longitudinal momentum goes to zero as $x \to 1$, leading to L_z=0. The wave function in Eq. (1) is general enough that it includes orbital angular momentum if $L_z = 0$. Relativistic effects could reduce A_1 because the small components of the Dirac wave function reduce the quark spin projection. However this effect also depends on the ratio of perpendicular to longitudinal momentum, so that it too does not affect our results for $x \to 1$.

Testing predictions as $x \to 1$ is a challenge for experiment. Current measurements indicate that A_{1p} may be increasing above the SU(6) value of $\frac{5}{9}$, but it is not clear that the trend around $x = 0.6$ is steep enough to reach 1 at $x = 1$. A_{1n} is positive in the new measurement at $x = 0.6$[1], but is still not close to 1. Experiments at higher x will be needed to test these $x \to 1$ limits. The experimental tests are complicated by a number of problems. It is difficult to achieve accurate asymmetry measurements at large x. The Q^2 dependence increases (beyond QCD evolution) as $x \to 1$. For A_{1n}, $\Delta d/d$, and the ratio F_{1n}/F_{1p}, unfolding nuclear effects from either deuteron or helium targets becomes more uncertain as $x \to 1$.

An important question is how high an x is high enough to test competing $x \to 1$ predictions for the ratios F_{1n}/F_{1p} and $\Delta d/d$. Because of the nuclear effects[10, 11], it is questionable if any x will be high enough to test whether F_{1n}/F_{1p} approaches $\frac{3}{7}$ or $\frac{1}{4}$. However, it may be easier to test whether $\Delta d/d$ can change sign and approach 1 as $x \to 1$. This ratio is close to -$\frac{1}{3}$ at x=0.6[1], with no indication of an upward trend. Planned experiments at x up to 0.8, should be high enough x to conclude that $\Delta d/d$ cannot become positive, or ever approach +1.

REFERENCES

1. X. Zheng *et. al.*, Phys. Rev. Lett. **92**,(2004) 012004.
2. J. Franklin, Phys. Lett. **587B** (2004) 211; **599B** (2004) 347(E).
3. J. Franklin, Phys. Rev. **184** (1968) 1807.
4. S. Capstick and N. Isgur, Phys. Rev. **D 34** (1986) 2809.
5. N. Isgur and G. Karl, Phys. Rev. **D 19** (1970) 2653; **D23** (1981) 817(E).
6. W. Rolnick, Phys. Rev. **D 25** (1982) 2439, **D 26** (1982) 1804(E).
7. Z. Dziembowski, Phys. Rev. **D 37** (1988) 768.
8. G. R. Farrar and D. R. Jackson, Phys. Rev. Lett. **35** (1975) 1416.
9. S. J. Brodsky, M. Burkardt, and I. Schmidt, Nucl. Phys. **B441** (1995) 197.
10. L. W. Whitlow *et al.*, Phys. Lett. **B282** (1992) 475.
11. W. Melnitchouk and A. W. Thomas, Phys. Lett. **377** (1996) 11.

DUALITY AND HIGHER TWISTS

Quark-hadron duality: higher twists and large-x structure functions

W. Melnitchouk

Jefferson Lab, 12000 Jefferson Avenue, Newport News, VA 23606, USA

Abstract. I review recent developments in the study of quark-hadron duality in inclusive electron scattering in the resonance-scaling transition region. Results on matrix elements of twist-4 operators extracted from moments of the spin-dependent g_1 structure function suggest that duality violating higher twists are small above $Q^2 \sim 1$ GeV^2. The systematics of the x dependence of local duality are analyzed within a quark model framework, and mechanisms are identified for spin-flavor symmetry breaking which underpin the behavior of structure functions at large x.

INTRODUCTION

The nature of the transition between quark and hadron degrees of freedom in QCD is one of the most fundamental problems in strong interaction physics. Assuming that QCD can ultimately describe the physics of hadrons, the transition from quarks and gluons to hadrons can be considered in principle trivial from the point of view of quark–hadron duality: as long as one has access to a complete set of states, it is immaterial whether physical quantities are calculated in terms of elementary quark or effective hadron degrees of freedom. In practice, truncations are of course unavoidable, and it is the consequences of working with incomplete sets of basis states that allows one to expose the dynamics underlying the quark–hadron transition.

The duality between quarks and hadrons reveals itself in most dramatic fashion in inclusive electron–nucleon scattering, $eN \to eX$. Here the inclusive nucleon structure function measured in the region dominated by low-lying nucleon resonances is observed to follow a global scaling curve describing the high energy data, to which the resonance structure function averages – a phenomenon known as Bloom-Gilman duality [1]. Recent high-precision data on the F_2 structure function from Jefferson Lab [2] have provided spectacular confirmation of this duality for each of the low-lying resonance regions, down to Q^2 values of ~ 1 GeV^2 or below.

More recent studies have explored the spin and flavor dependence of duality, as well as its workings in other reactions. Concurrently there has been considerable progress in the theoretical understanding of duality, with a number of model studies elucidating both the dynamical origins of duality and its phenomenological consequences.

In this talk I first review the basic elements of duality in inclusive electron scattering, and its formulation in terms of the operator product expansion (OPE) in QCD. Using the OPE, I describe how duality violations can be used to extract matrix elements of higher twist operators from recent data on the spin dependent g_1 structure function at

CP747, *Structure of the Nucleon at Large Bjorken x,* edited by C. Bourrely, Z.-E. Meziani, and J. Soffer

intermediate Q^2. The study of local duality (the equivalence of restricted resonance sums with the scaling function over a limited range of x) is illustrated within the quark model, and the conditions for the appearance of duality are explicitly identified.

DUALITY AND QCD

The standard method of analyzing structure functions in QCD is the operator product expansion. For large Q^2 the OPE allows the *moments* of a structure function $F(x,Q^2)$ to be expanded in inverse powers of the hard momentum scale, Q^2,

$$M^{(n)}(Q^2) \equiv \int_0^1 dx\, x^{n-2}\, F(x,Q^2) \tag{1}$$

$$= \sum_{i=2,4\ldots} \frac{A_i^{(n)}}{Q^{i-2}}, \tag{2}$$

where the expansion coefficients $A_i^{(n)}$ are matrix elements of operators with a specific twist (dimension – spin). The leading twist (twist-2) term, $A_2^{(n)}$, corresponds to scattering from free partons, and is responsible for the scaling (modulo QCD logarithmic corrections) of the structure functions. The higher twist terms $A_{i>2}^{(n)}$ involve multi-quark and mixed quark-gluon operators, and contain information on long-range, nonperturbative correlations between partons.

As pointed out by De Rújula et al. [3], in the OPE language the approximate independence on Q^2 of the moments is naturally attributed to the dominance of the twist-2 term, and suppression (or cancellation) of the higher twist contributions. This observation reveals two important practical applications of duality: if one finds empirically that higher twists are small down to some scale $Q^2_{\rm min}$, then one can extend leading twist analyses of structure function data to $Q^2_{\rm min}$. For the unpolarized F_2 structure function of the nucleon, $Q^2_{\rm min}$ has been found to be around 1 GeV^2 [2], which is lower than that used in standard parton distribution analyses. If, on the other hand, duality is found to be violated, and if the violations are not overwhelming, then the data can be used to extract matrix elements of higher twist operators. This provides unique information about the strength of nonperturbative, multi-parton interaction effects in the nucleon.

HIGHER TWISTS

In the context of global analyses of parton distributions, higher twist effects are often seen as unwelcome complications. However, higher twists contain valuable information on nucleon structure – no less fundamental than that contained in leading twists – and are therefore of tremendous interest in their own right.

Recently several analyses of moments of the spin dependent g_1 structure function of the nucleon have been performed with the aim of extracting so-called *color polarizabilities* of the nucleon, which are related to twist-4 matrix elements involving quark and

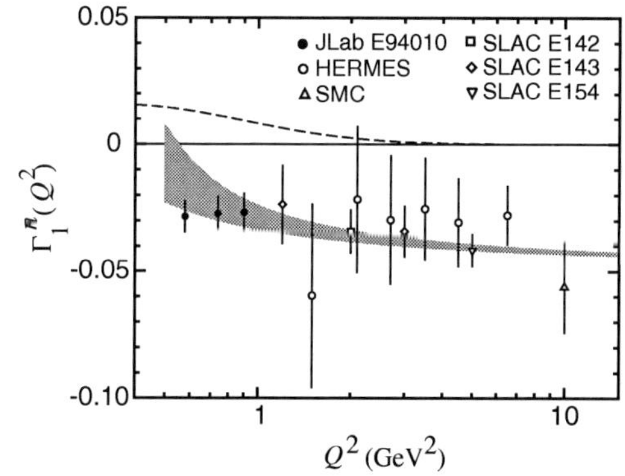

FIGURE 1. Lowest moment of the neutron g_1^n structure function [4]. The shaded band represents the uncertainty on the leading twist contribution due to α_s, and the dashed curve indicates the elastic contribution.

gluon fields. Defining the lowest ($n = 0$) moment of g_1 as

$$\Gamma_1(Q^2) = \int_0^1 dx\, g_1(x, Q^2) \,, \tag{3}$$

the OPE allows one to expand Γ_1 in a series in $1/Q^2$,

$$\Gamma_1(Q^2) \;=\; \mu_2 + \frac{\mu_4}{Q^2} + \frac{\mu_6}{Q^4} + \cdots \tag{4}$$

The leading, twist-2 term μ_2 is related to the spin carried by quarks in the nucleon, and can be expressed in terms of axial vector charges of the nucleon. The coefficient of the $1/Q^2$ correction is given by

$$\mu_4 \;=\; \frac{1}{9} M^2 \left(a_2 + 4d_2 + 4f_2\right) \,, \tag{5}$$

where a_2 is a (twist-2) target mass correction, d_2 is a twist-3 term related to the transverse g_2 structure function, and f_2 is a twist-4 term involving both quark and gluon fields

$$f_2\, M^2 S^\mu \;=\; \frac{1}{2} \sum_q e_q^2 \, \langle P,S|\, g\, \overline{\psi}_q\, \widetilde{G}^{\mu\nu} \gamma_\nu\, \psi_q \,|P,S\rangle \,. \tag{6}$$

Here ψ_q is a quark field, $\widetilde{G}^{\mu\nu} = \frac{1}{2}\varepsilon^{\mu\nu\alpha\beta} G_{\alpha\beta}$ is the dual gluon field strength tensor, g is the strong coupling constant, and P and S the momentum and spin vectors of the nucleon.

The $1/Q^2$ correction can be determined from Γ_1 data in the intermediate Q^2 region, where Q^2 is neither so large as to completely suppress the higher twists, nor so small as to render the twist expansion unreliable. The result of a recent reanalysis [4] of the world

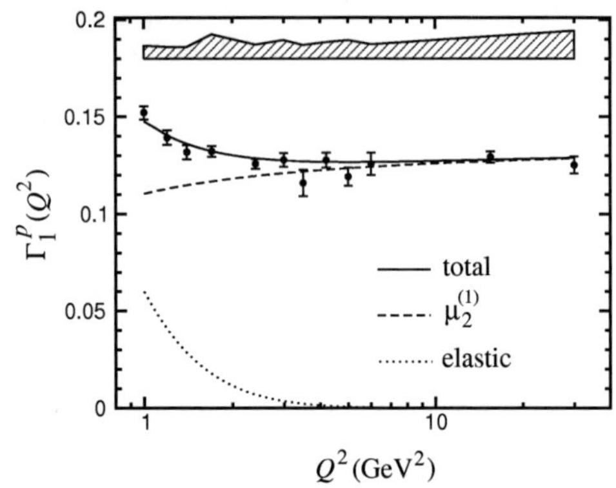

FIGURE 2. Lowest moment of the proton g_1^p structure function [5]. The error bars give statistical uncertainties only, while the systematic and low-x extrapolation errors are given by the shaded band. The shaded band represents the uncertainty on the leading twist contribution due to α_s, and the dashed curve indicates the elastic contribution.

data on the moment of the neutron g_1 structure functions is illustrated in Fig. 1, and for the proton in Fig. 2. Combining the moment data from various experiments is nontrivial since different analyses typically make use of different assumptions about extrapolations into unmeasured regions of kinematics. The structure function moments in Figs. 1 and 2 are therefore extracted using a single set of inputs and assumptions for *all* the data.

Fitting the neutron and proton data in Figs. 1 and 2, the extracted values for the f_2 matrix elements are

$$f_2^n = 0.034 \pm 0.043\,, \tag{7}$$
$$f_2^p = 0.039 \pm {}^{0.038}_{0.043}\,, \tag{8}$$

where the error includes statistical and (the more dominant) systematic uncertainties, as well as from the $x \to 0$ extrapolation and uncertainty in α_s at low Q^2. For the neutron, combining the $1/Q^2$ correction with the extracted $1/Q^4$ term, one finds that in fact the total higher twist contribution to Γ_1^n is almost exactly zero at $Q^2 = 1$ GeV^2 [4].

The twist-3 (d_2) and 4 (f_2) operators describe the response of the collective color electric and magnetic fields to the spin of the nucleon. Expressing these matrix elements in terms of the components of $\widetilde{G}^{\mu\nu}$ in the nucleon rest frame, one can relate d_2 and f_2 to color electric and magnetic polarizabilities. These are defined as [6, 7]

$$\chi_E\, 2M^2\vec{S} = \langle N|\, \vec{j}_a \times \vec{E}_a\, |N\rangle\,, \tag{9}$$
$$\chi_B\, 2M^2\vec{S} = \langle N|\, j_a^0\, \vec{B}_a\, |N\rangle\,, \tag{10}$$

respectively, where $j_a^\mu = -g\bar{\psi}\gamma^\mu t_a \psi$ is the quark current, t_a are color SU(3) matrices, and $\vec{E}_a$ and $\vec{B}_a$ are the color electric and magnetic fields, respectively. In terms of d_2 and

f_2 the color polarizabilities can be expressed as

$$\chi_E = \frac{2}{3}\left(2d_2 + f_2\right) , \quad \chi_B = \frac{1}{3}\left(4d_2 - f_2\right) . \tag{11}$$

With the above values for f_2, and the results for d_2^p from the global analysis in Ref. [5] and d_2^n from the SLAC E155 measurement [8], one finds

$$\chi_E^n = 0.033 \pm 0.029 \quad , \quad \chi_B^n = -0.001 \pm 0.016 , \tag{12}$$
$$\chi_E^p = 0.026 \pm 0.028 \quad , \quad \chi_B^p = -0.013 \mp 0.014 , \tag{13}$$

These results indicate that both the color electric and magnetic polarizabilities in the proton and neutron are relatively small, with the central values of the color electric polarizabilities being positive, and the color magnetic zero or slightly negative.

The small values of the higher twist corrections in polarized as well as unpolarized structure functions suggest that the long-range, nonperturbative interactions between quarks and gluons in the nucleon are not as dominant at $Q^2 > 1$ GeV^2 as one may have expected. This means that there are strong cancellations between nucleon resonances resulting in the dominance of the leading twist contribution to the moments. In order to see how such cancellations can take place, in the following we examine a model in which the resonance transitions can be evaluated exactly and the degree to which duality holds quantified.

LOCAL DUALITY

While duality for structure function moments can be analyzed in terms of the OPE, no such simple interpretation exists for the x dependence of the functions themselves. For this one must resort to theoretical & phenomenological models, which can be used to study how a scaling function can arise from a sum over resonances.

To understand the generation of a scaling function entirely out of resonances, each of which is described by form factors that fall rapidly with increasing Q^2, one must address the question of how coherent contributions ("square of sums of quark charges") can yield results consistent with incoherent scattering ("sum of squares of quark charges"). Close and Isgur [9] elucidated this problem by deriving the necessary conditions for duality to occur within the spin-flavor symmetric quark model (although the argument can be generalized to more complex systems). They found that for duality to hold at least one complete set of even and odd parity resonances must be summed over.

Table 1 gives the relative strengths of the contributions to the proton and neutron spin averaged and spin dependent structure functions from the $N \to N^*$ transition matrix elements, for the lowest even parity $\mathbf{56}^+$ and odd parity $\mathbf{70}^-$ representations of SU(6). The coefficients λ and ρ denote the relative strengths of the symmetric and antisymmetric contributions of the SU(6) ground state wave function, and the SU(6) limit corresponds to $\lambda = \rho$.

Summing over all of the states in the $\mathbf{56}^+$ and $\mathbf{70}^-$ multiplets gives rise to a neutron to proton ratio $R^{np} \equiv F_1^n/F_1^p = 2/3$, and polarization asymmetries $A_1^p \equiv g_1^p/F_1^p = 5/9$

TABLE 1. Relative strengths of electromagnetic $N \to N^*$ transitions in the SU(6) quark model [9, 10]. The coefficients λ and ρ denote the relative strengths of the symmetric and antisymmetric contributions of the SU(6) ground state wave function. The SU(6) limit corresponds to $\lambda = \rho$.

SU(6) rep$^{\rm n}$.	^{2}**8**[**56**$^+$]	^{4}**10**[**56**$^+$]	^{2}**8**[**70**$^-$]	^{4}**8**[**70**$^-$]	^{2}**10**[**70**$^-$]	total
F_1^p	$9\rho^2$	$8\lambda^2$	$9\rho^2$	0	λ^2	$18\rho^2+9\lambda^2$
F_1^n	$(3\rho+\lambda)^2/4$	$8\lambda^2$	$(3\rho-\lambda)^2/4$	$4\lambda^2$	λ^2	$(9\rho^2+27\lambda^2)/2$
g_1^p	$9\rho^2$	$-4\lambda^2$	$9\rho^2$	0	λ^2	$18\rho^2-3\lambda^2$
g_1^n	$(3\rho+\lambda)^2/4$	$-4\lambda^2$	$(3\rho-\lambda)^2/4$	$-2\lambda^2$	λ^2	$(9\rho^2-9\lambda^2)/2$

and $g_1^n/F_1^n = 0$, just as in the quark-parton model in which the structure functions are calculated in terms of partonic (rather than N^*) degrees of freedom.

While the SU(6) predictions for the structure functions hold approximately at $x \sim 1/3$, significant deviations are observed at larger x. It is important therefore to examine the conditions under which combinations of resonances can reproduce, via quark-hadron duality, the behavior of structure functions in the large-x region where SU(6) breaking effects are most prominent.

The most immediate breaking of the SU(6) duality could be achieved by varying the overall strengths of the coefficients for the **56**$^+$ and **70**$^-$ multiplets as a whole. However, since the cancellations of the $N \to N^*$ transitions for the case of g_1^n occur within each multiplet, a non-zero value of A_1^n can only be achieved if SU(6) is broken *within* each multiplet rather than *between* the multiplets. Some intuition is needed therefore on sensible breaking patterns within the supermultiplets.

If the mass difference between the nucleon and Δ is attributed to spin dependent forces, the energy associated with the symmetric part of the wave function will be larger than that of the antisymmetric component. A suppression of the symmetric (λ) configuration at large x will then give rise to a suppressed d quark distribution relative to u, which in turn leads to the famous neutron to proton ratio $R^{np} \to 1/4$ [11, 12].

On the other hand, duality implies that structure functions at large x are determined by transition form factors at high Q^2. At large enough Q^2 one expects these to be constrained by perturbative QCD, which predicts that photons predominantly couple to quarks with the same helicity as the nucleon. Since for massless quarks helicity is conserved, the helicity-3/2 cross section is expected to be suppressed relative to the helicity-1/2 cross section. The helicity-3/2 suppression scenario predicts that $A_1 \to 1$ for both protons and neutrons, and that the neutron to proton ratio $R^{np} \to 3/7$. This latter result is identical to that obtained in the classic quark level calculation of Farrar & Jackson [13] on the basis of perturbative QCD counting rules! Whether the $x \to 1$ behavior of structure functions follows the λ-suppression scenario or is governed by helicity conservation is being addressed experimentally at Jefferson Lab, where the 12 GeV energy upgrade will allow definitive tests of the properties of structure functions at large-x.

OUTLOOK

The discussion of duality in the simple quark model provides a clear illustration of how parton model results for structure functions can be replicated by explicit sums over nucleon resonances. More sophisticated models can of course be considered, and an important challenge will be to consistently include the effects of both resonances and the nonresonant background in the same framework.

Future experimental exploration of duality will focus on determining its flavor, spin and target dependence, as well as its workings when probed by the weak interaction, in neutrino–nucleon scattering. Another important avenue will be to explore duality in semi-inclusive reactions. Confirmation of duality here would open the way to an enormously rich semi-inclusive program in the preasymptotic regime, allowing unprecedented access to the spin and flavor distributions of the nucleon, especially at large x.

REFERENCES

1. Bloom, E. D., and Gilman, F. J., *Phys. Rev. Lett.*, **25**, 1140 (1970).
2. Niculescu, I. e. a., *Phys. Rev. Lett.*, **85**, 1186 (2000).
3. De Rújula, A., Georgi, H., and Politzer, H. D., *Annals Phys.*, **103**, 315 (1975).
4. Meziani, Z.-E. e. a., *[arXiv:hep-ph/0404066]*.
5. Osipenko, M. e. a., *[arXiv:hep-ph/0404195]*.
6. Stein, E. e. a., *Phys. Lett. B*, **353**, 107 (1995).
7. Ji, X., *[arXiv:hep-ph/9510362]*.
8. Abe, K. e. a., *Phys. Lett. B*, **493**, 19 (2000).
9. Close, F., and Isgur, N., *Phys. Lett. B*, **509**, 81 (2001).
10. Close, F., and Melnitchouk, W., *Phys. Rev. C*, **68**, 035210 (2003).
11. Close, F., and Thomas, A., *Phys. Lett. B*, **212**, 227 (1988).
12. Melnitchouk, W., and Thomas, A., *Phys. Lett. B*, **377**, 11 (1996).
13. Farrar, G., and Jackson, D., *Phys. Rev. Lett.*, **35**, 1416 (1975).

Quark-Hadron Duality: Experimental Status and Prospects

M.E. Christy

Hampton University, Hampton, VA 23668

Abstract.
The current experimental status of quark-hadron duality is discussed with particular emphasis on separated longitudinal and transverse structure functions. In addition, current and future experiments, which could help elucidate the nature of duality, are briefly discussed.

INTRODUCTION

Over the past several decades, Quantum Chromodynamics (QCD) has been established as the accepted theory of the strong interaction and has met with great success in describing the perturbative hard physics in many hadronic processes. However, the application of perturbative QCD (pQCD) is only possible for processes in which the long distance (non-perturbative) physics can be separated from the short distance (perturbative) physics. One such application is the accurate description of the logarithmic scaling violations observed in the nucleon F_2 structure function measured in inclusive lepton-nucleon scattering at large momentum transfer, Q^2. In addition, pQCD correctly predicts the vanishing of the longitudinal component (σ_L) of the deep inelastic scattering (DIS) inclusive cross section for $Q^2 \to \infty$. This vanishing of σ_L is required by helicity conservation if the exchanged virtual photon couples to *free* spin-$\frac{1}{2}$ constituents (the partons) in the nucleon, and this prediction was an early success of the quark-parton model.

While the applicability of pQCD to DIS at large enough Q^2 is guaranteed by asymptotic freedom, the long distance nature of QCD ensures that the constituents remain confined in the hadrons observed in the final state. At both small Q^2 and total hadronic mass, W^2, the cross section is dominated by the production of resonance hadronic states, which can not be described by scattering from single partons, and pQCD must give way to effective theories.

It was, therefore, surprising when it was first observed [1] that the SLAC data on the proton structure function $F_2(Q^2,x)$ measured in the nucleon resonance region (typically defined by $W^2 < 4\ \text{GeV}^2$) could be described, *on average*, by the DIS scaling curve with only the replacement of the Bjorken scaling variable, x, by the Nachtmann variable, $\xi = 2x/(1+\sqrt{1+4x^2/Q^2})$, to account for the finite nucleon mass. It seems reasonable to expect that this observed duality vetween the hadronic and scaling regimes must somehow be reflecting a fundamental relationship between asymptotic freedom and confinement. This 'quark-hadron' duality has recently been revisited [2, 3] in light of Jefferson Lab (JLab) high precision F_2 resonance region data at low Q^2. In these

CP747, *Structure of the Nucleon at Large Bjorken x,* edited by C. Bourrely, Z.-E. Meziani, and J. Soffer

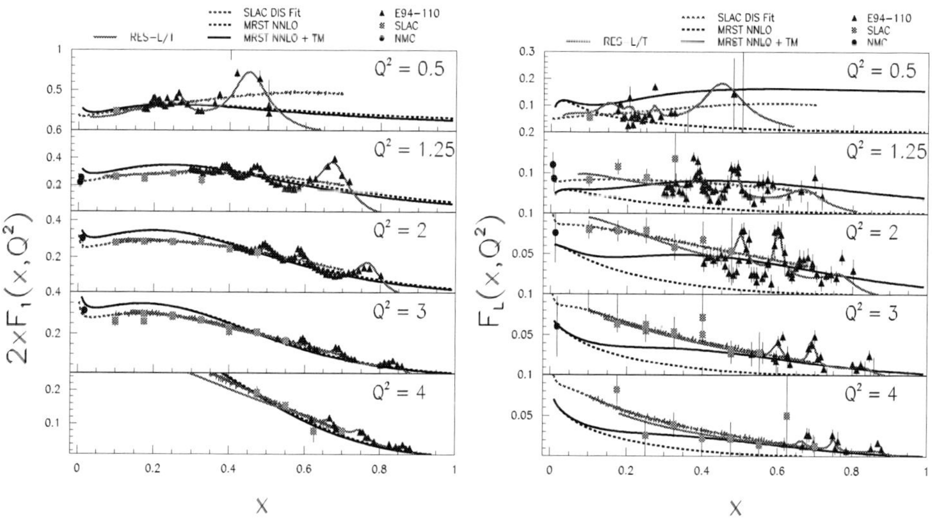

FIGURE 1. World data on proton L/T separated structure functions, $2xF_1$ (left) and F_L (right), for central Q^2 values of 0.5, 1.25, 2.0, 3.0, and 4.0 GeV^2. Fits to both the DIS and the current resonance region data are shown, as indicated by the legend. Also shown is the NNLO pQCD prediction based on the MRST2004 parton distributions both before and after including TM effects.

investigations it was found that even a 'local' averaging of F_2 (over only one or two resonances) produces a single scaling curve, and this phenomena is referred to as local duality.

Recently, experimental studies of quark-hadron duality have been extended to other nucleon structure functions. For instance, the Hermes Collaboration at HERA reported[4] evidence for quark-hadron duality in the proton spin asymmetry A_1. Such studies are just beginning and will be aided by the large body of data on resonance region spin structure functions which have either just become available or will be in the near future. Some examples for the proton are the data [5] on the spin structure function g_1 measured for $0.15 < Q^2 < 1.6$ GeV^2 in Jlab Hall B and the JLab Hall C measurements at higher Q^2 of g_1 and g_2, which are currently in the final stages of analysis. A discussion of the latter can found elsewhere in these proceedings [6].

Complementary to F_2 measurements are recent high precision measurements [7] of the separated transverse and longitudinal proton structure functions, $2xF_1$ and $F_L \equiv (1 + 4M^2x^2/Q^2)F_2 - 2xF_1$ in the resonance region. The results of these measurements are plotted in Figure 1 versus Bjorken x for central values of $Q^2 = 0.5$, 1.25, 2.0, 3.0, and 4.0 GeV^2. Also plotted are the fits to the longitudinal/transverse (L/T) separated data in both the DIS [8] and the resonance region [7], and these fits have been used to move the data to the central Q^2 values. It is observed from this data that the resonance region structure functions follow the DIS fit in both the transverse and longitudinal channels separately, as well as in the linear combination, F_2.

DUALITY AND THE OPERATOR PRODUCT EXPANSION

The first QCD description of the duality observed in F_2 was given by DeRujula, Georgi, and Politzer [9] in terms of the operator product expansion of the Cornwall-Norton moments of F_2, defined as $M_n^2(Q^2) \equiv \int_0^1 dx\ x^{n-2}\ F_2(x,Q^2)$. In the absence of power-like corrections due to parton correlations (HT effects) the Q^2 evolution of the F_2 moments is given solely by pQCD logarithmic scaling violations.

As can be seen in Figure 1, at large Q^2 the available x range for inelastic processes is dominated by DIS, while as Q^2 becomes small the resonance region covers an ever larger portion of the available x range. For the Q^2 dependence of the $n = 2$ moment to remain small, the resonances must follow the scaling curve, ie. HT must cancel on average. While this description frames the observation of duality in QCD terms, it does nothing to explain *why* the HT contributions to the moments are small. After all, clearly, the resonances themselves are the result of strong parton correlations. The question of why and over what local region in x (or more appropriately W) this cancellation takes place is an open one, and is the focus of much ongoing theoretical and experimental investigation.

The experimental $n = 2$ Cornwall-Norton moments of F_2, F_1, and F_L are shown in Figure 2, both with and without the contribution at $x = 1$ due to elastic scattering. Also shown is the NNLO pQCD calculation using the MRST2004 parton distribution functions and including the so-called target mass (TM) effects which account for kinematic power corrections resulting from the finite nucleon mass. It can be seen that the Q^2 dependence down to $\approx 2\text{GeV}^2$ is well described by pQCD, and in the case of F_2 and F_1 the description is still excellent down to values of less than 1 GeV^2 if the elastic contribution is excluded. Further investigations into the moments of the L/T separated structure functions and their analysis in terms of duality is ongoing.

EXPERIMENTAL PROSPECTS

Complementary to the proton data, experiments have been approved at JLab to measure both the resonance region L/T structure functions on the deuteron [10] and the F_2 structure function of the neutron [11]. The latter experiment (BoNuS) will measure the nearly free neutron structure function via the method of spectator tagging, and further details on this experiment can be found elsewhere [11] in these proceedings. By combining the results of these two experiments it should be possible to extract the L/T separated structure functions for the free neutron. This will allow for a comparison of how duality is made manifest in the proton and neutron, and will provide a stringent testing ground for different models of duality [12].

In addition to the unpolarized structure functions, the large body of data coming from JLab on the resonance region spin structure functions of both the proton and neutron will further help in the search for a fundamental understanding of duality. A discussion of the relevance and consequences of duality in the spin structure functions can be found in [13].

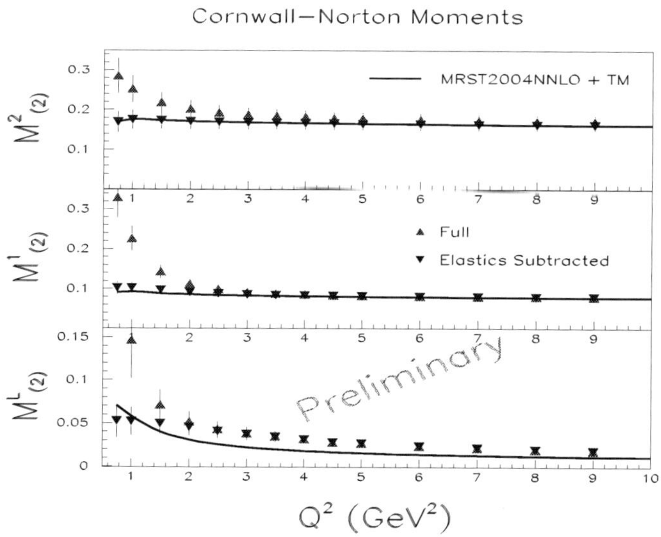

FIGURE 2. Experimental determination of the n=2 Cornwall-Norton moments of F_2 (top panel), F_1 (middle panel), and F_L (bottom panel). Also shown as the solid curve is the pQCD prediction based on the MRST2004 parton distributions including TM effects.

ACKNOWLEDGMENTS

This work was supported in part by research grants 0099540 and 9633750 from the National Science Foundation.

REFERENCES

1. Bloom, E. D., and Gilman, F. J., *Phys. Rev. D*, **4**, 2901 (1971).
2. I. Niculescu, *et.al.*, *Phys. Rev. Lett*, **85**, 1186–1189 (2000).
3. I. Niculescu, *et.al.*, *Phys. Rev. Lett*, **85**, 1182–1185 (2000).
4. A. Airapetian, *et al..*, *Phys. Rev. Lett*, **90**, 092002 (2003).
5. R. Fatemi, *et al..*, *Phys. Rev. Lett*, **91**, 222002 (2003).
6. Rondon, O. A., "Proton/Deuteron Measurements in Hall C," in *HiX2004: The 2nd International Workshop on the Structure of the Nucleon at Large Bjorken x, Marseille, France.*, 2004.
7. Y. Liang, *et al..*, *nucl-ex/0410027* (2004).
8. L. W. Whitlow, *et al..*, *Phys. Lett. B*, **250**, 193 (1990).
9. DeRujula, A., Georgi, H., and Politzer, H. D., *Ann. Phys.*, **103**, 315 (1977).
10. Christy, M. E., and Keppel, C. E., "Measurement of σ_L/σ_T on Deuterium in the Nucleon Resonance Region," in *JLab Proposal E02-109*, 2002.
11. Bueltmann, S., "The BONUS Experiment at JLab," in *HiX2004: The 2nd International Workshop on the Structure of the Nucleon at Large Bjorken x, Marseille, France.*, 2004.
12. Close, F. E., and Isgur, N., *Phys. Lett. B*, **509**, 81 (2001).
13. Carlson, C. E., and Mukhopadhyay, N. C., *Phys. Rev. D*, **58**, 094029 (1998).

Spin duality in the nucleon: Measurements at Jefferson Lab Hall A

Nilanga Liyanage [1]

University of Virginia, Charlottesville, VA, USA

Abstract. The Bloom-Gilman duality has been experimentally demonstrated for spin independent structure functions. Duality is observed when the smooth scaling curve at high momentum transfer is an average over the resonance bumps at lower momentum transfer, but at the same value of scaling variable x. Signs of quark-hadron duality for the spin Dependant structure function g_1 of the proton has been recently reported by the Hermes collaboration and the Jefferson lab Hall B collaboration. Jefferson Lab experiment 01-012 used the polarized ^3He target in Hall A for a precision extraction of the neutron spin structure function g_1^n and the virtual photon asymmetry A_1^n in the resonance region up to $Q^2 = 4(\mathrm{GeV/c})^2$. Data from E01-012 combined with Deep-Inelastic-Scattering data will provide a precision test of quark-hadron duality predictions for neutron spin structure functions. This will be one of the first precision tests of spin and flavor dependence of quark-hadron duality. The demonstration of duality for spin structure functions will enable the use of resonance data to study the nucleon spin structure in the very high x region.

Thirty years ago Bloom and Gilman [1] made the observation that the scaling curve seen at high momentum transfer is an accurate average over the resonance bumps at lower momentum transfer but at the same value of x. This duality between the resonance region, which is best described by constituent quark models, and the scaling region, which is well described by pQCD, hints a common origin for both regions. Several years after the observation of duality, De Rujula, Georgi and Politzer [2] suggested a framework based on the QCD operator product expansion (OPE) within which the averaging of the resonance bumps to the scaling curve can be interpreted in terms of the role of higher twists in DIS. In the OPE, the contributions to the QCD moments of the structure functions are organized according to the powers of $1/Q^2$. The leading terms are associated with free quark scattering and are responsible for the scaling of the structure functions. The $1/Q^2$ terms (higher twist terms) involve interactions between quarks and gluons and hence reflect elements of confinement dynamics. The experimental observation of duality for the unpolarized structure function F_2 of the proton indicates that the higher twist contributions to the proton F_2 moments are small or canceling on average, even in the low Q^2 regime where they should be maximized due to the $1/Q^2$ behavior. Recent data from Jefferson Lab Hall C [3] have further confirmed that Bloom-Gilman duality holds to a few percent level down to small values of Q^2. The striking agreement shown by these data for unpolarized structure functions between the resonance and the scaling regions raises the question whether duality holds for polarized

[1] This contribution was presented by J.-P.Chen

CP747, *Structure of the Nucleon at Large Bjorken x*, edited by C. Bourrely, Z.-E. Meziani, and J. Soffer

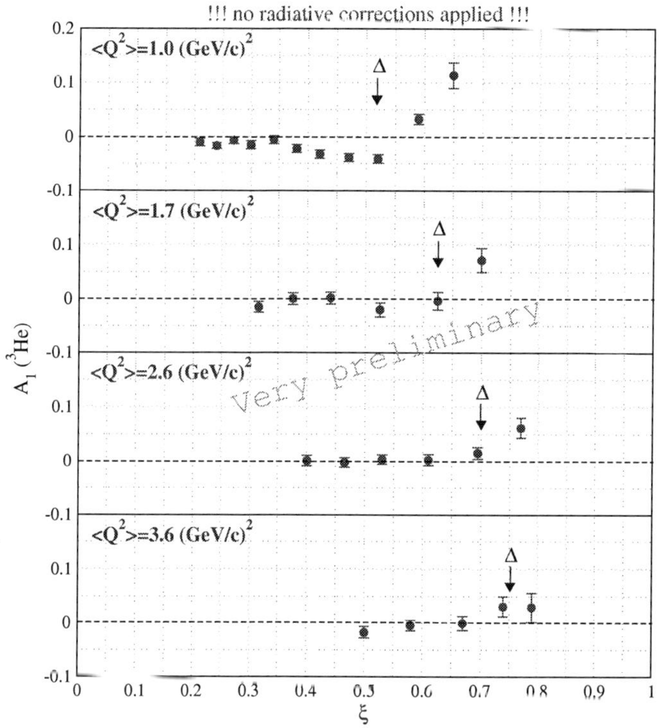

FIGURE 1. Preliminary results from E01-012 for A_1^{3He}. No radiative corrections have been applied yet and the error bars are statistical only.

structure functions as well. Hermes collaboration has recently reported observing global quark-hadron duality for the spin structure function g_1 of the proton [4]. Data from Jefferson Lab Hall B collaboration show the onset of quark hadron duality for g_1 of the proton and the deuteron [5, 6]. It should be noted that since g_1 of deuteron in the valence quark region is dominated by g_1 of the proton, spin duality observed in deuteron does not necessarily indicate spin duality for the neutron.

Jefferson Lab Hall A experiment E01-012 [7] used the Hall A polarized ^{3}He target for a precision extraction of the neutron spin structure function g_1^n and the virtual photon asymmetry A_1^n in the resonance region over a moderate Q^2 range (up to $Q^2 = 4(\mathrm{GeV/c})^2$). The results from this experiment, combined with Deep-Inelastic-Scattering data, will provide a precision test of quark-hadron (Bloom-Gilman) duality predictions for g_1^n and A_1^n.

E01-012 ran in January and February, 2003. We used the polarized beam and the polarized ^{3}He target to measure the inclusive $^3\vec{He}(e,e')X$ reaction. Both Hall A High Resolution spectrometers (HRS) were used in a symmetric configuration in electron detection mode. Three beam energies, 3 GeV, 4 GeV and 5 GeV were used with spectrometer angles of 25° and 32°. At each kinematic setting cross sections and asymmetries were measured with the target spin parallel and perpendicular to the electron beam respectively.

We have formed raw asymmetries at each setting. We are currently extracting absolute cross sections to calculate A_1 and g_1 directly from our data. In the first pass analysis however, $A_1^{^3He}$ was calculated from the measured asymmetries assuming a constant value of 0.18 for the longitudinal and transverse cross section ratio R. No radiative corrections have been applied yet and the error bars are statistical only. Figure 1 shows $A_1^{^3He}$ at the four Q^2 values as a function of the the Nachtmann scaling variable $\xi = 2x/(1+\sqrt{1+4M^2x^2/Q^2})^2$. The position of the $\Delta(1232)$ resonance is indicated by an arrow.

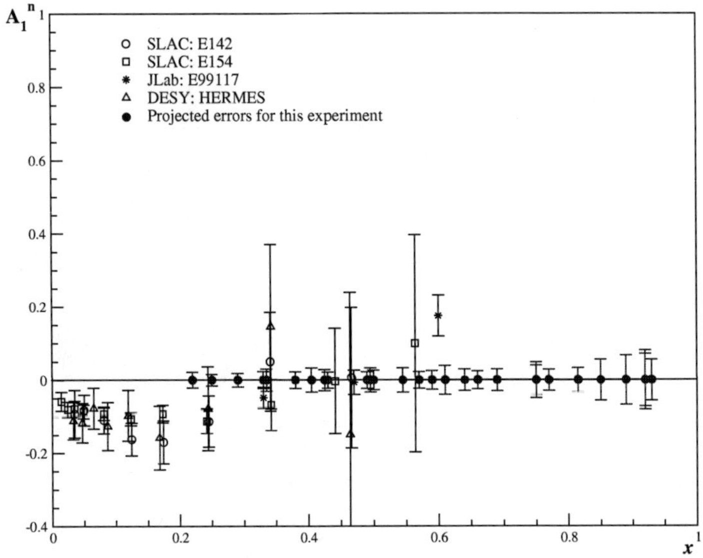

FIGURE 2. Available A_1^n DIS data and projected E01012 error bars.

The most noticeable feature of the plots is the negative contribution due to $\Delta(1232)$ at low Q^2. It has been noted that quark-hadron duality for spin structure functions is not expected in the Delta resonance region at this low Q^2. Our data seem to indicate that as the non-resonant background is starting to dominate the delta region with increasing Q^2, A_1 goes from negative to positive. Recent results from Jefferson Lab Experiment 99-117 [8] indicate that A_1^n in the DIS region also becomes positive at high x as expected by relativistic constituent quark models and by pQCD inspired models. Figure 2 shows the x range covered by E01-012 data, compared to the x range currently covered by the DIS data. As the plot indicates for $0.2 < x < 0.63$ duality can be tested with the current data. With the Jefferson Lab 12 GeV upgrade, the precision and the kinematic coverage of A_1 data in both DIS and resonance regions will improve significantly. This will enable

[2] The Nachtmann scaling variable takes into account the target mass corrections which become important at low values of Q^2. Note that for high Q^2, $\xi \to x$.

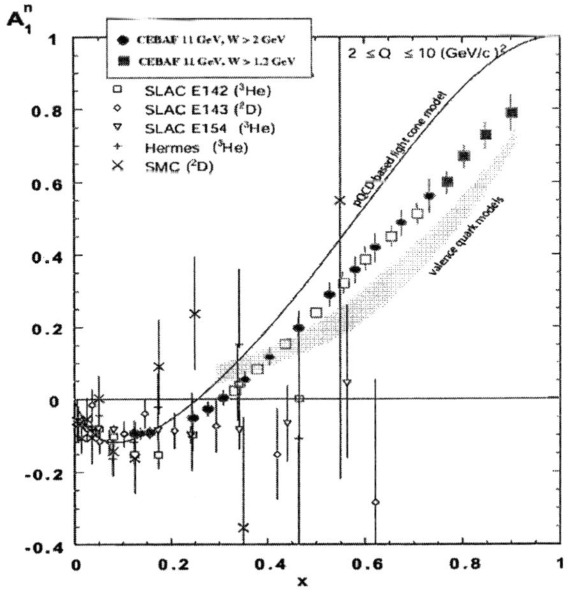

FIGURE 3. Projected A_1^n data in IS and resonance regions with Jefferson lab 12 GeV beam using the Hall A polarized ^3He target

us to extend the test of quark-hadron duality for spin structure function up to $x < 0.75$.

While it is interesting in its own right to test quark-hadron duality for spin structure functions, such a test can lead to very important applications. If duality is established between the resonance region and the DIS region in the overlap range of $x < 0.75$, we may be able to use the resonance data to study polarized quark distribution functions up to very high values of x. Figure 3 shows the projected x coverage for DIS and resonance regions with Jefferson lab 12 GeV beam.

REFERENCES

1. E. D. Bloom and F. J. Gilman, Phys. Rev. Lett. **25**, 1140 (1970); Phys. Rev. **D 4**, 2901 (1971).
2. De Rujula, H Georgi, and H. D. Politzer, Ann. Phys. (N.Y.) **103** 315(1977).
3. I. Niculescu *et al.*, Phys. Rev. Lett. **85**, 1186 (2000); I. Niculescu *et al.*, Phys. Rev. Lett. **85**, 1182 (2000).
4. A. Airapetian *et al.*, Phys. Rev. Lett. **90**, 092002 (2003).
5. Vipuli Dharmawardane, Ph.D Thesis, Old Dominion University (2004)
6. Yelena Prok, Ph.D Thesis, University of Virginia (2004).
7. Jefferson Lab Experiment E01-012, J.P Chen, S. Choi and N. Liyanage spokespeople.
8. X. Zheng *et al.* Phys. Rev. Lett. **92**, 012004 (2004) [arXiv:nucl-ex/0308011].

Higher Twists and the Measurements of g_2

W. Korsch

University of Kentucky, Lexington KY 40506

Abstract.
The 6 GeV accelerator at Jefferson Lab was used to perform precision studies of the neutron spin structure functions, $g_1^n(x,Q^2)$ and $g_2^n(x,Q^2)$. Using a dense polarized ^{3}He target allowed us to determine the lowest moment, $\Gamma_1^n(Q^2)$, for three different momentum-transfers (< 1 GeV^2) in the resonance region. Together with data from other laboratories we were able to perform a systematic analysis of higher twist effects. In addition, a first precision measurement of g_2^n outside the resonance region (W> 1.9 GeV) indicates clear evidence for higher twist contributions for Q^2 values less than 1 GeV^2.

INTRODUCTION

QCD, the quantum field theory of strong interactions, has been extremely successful in sharpening our understanding of high energy processes. Many properties of hadrons can be well described by the incoherent sum of weakly interacting point-like particles. However, at lower energies (scale of the nucleon mass) the application of QCD becomes nearly intractable, due to the non-Abelian nature of the theory. Here the point-like particles become highly correlated and interact strongly with each other. The transition from this partonic description to the hadronic description is strongly process dependent and quantitative results can only be obtained by experiment, and, in the future, by means of lattice QCD.

Jefferson Lab is an ideal environment to study this transition regime. The steady improvement in polarized target technology together with intense highly polarized electron beams allow us to perform such studies using spin-degrees of freedom. Spin sum rules and moments of spin structure functions together with the spin structure function g_2 itself are ideal observables to test QCD in this regime. The experiments described in the following sections were performed in Hall A using a polarized ^{3}He target [1].

HIGHER TWIST CONTRIBUTIONS TO $\Gamma_1^N(Q^2)$

The operator product expansion (OPE) in DIS is the commonly used framework to study the light-cone limit of the product of two electromagnetic current operators. The OPE relates the moments of the structure functions to an expansion in $1/Q^2$. For the lowest moment of the spin structure function $g_1(x,Q^2)$ this relation can be written as

CP747, *Structure of the Nucleon at Large Bjorken x,* edited by C. Bourrely, Z.-E. Meziani, and J. Soffer

$$\Gamma_1^N(Q^2) \equiv \int_0^1 dx\, g_1^N(x,Q^2) = \sum_{\tau=2,4,\ldots} \frac{\mu_\tau^N(Q^2)}{Q^{\tau-2}} = \mu_2^N(Q^2) + \frac{\mu_4^N(Q^2)}{Q^2} + \frac{\mu_6^N(Q^2)}{Q^4} + \ldots, \tag{1}$$

where the superscript N indicates a proton or a neutron. The μ_τ are the product of non-singular nucleon matrix elements and perturbatively calculable coefficient functions. Although the whole framework of the OPE is quite technical, some of the the individual terms have relatively simple physical interpretations. For example, the term $\mu_2(Q^2)$ incorporates the contribution of incoherent scattering off partons (including perturbative QCD corrections) and therefore dominates at large values of Q^2. The terms $\mu_{\tau>2}$ are due to coherent scattering off several (few) particles and are therefore a measure of quark-gluon and quark-quark correlations. The subscript of τ is called the "twist" of the operator.

In the limit of infinite Q^2, μ_2 for the spin structure function g_1 is given by:

$$\mu_2(Q^2 \to \infty) = \int_0^1 dx\, g_1(x) = \pm a_3 + a_8 + a_0 \tag{2}$$

where a_3, a_8, and a_0 are flavor triplet, octet, and singlet combinations of the hadronic currents, respectively. a_3 is related to the axial coupling constant, g_A, as extracted from neutron β-decay, a_8 can be accessed via the weak decay of hyperons, and, finally, a_0 is the singlet contribution of the quark spins to the spin of the nucleon. One assumes the validity of SU(3)$_f$ for the above relation. All these terms have been measured quite precisely (few percent level), i.e. the twist-2 part of g_1 is well known. The effect of higher twist contributions to Γ_1^N can be quantified using $\Delta\Gamma_1^N(Q^2) \equiv \Gamma_1^N(Q^2) - \mu_2(Q^2)$. The term containing μ_4 in Eq. 1 dominates the low Q^2 behavior of Γ_1^N. This twist-4 part can be written as: $\mu_4 = \frac{M^2}{9}(a_2 + 4d_2 + 4f_2)$ [2]. a_2 is a twist-2 target mass correction and is given by the second moment of g_1 [3]. A fit to the world data on g_1^n for $Q^2 >$ 5 GeV^2 results in a value of -0.0031(20) for a_2^n [4]. The term labeled d_2 is the twist-3 matrix element of g_2. For the neutron its value is 0.0079(48) [5]. The only unknown term is the twist-4 part, f_2. Figure 1 shows $\Delta\Gamma_1^n$ as a function of $1/Q^2$ with $\Delta\Sigma = 0.35$. To ensure consistency between the data sets, the low x extrapolation was performed using the parameterization suggested by Bianci and Thomas [6] for invariant masses up to $\approx$ 33 GeV.

From the two-parameter fit to the data in Fig. 1 we find: $f_2^n = 0.034 \pm 0.043$ and $\mu_6^n/M^4 = -0.019 \pm 0.017$ at a Q^2 value of 1 GeV^2 with the total uncertainties included [4]. This result is quite surprising, indicating that higher twist effects are consistent with zero down to Q^2 values of 1 GeV^2 or even lower. The lowest moment of the neutron, Γ_1^n is dominated by leading twist operators down to very low values of Q^2.

HIGHER TWIST CONTRIBUTIONS TO $G_2^N(Q^2)$

The spin structure function g_2^N is a very interesting and unique structure function. First of all, unlike g_1^N, it does not have a simple interpretation in the parton model. It is a higher

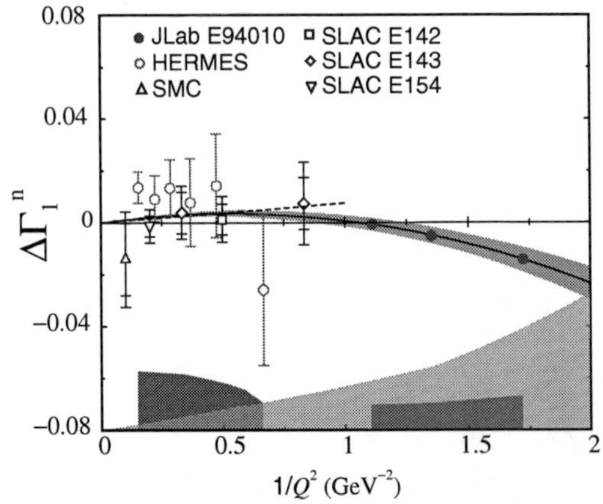

FIGURE 1. $\Delta\Gamma_1^n = \Gamma_1^n - \mu_2^n$ as a function of $1/Q^2$. The solid curve is a two-parameter fit (f_2^n and μ_6^n) to the data with $Q^2 > 0.5$ GeV^2 and the dashed curve is a one-parameter fit (f_2^n) to data with $Q^2 > 1$ GeV^2. The shaded band around the solid line is due to statistical uncertainties and the light band at the bottom corresponds to the total uncertainty.

twist structure function where more than a single parton is involved in the scattering process. Twist-3 contributions due to quark-gluon correlations are not suppressed by $1/Q$ in this structure function [7]. Wandzura and Wilczek showed that the twist-2 part of $g_2(x,Q^2)$ is completely determined by the twist-2 part of $g_1(x,Q^2)$ [8]:

$$g_2^{WW}(x,Q^2) = -g_1(x,Q^2) + \int_{x'}^{1} dx' \, \frac{g_1(x',Q^2)}{x'} \tag{3}$$

A fit to precision data of g_1 at high values of Q^2 together with a standard DGLAP evolution in NLO can be used to calculate g_2^{WW} down to Q^2 values of 0.5 GeV^2. Figure 2 shows the preliminary results of $g_2^n(Q^2)$ for the neutron at $x \approx 0.2$ as determined by experiment E97-103. This experiment was performed in Hall A using the two standard high resolution spectrometers (HRS) and the polarized ^{3}He target. Details of the experimental setup are described in [1, 9]. The figure clearly shows that g_2^n is not zero in this kinematical range. For Q^2 values above 1 GeV^2 the higher twist effects are small or even consistent with zero. However, at lower Q^2 a deviation from the twist-2 expectation is visible. The difference between the measured values of g_2^n and the twist-2 part is positive which is consistent with the value of the twist-3 matrix element, d_2^n (see above), extracted by the E155(x) [5] and E99-117 [1] collaborations.

SUMMARY

High precision studies of the lowest moment of the neutron, $\Gamma_1^n(Q^2)$, as well as a measurement of the Q^2 dependence of the spin structure function $g_2(x,Q^2)$ at $x \approx 0.2$ have been performed in Hall A at Jefferson Lab. The results yielded unprecedented

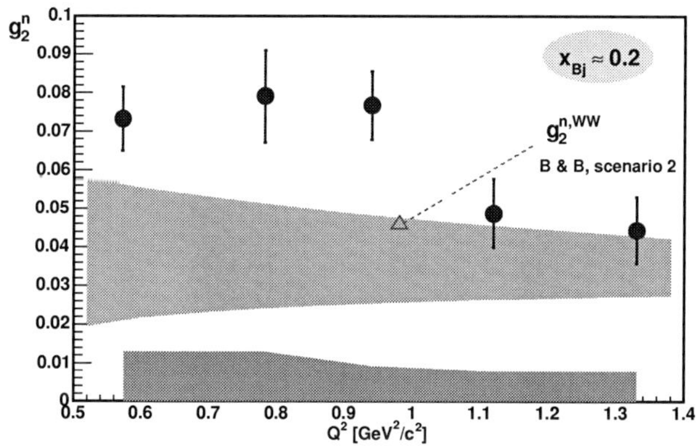

FIGURE 2. $g_2^n(Q^2)$ as a function of Q^2 at $x \approx 0.2$. The shaded band in the middle is the twist-2 expectation for g_2^n and was calculated using the NLO parameterizations of the polarized parton distribution functions from Blümlein and Böttcher [10]. The band at the bottom corresponds to the total systematic uncertainty. The data are *preliminary*.

statistical uncertainties and allowed us to search for higher twist contributions to the moment and the structure function. In both cases higher twist effects are consistent with zero down to Q^2 values of ≈ 1 GeV^2. Higher twist effects in g_2^n might become important below $Q^2 = 1$ GeV^2. These studies will continue when the energy upgrade to 12 GeV at Jefferson Lab is accomplished.

ACKNOWLEDGMENTS

The results presented here were obtained with strong support of the EG1b, Hall A, and the Polarized ^{3}He Collaborations. This work was supported by the US Department of Energy (DOE) under contracts DE-AC05-84ER40150 and DE-FG02-99ER41101, the US National Science Foundation (NSF), and the DOE/EPSCoR program.

REFERENCES

1. X. Zheng *et al.*, nucl-ex/0405006.
2. X. Ji and P. Unrau, Phys. Lett. B **333**, 228 (1994); E. Stein *et al.*, Phys. Lett. B **353**, 107 (1995); X. Ji and W. Melnitchouk, Phys. Rev. D **56**, 1 (1997).
3. X. Ji, Phys. Lett. B **309**, 187 (1993).
4. Z.-E. Meziani *et al.*, hep-ph/0404066.
5. P. L. Anthony *et al.*, Phys. Lett. B **553**, 18 (2003).
6. N. Bianchi and E. Thomas, Phys. Lett. B **450**, 439 (1999).
7. R. L. Jaffe, Comments Nucl. Part. Phys. **19**, 239 (1990).
8. S. Wandzura and F. Wilcczek, Phys. Lett. B **72**, 195 (1977).
9. K. Kramer, Ph. D. thesis, College of William and Mary, 2003.
10. J. Blümlein and H. Böttcher, Nucl. Phys. B **636**, 225 (2002).

Quark-hadron duality and higher twist contributions in structure functions

A. Fantoni*, N. Bianchi* and S. Liuti†

*Laboratori Nazionali dell'INFN, Via E.Fermi 40, 00044 Frascati (RM) Italy
†University of Virginia, Charlottesville, Virginia 22901, USA

Abstract. The quark-hadron duality is studied in a systematic way for polarized and unpolarized structure functions, by taking into account all the available data in the resonance region. In both cases, a precise perturbative QCD based analysis to the integrals of the structure functions in the resonance region has been done: non perturbative contributions have been disentangled, the higher twist contributions have been evaluated and compared with the ones extracted in the DIS region. A different behavior for the unpolarized and polarized structure functions at low Q^2 has been found.

The structure and the interaction of hadrons is generally described by two different but complementary approaches: the quark-gluon context at high energy, where the quarks are asymptotically free, and the meson and baryon description at low energy, where the effects of the confinement are large. In some specific cases where the natural description in terms of hadrons should be applied, the quark-gluon description can be successfully also used. This evidence is called quark-hadron duality and it was introduced by Bloom and Gilman who noted a relationship between the nucleon resonance region and the deep-inelastic-scattering (DIS) one [1].

A quantitative analysis of the Q^2 dependence of quark-hadron duality in both polarized and unpolarized ep scattering is presented. All current data in the resonance region, $1 \leq W^2 \leq 4$ GeV^2, have been taken into account. For the unpolarized case it has been used the data obtained at Jefferson Lab in the range $0.3 \leq Q^2 \leq 5$ GeV^2 [2], and the data from SLAC ([3] and references therein) for $Q^2 \geq 4$ GeV^2. For the polarized case there are only few experimental data in the resonance region. One set is part of the E143 data [4], and it corresponds to $Q^2 = 0.5$ and 1.2 GeV^2. Another set is the one from HERMES [5, 6] in the range $1.2 \leq Q^2 \leq 12$ GeV^2.

The full procedure of the analysis is described in [7]. The quark-hadron duality in DIS is studied by considering the ratio of the integrals of the structure functions integrated in a defined x-range, corresponding to the W range of the resonance region. The structure function in the numerator is evaluated using the experimental data in the resonance region, while the one at the denominator is calculated from parametrizations that reproduce the DIS behavior of the data at large Q^2. The ratios R_{unpol} and R_{pol} have been calculated in unpolarized and polarized cases, respectively.

In order to understand the nature of the remaining Q^2 dependence that cannot be described by NLO pQCD evolution, the effect of target mass corrections and large x resummation have been studied. The analysis was performed by using x as an integration variable, which avoids the ambiguities associated to the usage of other *ad hoc* kinemati-

CP747, *Structure of the Nucleon at Large Bjorken x,* edited by C. Bourrely, Z.-E. Meziani, and J. Soffer

cal variables. Standard input parametrizations with initial scale $Q_o^2 = 1$ GeV^2 have been used. Once both effects have been subtracted from the data, and assuming the validity of the twist expansion, one can interpret any remaining discrepancy of the ratio from unity in terms of higher twist.

The Target Mass Corrections (TMC) are necessary to take into account the finite mass of the initial nucleon. They are corrections to the leading twist (LT) part of the unpolarized structure function F_2.

Large x Resummation (LxR) effects arise formally from terms containing powers of $\ln(1-z)$, z being the longitudinal variable in the evolution equations, that are present in the Wilson coefficient functions $C(z)$. The logarithmic terms in $C_{NS}(z)$ become very large at large x, and they need to be resummed to all orders in α_S [8].

All the effects are summarized in Fig.1, where the ratio between the resonance region and the 'DIS' one is reported for the unpolarized and for the polarized case: the numerator is obtained from the experimental data, while the denominator includes the different components of the present analysis, one by one.

For unpolarized scattering it has been found that TMC and LxR diminish considerably the space left for HT contributions. The contribution of TMC is large at the largest values of Q^2 because these correspond also to large x values. Moreover, the effect of TMC is larger than the one of LxR. Similarly, in polarized scattering the inclusion of TMC and LxR decreases the ratio R_{pol}^{LT}. However, in this case these effects are included almost completely within the error bars. Clearly, duality is strongly violated at $Q^2 < 1.7$ GeV^2. The present mismatch between the unpolarized and polarized low Q^2 behavior might indicate that factorization is broken differently for the two processes, and that the universality of quark descriptions no longer holds. The discrepancy from unity of the ratios already presented is interpreted in terms of HTs. In Figs. 2,3 the question of the size of the HT corrections is addressed explicitely. For F_2, they are defined as:

$$H(x,Q^2) = Q^2\left(F_2^{\mathrm{res}}(x,Q^2) - F_2^{\mathrm{LT}}\right) \; ; \; C_{HT}(x) = \frac{H(x,Q^2)}{F_2^{pQCD}(x/Q^2)} = Q^2 \frac{F_2^{\mathrm{res}}(x,Q^2) - F_2^{LT}}{F_2^{\mathrm{LT}}} \qquad (1)$$

A similar expression is assumed for g_1. C_{HT} is the so-called factorized form obtained by assuming that the Q^2 dependences of the LT and of the HT parts are similar and therefore they cancel out in the ratio. Although the anomalous dimensions of the HT part could in principle be different, such a discrepancy has not been found so far in accurate analyses of DIS data. The HT coefficient, C_{HT} has been evaluated for the three cases listed also in Fig.1, namely with respect to the NLO pQCD calculation, to NLO+TMC and to NLO+TMC+LxR. The values of $1 + C_{HT}/Q^2$ are plotted in Fig.2 (left panel) as a function of the average value of x for each spectrum. One can see that the NLO+TMC+LxR analysis yields very small values for C_{HT} in the whole range of x. Furthermore, the extracted values are consistent with the ones obtained in Ref.[9] using a different method, however the present extraction method gives more accurate results.

In the polarized case (right panel) the HTs are small within the given precision, for $Q^2 > 1.7$ GeV^2, but they appear to drop dramatically below zero for lower Q^2 values. The inclusion of TMC and LxR renders these terms consistent with zero at the larger Q^2 values, but it does not modify substantially their behavior at lower Q^2. From a

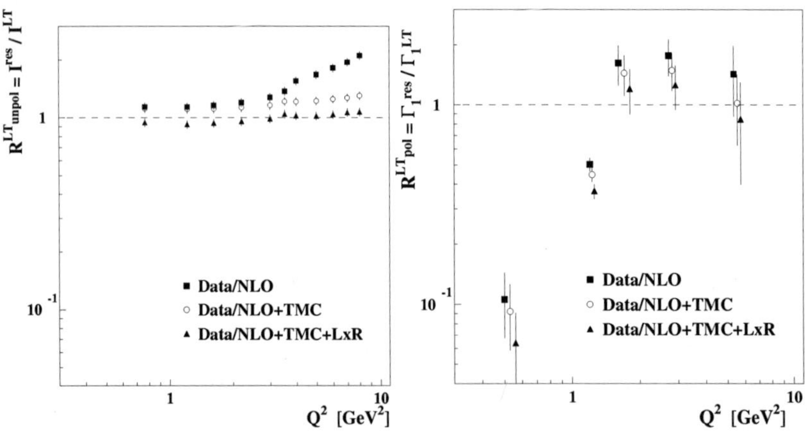

FIGURE 1. Ratio between the integrals of the measured structure functions and the calculated ones plotted as a function of Q^2 for the unpolarized case (left) and the polarized one (right). The calculation includes one by one the effects of NLO pQCD.

comparison with results of ratio including phenomenological parametrizations [7] that includes some extra non perturbative behaviors it's possible to see that their effect seems not be large.

As mentioned in [10], an accurate extraction of the Q^2 dependence is fundamental. The results shown in [7] have been extendend with the recent results available in literature for the unpolarized and polarized case. In Fig. 3, the higher twist coefficients of the present extraction in the resonance region are compared with all existing results of HT coefficients calculated in the DIS region. For the unpolarized case there is the BCDMS evaluation [12] (already shown in [7]) and the new MRST calculation [11]. The HT coefficients have been calculated following the factorization formula, displayed in Eq. 1, which can be expressed as $F_2^{LT+HT} = F_2^{LT} \cdot (1 + C(x)/Q^2)$. For the polarized one the only data available [13] are using the additive formula, for which $F_2^{LT+HT} = F_2^{LT} + H(x)/Q^2$ In the expression of $C(x)$ and $H(x)$ there is no Q^2 dependence hidden. A different behavior for the unpolarized and polarized HT terms is evident. In details, for the unpolarized case in the region of high x there is a big discrepancy between the HT terms in the resonance region ($C_{res}(x)$) and in the DIS region ($C_{DIS}(x)$). In the polarized case, at high x this comparison is little bit complicated, due to the fact that there is only one point with x >0.6 in the resonance region ($H_{res}(x)$) and no value for the DIS region ($H_{DIS}(x)$).

REFERENCES

1. E.D. Bloom and F.J. Gilman, Phys. Rev. Lett. **25**, (1970) 1140; Phys. Rev. D **4**, (1971) 2901.
2. I. Niculescu *et al.*, Phys. Rev. Lett. **85**, (2000) 1186.
3. L.W. Whitlow *et al.*, Phys. Lett. B **282**, (1992) 475.
4. E143 Coll., K. Abe *et al.*, Phys. Rev. D **58**, (1998) 112003.
5. HERMES Coll., A. Airapetian *et al.*, Phys. Rev. Lett. **90**, (2003) 092002.

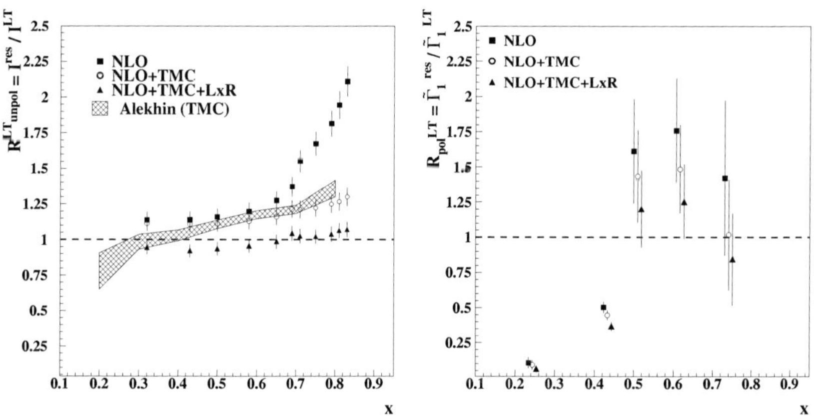

FIGURE 2. HT coefficients extracted in the resonance region according to Eq.(1). Shown in the figure is the quantity $1 + C_{HT}(x)/Q^2$. For comparison the values for HT coefficients obtained in ref. [14] using DIS data and the effect of TMC are shown. The left (right) panel refers to the unpolarized (polarized) case.

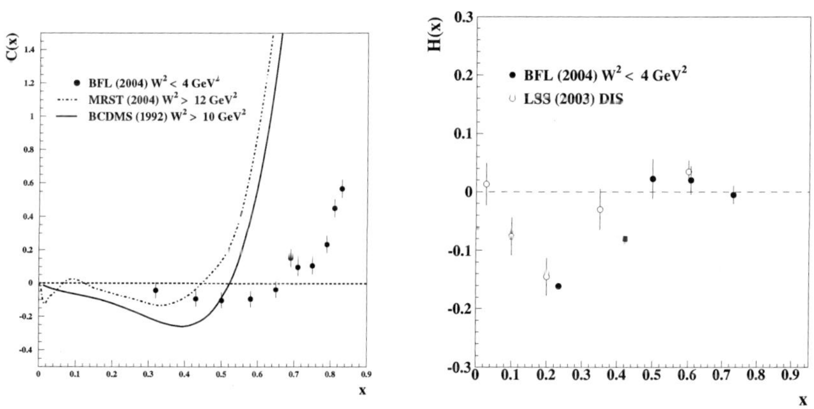

FIGURE 3. HT coefficients for the unpolarized (left) and polarized (right) case. The full points represent the values in the resonance region, while the empty points and the curves are related to the DIS region.

6. A. Fantoni, Eur. Phys. J **A17**, (2003) 385.
7. N. Bianchi, A. Fantoni & S. Liuti, Phys. Rev. D **69**, (2004) 014505.
8. R. G. Roberts, Eur. Phys. J **C10**, (1999) 697.
9. S. Liuti, R. Ent, C.E. Keppel & I. Niculescu, Phys. Rev. Lett. **89**, (2002) 162001.
10. A. Fantoni, Eur. Phys. J **A**, (2005) to be published.
11. A.D. Martin *et al.*, hep-ph/0407311.
12. M. Virchaux & A. Milsztajn, Phys. Lett. B **274**, (1992) 221.
13. E. Leader *et al.*, Phys. Rev. D **67**, (2003) 074017.
14. S.I. Alekhin, Phys. Rev. D **68**, (2003) 014002; arXiv:hep-ph/0212370.

SEMI-INCLUSIVE
DEEP INELASTIC SCATTERING

Semi-Inclusive DIS and Transversity

Philip G. Ratcliffe

Dipartimento di Fisica e Matematica, Università degli Studi dell'Insubria
and
Istituto Nazionale di Fisica Nucleare, Sezione di Milano

Abstract. A review is presented of some aspects of semi-inclusive deeply inelastic scattering and transversity. In particular, the role of k_T-dependent and higher-twist (or multi-parton) distributions in generating single-spin asymmetries is discussed.

INTRODUCTION

A large part of the material presented here is gleaned from [1, 2], where the coverage is more complete. Therefore, much credit and thanks go to my two collaborators: Enzo Barone and Alessandro Drago. Further (less condensed and more complete) conference reviews may also be found in other proceedings [3, 4]. Finally, the space available forces a limited choice of topics and there are, unfortunately, many I cannot even mention.

By way of motivation for the interest, let me recall that transversity is the last piece in the partonic jig-saw puzzle of the hadron. A number of experiments aim at its study: HERMES, COMPASS and the RHIC spin programme. And while the QCD theory describing transversity is now solid, transverse-spin effects are notoriously surprising; consider, *e.g.*, the unexpectedly large single-spin asymmetries (SSA's).

Transversity then is the third and final leading twist ($\tau = 2$) partonic distribution function. Now, it is important to distinguish between partonic distributions: $q(x)$, $\Delta q(x)$, $\Delta_T q(x)$ *etc.* and DIS structure functions: F_1, F_2, g_1, g_2 *etc.* In both the unpolarised and helicity-dependent cases at leading twist there is a simple correspondence between the two: DIS structure functions are little more than weighted sums of partonic distributions (or densities). However, in the transverse-spin case, firstly, there is no DIS transversity structure function and, secondly, g_2 does not correspond to a partonic density.

The parton-model description provides a simple probabilistic view of hadron structure (herein I shall use f_1, g_1 and h_1 generically):

$f_1(x)$ or $q(x)$ represent the probability of finding a given parton type with light-cone momentum fraction x inside a given parent hadron;

$g_1(x)$ or $\Delta q(x)$ the same but weighted by parton helicity relative to parent helicity;

$h_1(x)$ or $\Delta_T q(x)$ weighted by transverse-spin projection relative to parent transverse-spin direction.

Note, however, $h_1(x)$ does not measure quark transverse polarisation; g_2 has this role.

CP747, *Structure of the Nucleon at Large Bjorken x,* edited by C. Bourrely, Z.-E. Meziani, and J. Soffer

Turning now to SSA's, they generically reflect correlations of the form $\vec{s}\cdot(\vec{p}\wedge\vec{k})$, where $\vec{s}$ is a particle spin vector, $\vec{p}$ and $\vec{k}$ are initial/final particle/jet momenta; for example, $\vec{s}$ might be a target polarisation vector (transverse), $\vec{p}$ a beam direction and $\vec{k}$ a final-state particle direction. Thus, spins involved in SSA's are typically transverse (however, there are exceptions). Transforming the spin basis from transversity to helicity such an asymmetry takes on the schematic form, using $|\uparrow/\downarrow\rangle = \frac{1}{\sqrt{2}}[|+\rangle \pm i|-\rangle]$,

$$\mathscr{A}_N \sim \frac{\langle\uparrow|\uparrow\rangle - \langle\downarrow|\downarrow\rangle}{\langle\uparrow|\uparrow\rangle + \langle\downarrow|\downarrow\rangle} \sim \frac{2\,\mathrm{Im}\,\langle+|-\rangle}{\langle+|+\rangle + \langle-|-\rangle}. \tag{1}$$

The appearance of both $|+\rangle$ and $|-\rangle$ in the numerator indicates a spin-flip amplitude. Indeed, interference between a spin-flip and a non-flip amplitude, with a relative phase difference, is necessary. It was realised early on [5] that in the Born approximation and massless (or high-energy) limit a gauge theory, such as QCD, cannot furnish either requirement: fermion helicity is conserved and tree diagrams are real. This naturally led to the claim [5] that "... *observation of significant polarizations in the above reactions would contradict either QCD or its applicability.*"

Now, although large experimental asymmetries were found, QCD survived! A way out was discovered [6] by considering the three-parton correlators involved in g_2: the relevant mass scale when considering helicity flip is not the quark mass but a hadronic mass; and the pseudo two-loop nature of the diagrams leads to an imaginary part in certain regions of partonic phase space. However, it took some time before progress was made and the richness of the available structure was fully exploited (see, *e.g.*, [7]).

A BRIEF HISTORY OF TRANSVERSITY

Quark *transversity* (the concept though not the term[1]) was introduced by Ralston and Soper [9] in the Drell–Yan (DY) process. An important clarification of transversity, the role of chirally-odd parton distributions and the general twist classification was provided by Jaffe and Ji in [10][2], see Table 1. As twist runs from 2 to 4, the number of "bad" light-cone components runs from 0 to 2.

Table 1. The twist classification (up to $\tau = 4$) for the various parton distributions, including spin dependence (both transverse and longitudinal).

twist	2	3	4
unpolarised	f_1	e	f_4
longitudinal	g_1	h_L	g_3
transverse	h_1	g_T	h_3

The leading-order (LO) anomalous dimensions for transversity were first calculated very early on in [12], which went unnoticed, and later re-calculated in [13]. They were also calculated (as part of the g_2 evolution) in [14–17]. The next-to-leading-order (NLO) DY coefficient functions were calculated in [18, 19] while the NLO anomalous dimensions were calculated in [20–22]. The effects of evolution have been studied by many authors—for more details and a general review see, *e.g.*, [1].

[1] For very early use of the term *transversity*, see [8].
[2] The term *transversity*, following [11], was also suggested to distinguish it from transverse spin.

TRANSVERSE-SPIN BASICS

Transversity is one of three twist-two structures:

$$\begin{aligned} q(x) &= \int \frac{d\xi^-}{4\pi} e^{ixP^+\xi^-} \langle PS|\overline{\psi}(0)\gamma^+\psi(0,\xi^-,\vec{0}_\perp)|PS\rangle, \\ \Delta q(x) &= \int \frac{d\xi^-}{4\pi} e^{ixP^+\xi^-} \langle PS|\overline{\psi}(0)\gamma^+\gamma_5\psi(0,\xi^-,\vec{0}_\perp)|PS\rangle, \\ \Delta_T q(x) &= \int \frac{d\xi^-}{4\pi} e^{ixP^+\xi^-} \langle PS|\overline{\psi}(0)\gamma^+\gamma^1\gamma_5\psi(0,\xi^-,\vec{0}_\perp)|PS\rangle. \end{aligned} \tag{2}$$

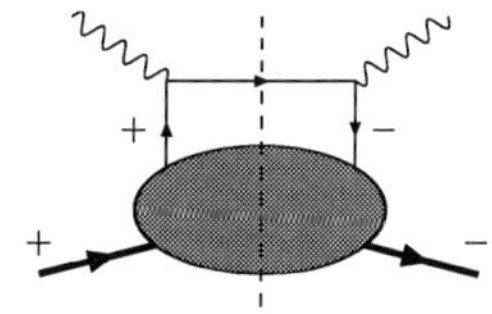

Figure 1. Chirality-flip DIS diagram forbidden by helicity conservation.

Here the γ_5 matrix signals spin dependence while the extra γ^1 matrix in $\Delta_T q(x)$ signals chirality flip; this last precludes transversity contributions in DIS, see Fig. 1.

QCD and electroweak vertices conserve quark chirality. Thus, charged-current interactions exclude transversity, since only a single chirality interacts. Note though that chirality flip is not a problem if the quarks connect to different hadrons (as in DY). However, a caveat to measuring transversity in DY is that the azimuth of the lepton pair be left unintegrated. The same observations lead to another important consequence: the LO QCD evolution of transversity is of the non-singlet type, see Fig. 2.

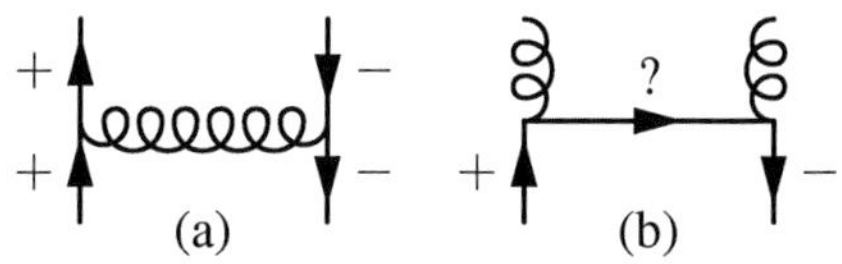

Figure 2. (a) Evolution kernel for transversity. (b) Disallowed gluon–fermion mixing kernel.

TRANSVERSITY: MODELS AND QCD

Having opposite charge-conjugation properties, Δq and $\Delta_T q$ are not simply related. Decomposing Δq as $\Delta q^{\rm NS} + \Delta q^{\rm S}$, one might imagine $\Delta_T q \simeq \Delta q^{\rm NS}$. In the non-relativistic quark model this is the case. However, in a relativistic model the lower quark wave-function components spoil the identity and, *e.g.*, the MIT bag gives [10]:

$$\Delta q^{\rm NS} = c \int r^2 dr \left(f^2 - \tfrac{1}{3}g^2\right) \quad \text{while} \quad \Delta_T q = c \int r^2 dr \left(f^2 + \tfrac{1}{3}g^2\right), \tag{3}$$

where f, g (the upper, lower components) contribute differently due to the extra γ_0.

By considering hadron–parton amplitudes Soffer [23] derived an intriguing bound: $|\Delta_T q(x)| \leq q_+(x)$ or $2|\Delta_T q(x)| \leq q(x) + \Delta q(x)$. In QCD the question arises as to the effects of evolution on this bound [24]: its maintenance has been checked explicitly to LO in [25], to NLO in [26] and discussed on more general grounds in [27]; experimental verification thus becomes an important test.

The LO (non-singlet) DGLAP quark–quark splitting functions are

$$\Delta P_{qq}^{(0)}(x) = P_{qq}^{(0)}(x) = C_{\rm F} \left[\frac{1+x^2}{1-x}\right]_+ \qquad \text{(due to helicity conservation)}, \tag{4a}$$

$$\Delta_T P_{qq}^{(0)}(x) = P_{qq}^{(0)}(x) - C_{\rm F}(1-x). \tag{4b}$$

Thus, for both $P_{qq}^{(0)}$ and $\Delta P_{qq}^{(0)}$ the first moments vanish (leading to conservation laws and sum rules). The same is not true for $\Delta_T P_{qq}$ and so, there are no transversity sum rules.

The importance of this difference has been studied both at LO and NLO. The sign indicates that quark transverse polarisation decreases, as compared to longitudinal polarisation. This is both bad news, since transversity effects therefore die at very high energies (though only very slowly), and good news, since evolution effects are stronger and therefore more measurable (*i.e.*, they are good for testing QCD).

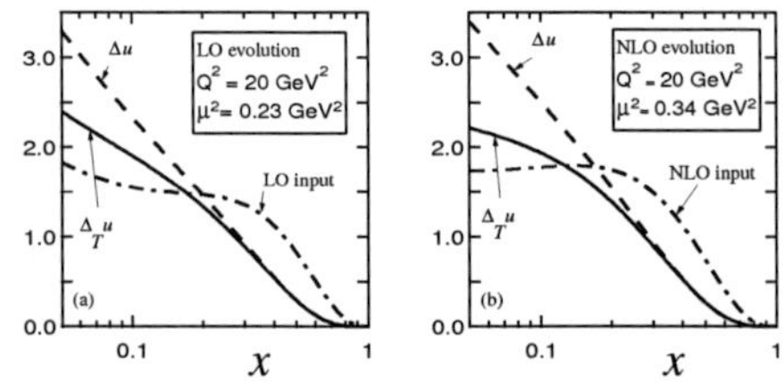

Figure 3. The Q^2-evolution of $\Delta_T u(x,Q^2)$ and $\Delta u(x,Q^2)$ at (a) LO and (b) NLO, from [20].

The chirally-odd nature of transversity means that at least two hadrons are needed to probe transversity: $p^\uparrow p^\uparrow \to \mu^+\mu^- X$ [9, 28]; $ep^\uparrow \to e'\pi X$ [29–31]; $pp^\uparrow \to \Lambda^\uparrow X$ [32]; $ep^\uparrow \to \Lambda^\uparrow X$ [33]; $ep^\uparrow \to e'\pi^+\pi^- X$ [30, 34, 35] *etc.* There are thus two basic categories: double- and single-spin asymmetries. I shall now examine the latter more closely.

SINGLE-SPIN ASYMMETRIES

SSA's can be generated in various ways: higher-twist, k_T-dependent distribution or fragmentation functions; or interference and vector-meson fragmentation functions. Consider single-hadron production with a transversely polarised beam or target:

$$A^\uparrow(P_A) + B(P_B) \to h(P_h) + X, \tag{5}$$

where A is transversely polarised and the unpolarised (or spinless) hadron h is produced at large transverse momentum; PQCD is thus applicable. One measures the SSA:

$$\mathscr{A}_T^h = \frac{d\sigma(\vec{S}_T) - d\sigma(-\vec{S}_T)}{d\sigma(\vec{S}_T) + d\sigma(-\vec{S}_T)}. \tag{6}$$

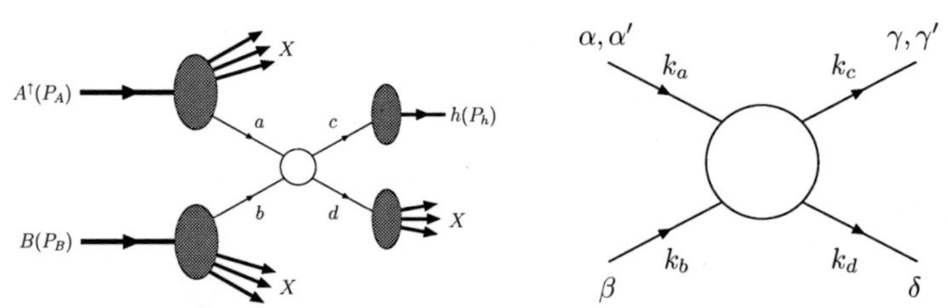

Figure 4. Left: a schematic representation of single-hadron production with a transversely polarised beam or target. Right: the hard partonic scattering amplitude $\mathscr{M}_{\alpha\beta\gamma\delta}$ ($\alpha \ldots \delta$ are Dirac indices).

According to the factorisation theorem, the differential cross-section for semi-inclusive production may be written formally as

$$\mathrm{d}\sigma = \sum_{abc} \sum_{\alpha\alpha'\gamma\gamma'} \rho^a_{\alpha'\alpha} f_a(x_a) \otimes f_b(x_b) \otimes \mathrm{d}\hat{\sigma}_{\alpha\alpha'\gamma\gamma'} \otimes \mathscr{D}^{\gamma'\gamma}_{h/c}(z), \tag{7}$$

where f_a (f_b) is the density of parton type a (b) in hadron A (B), $\rho^a_{\alpha\alpha'}$ is the spin density matrix of parton a, and $\mathscr{D}^{\gamma\gamma'}_{h/c}$ is the fragmentation matrix for parton c into final hadron h. The elementary cross-section is ($\mathscr{M}_{\alpha\beta\gamma\delta}$ is the hard partonic scattering amplitude)

$$\left(\frac{\mathrm{d}\hat{\sigma}}{\mathrm{d}\hat{t}}\right)_{\alpha\alpha'\gamma\gamma'} = \frac{1}{2}\sum_{\beta}\left(\frac{\mathrm{d}\hat{\sigma}}{\mathrm{d}\hat{t}}\right)_{\alpha\alpha'\beta\beta'\gamma\gamma'} = \frac{1}{16\pi\hat{s}^2}\frac{1}{2}\sum_{\beta\delta}\mathscr{M}_{\alpha\beta\gamma\delta}\mathscr{M}^*_{\alpha'\beta\gamma'\delta}. \tag{8}$$

The off-diagonal elements of $\mathscr{D}^{\gamma\gamma'}_{h/c}$ vanish, *i.e.*, $\mathscr{D}^{\gamma\gamma'}_{h/c} \propto \delta_{\gamma\gamma'}$ when the hadron produced is unpolarised. Helicity conservation then implies $\alpha = \alpha'$, thus precluding any dependence on the polarisation of hadron A. Either intrinsic quark transverse motion, or higher-twist effects can circumvent this conclusion.

Quark intrinsic transverse motion can generate SSA's in three different ways (always necessarily as T-odd effects):

1. $\vec{\kappa}_T$ in the final hadron h allows $\mathscr{D}^{\gamma\gamma'}_{h/c}$ to be non-diagonal (fragmentation level).
2. $\vec{k}_T$ in hadron A requires $f_a(x_a)$ be replaced by $\mathscr{P}_a(x_a,\vec{k}_T)$, which may depend on the spin of A (distribution level).
3. $\vec{k}'_T$ in hadron B requires $f_b(x_b)$ be replaced by $\mathscr{P}_b(x_b,\vec{k}'_T)$. The transverse spin of b in the unpolarised B may then couple to the transverse spin of a (distribution level).

The three mechanisms are: 1. Collins [29], 2. Sivers [36]; and 3. an effect studied in by Boer in [37]. All such intrinsic-$\vec{\kappa}_T$, -$\vec{k}_T$, or $\vec{k}'_T$ effects are T-odd; they require initial- or final-state interactions. Note that when quark transverse motion is included, the QCD factorisation theorem is not completely proven, but see recent work in [38].

The Collins mechanism exploits intrinsic quark motion inside the produced hadron h. Thus, assuming factorisation to be valid, the cross-section difference is

$$E_h\frac{\mathrm{d}^3\sigma(\vec{S}_T)}{\mathrm{d}^3\vec{P}_h} - E_h\frac{\mathrm{d}^3\sigma(-\vec{S}_T)}{\mathrm{d}^3\vec{P}_h} = -2\,|\vec{S}_T|\sum_{abc}\int\mathrm{d}x_a\int\mathrm{d}x_b\int\mathrm{d}^2\vec{\kappa}_T\,\frac{1}{\pi z}$$
$$\times\,\Delta_T f_a(x_a)\,f_b(x_b)\,\Delta_{TT}\hat{\sigma}(x_a,x_b,\vec{\kappa}_T)\,\Delta^0_T D_{h/c}(z,\vec{\kappa}^2_T), \tag{9}$$

where $\Delta_{TT}\hat{\sigma}$ is a partonic spin-transfer differential cross-section difference (see, *e.g.*, [1]). The Sivers effect relies on T-odd k_T-dependent partonic distribution functions and predicts SSA's of the form

$$E_h\frac{\mathrm{d}^3\sigma(\vec{S}_T)}{\mathrm{d}^3\vec{P}_h} - E_h\frac{\mathrm{d}^3\sigma(-\vec{S}_T)}{\mathrm{d}^3\vec{P}_h} = |\vec{S}_T|\sum_{abc}\int\mathrm{d}x_a\int\mathrm{d}x_b\int\mathrm{d}^2\vec{k}_T\,\frac{1}{\pi z}$$
$$\times\,\Delta^T_0 f_a(x_a,\vec{k}^2_T)\,f_b(x_b)\,\frac{\mathrm{d}\hat{\sigma}(x_a,x_b,\vec{k}_T)}{\mathrm{d}\hat{t}}\,D_{h/c}(z), \tag{10}$$

where $\Delta_0^T f$ (related to $f_{1T}^{\perp}$) is a T-odd distribution. Finally, the effect studied by Boer [37] gives rise to an asymmetry involving another T-odd k_T-dependent distribution $\Delta_T^0 f$ (related to $h_1^{\perp}$) and a partonic initial-spin correlation differential cross-section $\Delta_{TT}\hat{\sigma}'$:

$$E_h \frac{\mathrm{d}^3\sigma(\vec{S}_T)}{\mathrm{d}^3\vec{P}_h} - E_h \frac{\mathrm{d}^3\sigma(-\vec{S}_T)}{\mathrm{d}^3\vec{P}_h} = -2|\vec{S}_T| \sum_{abc} \int \mathrm{d}x_a \int \mathrm{d}x_b \int \mathrm{d}^2\vec{k}'_T \, \frac{1}{\pi z} \times \Delta_T f_a(x_a)\, \Delta_T^0 f_b(x_b, \vec{k}'^2_T)\, \Delta_{TT}\hat{\sigma}'(x_a, x_b, \vec{k}'_T)\, D_{h/c}(z). \quad (11)$$

It was shown in [39] that non-vanishing SSA's can also be obtained in PQCD by resorting to the gluonic poles present in higher-twist diagrams involving qqg correlators. Such asymmetries were evaluated in [7], where direct photon production was studied, and for single hadron production later in [40]. This program was extended to cover chirally-odd contributions in [41]. In the general case one has

$$\mathrm{d}\sigma = \sum_{abc} \Big\{ G_F^a(x_a, y_a) \otimes f_b(x_b) \otimes \mathrm{d}\hat{\sigma} \otimes D_{h/c}(z) + \Delta_T f_a(x_a) \otimes E_F^b(x_b, y_b) \otimes \mathrm{d}\hat{\sigma}' \otimes D_{h/c}(z) + \Delta_T f_a(x_a) \otimes f_b(x_b) \otimes \mathrm{d}\hat{\sigma}'' \otimes D_{h/c}^{(3)}(z) \Big\}. \quad (12)$$

The first term does *not* contain transversity and is the chirally-even mechanism studied in [40]; the second *does* contain transversity and is the chirally-odd contribution analysed in [41]; and the third contains transversity and a twist-3 fragmentation function $D_{h/c}^{(3)}$.

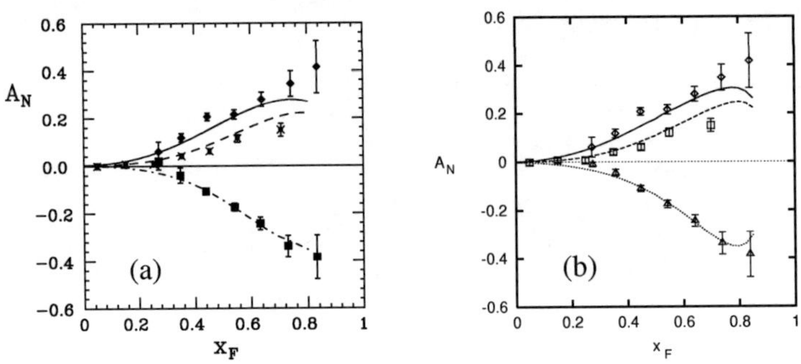

Figure 5. Single-spin asymmetries Assuming (a) only Sivers [42] and (b) only Collins [43] mechanisms.

Models inspired by the previous possible (k_T-dependent) contributions compare well to data in [42, 43]. Indeed, phenomenological fits based on either the Sivers or Collins mechanisms work equally well and are thus impossible to distinguish at present, see Fig. 5. The calculations based on three-parton correlators [7] are opaque, involving many diagrams, complex momentum flow, colour and spin structure. One sees that unfortunately the current knowledge at both the theoretical and experimental level do not permit a clear and concise description of these phenomena.

However, the twist-3 correlators obey constraining relations with k_T-dependent densities and they also exhibit a novel factorisation property. It is this simplification that I shall now describe in more detail. The twist-3 diagrams involving 3-parton correlators supply an imaginary part via a pole term [6] (spin-flip is implicit).

The $i\varepsilon$ propagator prescription (for -•- in Fig. 6), leads to an imaginary contribution for $k^2 \to 0$:

$$\frac{1}{k^2 \pm i\varepsilon} = \mathrm{I\!P}\frac{1}{k^2} \mp i\pi\delta(k^2), \tag{13}$$

For a gluon, with momentum $x_g p$, inserted into an initial or final external line p', one has $k = p' - x_g p$, and thus $x_g \to 0$. This can be shown systematically for all poles (gluon and fermion), *i.e.*, on all external legs with all possible insertions [44].

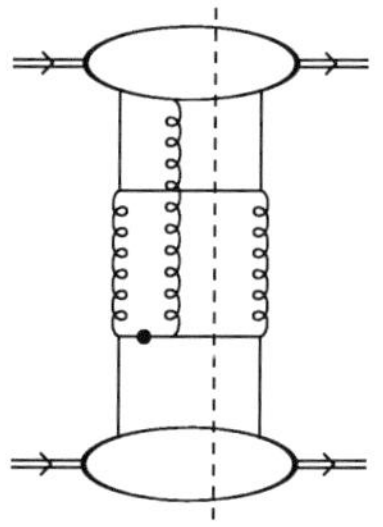

Figure 6. Example of a dominant higher-twist diagram in the large-N_c limit.

This still generates very complex structures; there are many possible insertions, with contributions of different sign and momentum dependence. The colour structure of the various diagrams is also very different. In all cases examined just one diagram dominates in the large-N_c limit (see Fig. 6), *i.e.*, other insertions into external legs are suppressed by $1/N_c^2$. Assuming that similar simplifications can also be found for the k_T-dependence, then by relating the higher-twist to the k_T-dependent mechanisms via the equations of motion, unique predictions may be possible for azimuthal SSA's. A study along these lines has indeed already been made for DY in [45]; predictions were, however, found *not* to be unique there. The k_T-dependence and higher-twist connections have also been exploited in [46].

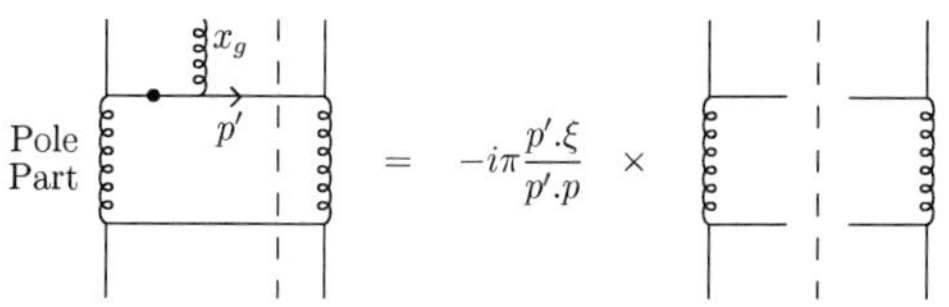

Figure 7. A graphical representation of the higher-twist pole-factorisation mechanism.

SUMMARY AND CONCLUSIONS

Transversity is now considered equally important an aspect of nucleon structure as the other two leading-twist parton densities and a complete description of the nucleon requires its understanding. Theoretically, all the pieces of the PQCD puzzle (up to NLO) are in place and lattice calculations indicate transversity to be sizable. On the experimental side, unfortunately, there are as yet no real data. However, the future is promising and before long we shall start to harvest interesting results. The phenomenology, while not dissimilar to the other leading-twist structures, has interesting peculiarities: evolution is non-singlet (so analysis should be cleaner), spin-half objects can contain gluonic transversity but it is not accessible via the usual partonic hard-scattering processes (this should perhaps be examined) and there is no associated sum rule (thus QCD evolution is faster). This all suggests that transversity could, in principle, allow clean extraction of α_s from the evolution fits to the scale violating Q^2 variation (*cf.* unpolarised).

SSA's have progressed from having essentially no (PQCD) theory to almost too much! Hopefully, the multitude of mechanisms can be reduced to just a few simple terms: experiment can eliminate some possibilities if null results are obtained; relationships between three-parton correlators and k_T-dependent densities should show up equivalences between apparently different phenomenological models; while pole-factorisation and the large-N_c limit can simplify calculations and allow a clear pattern to emerge.

REFERENCES

1. V. Barone, A. Drago and P.G. Ratcliffe, *Phys. Rep.* **359** (2002) 1.
2. V. Barone and P.G. Ratcliffe, *Transverse Spin Physics* (World Sci., 2003).
3. P.G. Ratcliffe, in Proc. of the *Advanced Studies Institute on Symmetries and Spin—Praha–SPIN 2002* (Prague, July 2002), eds. M. Finger *et al.*; *Czech. J. Phys.* **53** (2003) B27; the *XV Int. Symp. on High-Energy Spin Physics—SPIN 2002* (Upton/Danvers, Sept. 2002), eds. Y.I. Makdisi, A.U. Luccio and W.W. Mackay; *AIP Conf. Proc.* **675** (2003) 176; the *2nd. Int. Workshop on QCD* (Conversano, June 2003), eds. P. Colangelo *et al.*; *eConf* **C030614** (2003) 151.
4. X. Artru, in Proc. of the *10th. Rhodanien Seminar: The Spin in Physics* (Torino, Mar. 2002), eds. M. Anselmino, F. Mila and J. Soffer (Frontier Group, 2002), p. 115.
5. G.L. Kane, J. Pumplin and W. Repko, *Phys. Rev. Lett.* **41** (1978) 1689.
6. A.V. Efremov and O.V. Teryaev, *Phys. Lett.* **B150** (1985) 383.
7. J. Qiu and G. Sterman, *Phys. Rev. Lett.* **67** (1991) 2264; *Nucl. Phys.* **B378** (1992) 52.
8. A. Kotanski, *Acta Phys. Pol.* **B29** (1966) 127; **B30** 629.
9. J. Ralston and D.E. Soper, *Nucl. Phys.* **B152** (1979) 109.
10. R.L. Jaffe and X.-D. Ji, *Nucl. Phys.* **B375** (1992) 527.
11. G.R. Goldstein and M.J. Moravcsik, *Annals Phys.* **195** (1989) 213.
12. F. Baldracchini, N.S. Craigie, V. Roberto and M. Socolovsky, *Fortschr. Phys.* **30** (1981) 505.
13. X. Artru and M. Mekhfi, *Z. Phys.* **C45** (1990) 669.
14. J. Kodaira, S. Matsuda, K. Sasaki and T. Uematsu, *Nucl. Phys.* **B159** (1979) 99.
15. I. Antoniadis and C. Kounnas, *Phys. Rev.* **D24** (1981) 505.
16. A.P. Bukhvostov, É.A. Kuraev and L.N. Lipatov, *Sov. J. Nucl. Phys.* **38** (1983) 263.
17. P.G. Ratcliffe, *Nucl. Phys.* **B264** (1986) 493.
18. W. Vogelsang and A. Weber, *Phys. Rev.* **D48** (1993) 2073.
19. A.P. Contogouris, B. Kamal and Z. Merebashvili, *Phys. Lett.* **B337** (1994) 169.
20. A. Hayashigaki, Y. Kanazawa and Y. Koike, *Phys. Rev.* **D56** (1997) 7350.
21. S. Kumano and M. Miyama, *Phys. Rev.* **D56** (1997) R2504.
22. W. Vogelsang, *Phys. Rev.* **D57** (1998) 1886.
23. J. Soffer, *Phys. Rev. Lett.* **74** (1995) 1292.
24. G.R. Goldstein, R.L. Jaffe and X.-D. Ji, *Phys. Rev.* **D52** (1995) 5006.
25. V. Barone, *Phys. Lett.* **B409** (1997) 499.
26. O. Martin, A. Schäfer, M. Stratmann and W. Vogelsang, *Phys. Rev.* **D57** (1998) 3084.
27. C. Bourrely, J. Soffer and O.V. Teryaev, *Phys. Lett.* **B420** (1998) 375.
28. J.L. Cortes, B. Pire and J.P. Ralston, *Z. Phys.* **C55** (1992) 409.
29. J.C. Collins, *Nucl. Phys.* **B396** (1993) 161.
30. X.-D. Ji, *Phys. Rev.* **D49** (1994) 114.
31. L.P. Gamberg, G.R. Goldstein and K.A. Oganessyan, in Proc. of the *11th. Int. Workshop on Deep Inelastic Scattering—DIS 2003* (St. Petersburg, Apr. 2003); hep-ph/0309137.
32. D. de Florian, J. Soffer, M. Stratmann and W. Vogelsang, *Phys. Lett.* **B439** (1998) 176.
33. M. Anselmino, D. Boer, U. D'Alesio and F. Murgia, *Phys. Rev.* **D65** (2002) 114014.
34. J.C. Collins, S.F. Heppelmann and G.A. Ladinsky, *Nucl. Phys.* **B420** (1994) 565.
35. R.L. Jaffe, X.-M. Jin and J. Tang, *Phys. Rev.* **D57** (1998) 5920.
36. D. Sivers, *Phys. Rev.* **D41** (1990) 83.
37. D. Boer, *Phys. Rev.* **D60** (1999) 014012.
38. X. Ji, J.-P. Ma and F. Yuan, hep-ph/0404183.
39. A.V. Efremov and O.V. Teryaev, *Yad. Fiz.* **36** (1982) 242; *transl.*, *Sov. J. Nucl. Phys.* 140.
40. J. Qiu and G. Sterman, *Phys. Rev.* **D59** (1998) 014004.
41. Y. Kanazawa and Y. Koike, *Phys. Lett.* **B478** (2000) 121; *ibid.* **B490** (2000) 99.
42. M. Anselmino and F. Murgia, *Phys. Lett.* **B442** (1998) 470.
43. M. Anselmino, M. Boglione and F. Murgia, *Phys. Rev.* **D60** (1999) 054027.
44. P.G. Ratcliffe, *Eur. Phys. J.* **C8** (1999) 403.
45. J.-P. Ma and Q. Wang, hep-ph/0310245.
46. A. Bacchetta, P.J. Mulders and F. Pijlman, hep-ph/0405154.

Transversity and SIDIS at Jefferson Lab

Jen-Chieh Peng*, Jian-Ping Chen†, Xiaodong Jiang** and Lingyan Zhu*

*University of Illinois at Urbana-Champaign, Urbana, Illinois, 61801, U.S.A.
†Jefferson Lab, Newport News, Virginia, 23606, U.S.A.
**Rutgers University, Piscataway, New Jersey, 08855, U.S.A.

INTRODUCTION

Remarkable progress in the study of spin-dependent structure functions has been made since the discovery of the "proton spin puzzle" in the late 1980's. Very active spin-physics programs have been pursued at many facilities including SLAC, CERN, HERA, JLab, and RHIC. In recent years, new experimental tools such as semi-inclusive polarized DIS, polarized proton-proton collision, and deeply exclusive reactions have been employed to address various aspects of the proton spin. In this paper, we present the plans at the Jefferson Lab to use the semi-inclusive DIS (SIDIS) reactions to measure various unpolarized and polarized nucleon parton distribution and fragmentation functions. In addition, the prospect of probing the flavor structure of the kaon fragmentation function is discussed.

TRANSVERSITY

The nucleon unpolarized and polarized quark distributions, $q(x,Q^2)$ and $\Delta q(x,Q^2)$, are now rather well known. In contrast, a third quark distribution, called transversity, is the remaining twist-2 distribution which is practically unmeasured. This helicity-flip quark distribution, $\delta q(x,Q^2)$, can be described in quark-parton model as the net transverse polarization of quarks in a transversely polarized nucleon [1]. The transversity distributions have many interesting theoretical aspects, related to general properties of QCD as well as to the structure of the nucleon:

- In the non-relativistic quark model, where boosts and rotations commute, the transversity distributions are identical to the longitudinally polarized distributions. The transversity distributions provide a detailed measure of the relativistic nature of the quarks inside the nucleon.
- The transversity distribution is expected to have a valence like behavior. Since the transversity of gluons in a nucleon does not exist, the transversity distribution is expected to follow a simple evolution as a flavor non-singlet quantity [2].
- The transversity obeys some important inequalities. The first, $|h_1^q| \leq f_1^q$, follows from its interpretation as a difference of probabilities. The second (Soffer's bound)

CP747, *Structure of the Nucleon at Large Bjorken x*, edited by C. Bourrely, Z.-E. Meziani, and J. Soffer

has its origins in the positivity of helicity amplitudes [3]: $|h_1^q| \leq \frac{1}{2}(f_1^q + g_1^q)$.

- The lowest moment of h_1^q measures a simple local operator analogous to the axial charge, known as the "tensor charge".

Due to the chiral-odd nature of the transversity distribution, it can not be measured in inclusive DIS experiments. In order to measure $\delta q(x, Q^2)$, an additional chiral-odd object is required. For example, the double spin asymmetry, A_{TT}, for Drell-Yan cross section in transversely polarized pp collision, is sensitive to transversity since $A_{TT} \sim \sum_i e_i^2 \delta q_i(x_1) \delta \bar{q}_i(x_2)$. Such a measurement could be carried out at RHIC [4], although the anticipated effect is small, on the order of $1-2\%$.

Several other methods for measuring transversity have been proposed for semi-inclusive DIS. In particular, Collins suggested [5] that a novel chiral-odd fragmentation function (Collins function) in conjunction with the chiral-odd transversity distribution would lead to a single-spin azimuthal asymmetry in semi-inclusive pion production. Results on target single spin azimuthal asymmetries in SIDIS reactions from SMC [6] and HERMES [7] offered the first glimpse of possible effects caused by the transversity distribution.

The semi-inclusive DIS is a powerful tool for probing various parton distribution and fragmentation functions. For producing a spin-zero meson, the SIDIS differential cross section at the leading order is [8, 9]:

$$
\begin{aligned}
d^6\sigma^{LO} &= \frac{d\sigma^{\ell+N\to\ell'+h+X}}{dxdyd\phi^\ell dzd^2P_{h\perp}} = \frac{4\pi\alpha^2 sx}{Q^4}\times \\
&\Big\{ \left[1+(1-y)^2\right] \sum_{q,\bar{q}} e_q^2 f_1^q(x) D_1^q(z, P_{h\perp}^2) \\
&+(1-y)\frac{P_{h\perp}^2}{4z^2 M_N M_h} \cos(2\phi_h^\ell) \sum_{q,\bar{q}} e_q^2 h_1^{\perp(1)q}(x) H_1^{\perp q}(z, P_{h\perp}^2) \\
&-|S_L|(1-y)\frac{P_{h\perp}^2}{4z^2 M_N M_h} \sin(2\phi_h^\ell) \sum_{q,\bar{q}} e_q^2 h_{1L}^{\perp(1)q}(x) H_1^{\perp q}(z, P_{h\perp}^2) \\
&+|S_T|(1-y)\frac{P_{h\perp}}{zM_h} \sin(\phi_h^\ell + \phi_S^\ell) \sum_{q,\bar{q}} e_q^2 h_1^q(x) H_1^{\perp q}(z, P_{h\perp}^2) \\
&+|S_T|\left(1-y+\frac{1}{2}y^2\right)\frac{P_{h\perp}}{zM_N} \sin(\phi_h^\ell - \phi_S^\ell) \sum_{q,\bar{q}} e_q^2 f_{1T}^{\perp(1)q}(x) D_1^q(z, P_{h\perp}^2) \\
&+|S_T|(1-y)\frac{P_{h\perp}^3}{6z^3 M_N^2 M_h} \sin(3\phi_h^\ell - \phi_S^\ell) \sum_{q,\bar{q}} e_q^2 h_{1T}^{\perp(2)q}(x) H_1^{\perp q}(z, P_{h\perp}^2) \\
&+\lambda_e |S_L| y(1-\frac{1}{2}y) \sum_{q,\bar{q}} e_q^2 g_1^q(x) D_1^q(z, P_{h\perp}^2) \\
&+\lambda_e |S_T| y(1-\frac{1}{2}y) \frac{P_{h\perp}}{zM_N} \cos(\phi_h^\ell - \phi_S^\ell) \sum_{q,\bar{q}} e_q^2 g_{1T}^{(1)q}(x) D_1^q(z, P_{h\perp}^2) \Big\}, \qquad (1)
\end{aligned}
$$

where x, y, z, Q^2 are the familiar kinematic variables. $|S_L|$ and $|S_T|$ are the longitudinal and transverse target spin components with respect to the virtual photon direction, and λ_e is the electron helicity. Azimuthal angles are defined relative to the lepton plane, e.g. $\phi_h^\ell = \phi_h - \phi^\ell$ is the angle between the hadron plane ($\vec{P}_h \wedge \vec{q}$) and the lepton plane. $\phi_S^\ell = \phi_S - \phi^\ell$ is the angle between the $\vec{S} \wedge \vec{q}$ plane and the lepton plane. $P_{h\perp}$ is the component of the hadron's momentum perpendicular to the virtual photon direction.

In Eq. 1, $f_1^q(x), g_1^q(x)$ and $h_1^q(x)$ are the helicity-averaged, helicity-dependent, and transversity distribution functions, respectively. These three functions survive the integration of intrinsic quark transverse momentum p_T in the initial nucleon. The other p_T-dependent distributions are $f_{1T}^{\perp q}$, $g_{1T}^{\perp q}$, $h_1^{\perp q}$, $h_{1L}^{\perp q}$ and $h_{1T}^{\perp q}$. They only survive the p_T integration after weighting with $(p_T^2/2M_N^2)^n$. Such weighting is indicated by a superscript (n) in Eq. 1. D_1^q and $H_1^{\perp q}$ are the regular fragmentation function and the chiral-odd, T-odd Collins fragmentation function, respectively.

Eq. 1 shows that SIDIS can be used to extract all eight of the twist-2 structure functions with a judicial selection of beam helicity (λ_e), target polarization (S_L, S_T), and the detection of azimuthal angular dependences (ϕ_h^l and ϕ_s^l).

The HERMES collaboration recently reported [7] observation of single-spin azimuthal asymmetry for charged and neutral hadron electroproduction. Using unpolarized positron beam on longitudinally polarized hydrogen and deuterium targets, the cross section was found to have a sinϕ_h^l dependence. Although a longitudinally polarized target was used in the HERMES experiment, S_T is nonzero with a mean value of ≈ 0.15. Eq. 1 shows that a sinϕ_h^l dependence can arise from the transversity plus collins fragmentation function (the $\sin(\phi_h^l + \phi_s^l)$ term), as well as from $f_{1T}^{\perp}$, the chiral-odd T-odd Sivers distribution function [10] (the $\sin(\phi_h^l - \phi_s^l)$ term). The Collins fragmentation function represents a correlation between the quark's transverse spin and the transverse momentum of the leading hadron formed in the fragmentation process. The Sivers distribution function reflects the correlation between the quark's transverse spin and its transverse momentum within the nucleon. For a longitudinally polarized target ($\phi_s^l = 0$) the Collins and the Sivers mechanisms can not be distinguished.

If the azimuthal asymmetry observed by HERMES is indeed caused by the nucleon transversity, a much larger asymmetry is expected for a transversely polarized target. The HERMES [11] and COMPASS [12] collaborations have collected polarized SIDIS using transversely polarized hydrogen and LiD targets, respectively. A simultaneous fit to $\sin(\phi_h^l + \phi_s^l)$ and $\sin(\phi_h^l - \phi_s^l)$ dependences was applied to the HERMES data to extract the "Collins moments" and the "Sivers moments", respectively. Nonzero Collins moments were clearly observed in the HERMES data with positive azimuthal asymmetry for π^+ and negative azimuthal asymmetry for π^-. The unexpectedly large magnitude for the azimuthal asymmetry of π^- relative to π^+ seems to suggest that the disfavored Collins function is of comparable magnitude, but opposite sign, to the favored Collins function. Furthermore, the observation of the $\sin(\phi_h^l - \phi_s^l)$ moments shows that the Sivers function is nonzero and indeed contribute to the azimuthal asymmetry in SIDIS. Preliminary result [12] from the COMPASS 2002 data shows that the $\sin(\phi_h^l + \phi_s^l)$ azimuthal asymmetry is consistent with zero for transversely polarized LiD target. A factor of 4 increase in statistics is expected for the COMPASS 2002-2004 data.

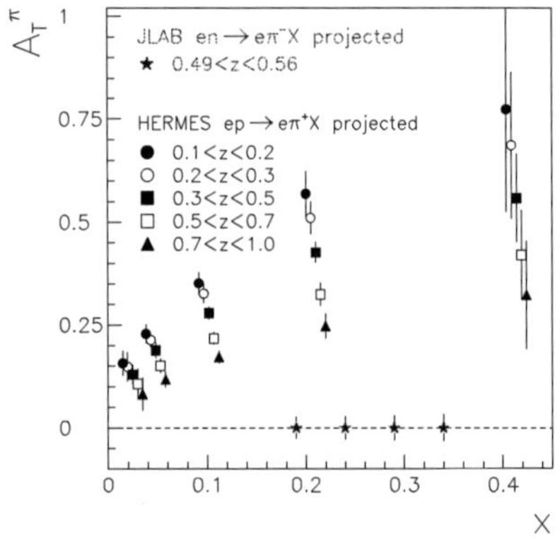

FIGURE 1. Expected statistical precision of this experiment in comparison with HERMES run-II projected precision on a proton target [14].

In order to confront theoretical predictions on the transversity distribution, precise SIDIS data on both proton and neutron covering a wide range of kinematics is required. A recently approved JLab proposal [13] plans to measure the single spin asymmetry of the $\vec{n}(e,e'\pi^-)X$ reaction on a transversely polarized ^{3}He target. The goal of this experiment is to provide the first measurement on the neutron transversity, complementary to the ongoing HERMES and COMPASS measurements on the proton and deuteron. This experiment focuses on the valence quark region, $x = 0.19 \sim 0.34$, at $Q^2 = 1.77 \sim 2.73$ GeV^2. This kinematics is comparable to the HERMES measurement. The variation of single spin asymmetry relative to x and z will isolate effect of parton distribution from that of fragmentation. Data from this experiment, when combined with data from HERMES and COMPASS, will provide powerful constraints on the transversity distributions on both u–quark and d-quark in the valence region.

The experiment will use 6 GeV electron beam with the Hall A left-side high resolution spectrometer (HRS$_L$) situated at 16° as the hadron arm, and the BigBite spectrometer located at 30° beam-right as the electron arm. A Rb spin-exchange ^{3}He target of 50 mg/cm^2 and ~42% transverse polarization with tunable direction will be used. By rotating the target polarization direction, the coverage in ϕ_s^l is increased, hence facilitating the separation of the Collins and the Sivers effects. Figure 1 shows the expected statistical precision of this experiment for $\vec{n}(e,e'\pi^-)X$ single spin asymmetry together with the projected precision[14] of HERMES run–II of the $\vec{p}(e,e'\pi^+)X$ reaction.

The $h_1^\perp$ quark distributions

In addition to the Svers function, the other T-odd quark distribution is the $h_1^\perp(x)$ function. $h_1^\perp$ is interpreted as the distribution of a transversely polarized quark with nonzero transverse momentum inside an unpolarized nucleon. Eq. 1 shows that the $h_1^\perp(x)$

can be extracted by isolating the $\cos(2\phi_h^l)$ azimuthal dependence. It would be of great interest if existing and future SIDIS experiments can shed some light on the T-odd $h_1^{\perp}(x)$ quark distributions.

$h_1^{\perp}$ can also lead to a $\cos 2\phi$ dependence in unpolarized Drell-Yan process [15]. The azimuthal angle in this case is the angle between l^- and the plane formed by the beam and the virtual photon. As a time-odd quantity, $h_1^{\perp}$ in the Drell-Yan process is predicted [16] to have an opposite sign relative to the $h_1^{\perp}$ in the SIDIS process. However, this prediction remains to be confirmed by experiments. A large $\cos 2\phi$ dependence was observed in pion-induced Drell-Yan process, and a first attempt to extract $h_1^{\perp}$ information from these data has been made [15]. The $\cos 2\phi$ dependence has never been extracted for the proton-induced Drell-Yan data. The recent FNAL E866 experiment has collected $\sim 2.5 \times 10^5$ $p+p$ and $p+d$ Drell-Yan events using 800 GeV/c proton beams. These data are being analysed to determine the $\cos 2\phi$ dependence. A comparison of proton Drell-Yan data with the pion Drell-Yan data could provide additional light on the role of $h_1^{\perp}$ on the $\cos 2\phi$ dependence in the Drell-Yan process.

Other SIDIS initiaves at JLab

A new JLab Hall-C approved proposal [17] plans to measure spin asymmetries in $\vec{e}\vec{p} \to e'hX$ and $\vec{e}\vec{d} \to e'hX$ SIDIS reactions. Taking advantage of the high flux electron beam and a large-acceptance BETA spectrometer for electron detection, this experiment aims at a high statistics measurement of $\Delta u, \Delta d, \Delta\bar{u}, \Delta\bar{d}$ over the $0.1 < x < 0.45$ range. This experiment could address the important unresolved issue on the flavor structure of polarized sea-quark distributions. Recent polarized SIDIS results from HERMES have already provided a first measurement of the $\Delta\bar{u}$ and $\Delta\bar{d}$ [18].

Another new JLab proposal [19] plans to measure unpolarized $(e,e'\pi^{\pm})$ and $(e,e'K^{\pm})$ reactions. The pion SIDIS measurement aims at a determination of $\bar{d}-\bar{u}$ with better statistical accuracy than the existing HERMES data [20]. The kaon SIDIS measurement is motivated by a search for the $\bar{u},\bar{d}$ asymmetry in the kaon fragmentation function. The interest for such a study is discussed in the next Section.

FLAVOR ASYMMETRY OF FRAGMENTATION FUNCTIONS

The $\bar{d}/\bar{u}$ asymmetry in the proton is now well established [21]. It is of interest to examine the flavor structure of sea quark distributions in other hadrons. Alberg et al. [22] have considered the sea-quark distributions in Σ^+. The meson-cloud model implies a $\bar{d}/\bar{u}$ asymmetry in the Σ^+ even larger than that of the proton. The opposite effect, however, is expected from SU(3) symmetry. A Drell-Yan experiment using a Σ^+ hyperon beam has been considered [22] for such study, but appears quite challenging.

An alternative area of investigation is in the meson sector. A flavor asymmetry in the π^+ sea would show up as $u_{sea} \neq d_{sea}$, which is equivalent to $\bar{u} \neq d$. However, the charge-conjugation symmetry and isospin symmetry dictate that $\bar{u} = d$ in the π^+. Therefore, the sea quark distributions in charged pions are u/d symmetric. This is readily understood in the Pauli-blocking picture. The presence of the u and $\bar{d}$ valence quarks in π^+ would equally block the $u\bar{u}$ and the $\bar{d}d$ sea.

The situation is quite different for charged kaons. One anticipates an $\bar{d}/\bar{u}$ asymmetry based on the Pauli-blocking picture. The presence of u quark in K^+ would inhibit the $\bar{u}$ sea relative to the $\bar{d}$ sea in K^+. Therefore, one expects $\bar{d} > \bar{u}$ in K^+, analogous to the situation for proton. Unfortunately, the putative $\bar{u}/\bar{d}$ flavor asymmetry in K^+ is not accessible in DIS experiments.

A promising avenue for probing the possible $\bar{u}/\bar{d}$ asymmetry in K^+ is the measurement of kaon fragmentation functions. The sea-quark flavor asymmetry would imply that $\bar{u}$ and $\bar{d}$ would have unequal probability to hadronize into a K^+, namely, $D_{\bar{u}}^{K^+} \neq D_{\bar{d}}^{K^+}$. A possible evidence for such asymmetry might already exist. Kaon fragmentation functions have been extracted by Kniehl et al. [23] who fitted the hadron production data from e^+e^- annihilation. They obtain the following surprising result:

$$\int_{0.05}^{1} dx\, x\, D_{u,s}^{K^\pm}(x, Q_0^2) = 0.19, \quad \int_{0.05}^{1} dx\, x\, D_{u,s}^{K^\pm}(x, Q_0^2) = 0.25 \tag{2}$$

This result shows that the d quark, which is a sea quark for $K^\pm$, has a higher probability to fragment into $K^\pm$ than for the valence u and s quarks. This surprising result might reflects a large $\bar{d}/\bar{u}$ flavor asymmmtry in the kaon sea. Clearly, additional data on kaon fragmentation function would be of great interest. The JLab proposal [19] plans to measure kaon SIDIS to examine possible flavor asymmetry of the kaon fragmentation functions.

REFERENCES

1. Barone, V., Drago, A., and Ratcliffe, P., *Phys. Rep.*, **359**, 1 (2002).
2. Bourrely, C., Soffer, J., and Teryaev, O. V., *Phys. Lett.*, **B420**, 375 (1998).
3. Soffer, J., *Phys. Rev. Lett.*, **74**, 1292 (1995).
4. Bunce, G., et al., *Ann. Rev. Nucl. Part. Sci.*, **50**, 525 (2000).
5. Collins, J., *Nucl. Phys.*, **B396**, 161 (1993).
6. Bravar, A., et al., *Nucl. Phys. Proc. Suppl.*, **79**, 520 (1999).
7. Airapetian, A., et al., *Phys. Lett.*, **B562**, 182 (2003).
8. Mulders, P., and Tangerman, R. D., *Nucl. Phys.*, **B461**, 197 (1996).
9. Boer, D., and Mulders, P., *Phys. Rev.*, **D57**, 5780 (1998).
10. Sivers, D. W., *Phys. Rev.*, **D41**, 83 (1990).
11. Airapetian, A., et al., *hep-ex/0408013* (2004).
12. Bressan, A., *Spin2004 Proceedings* (2004).
13. Chen, J.-P., Jiang, X., Peng, J., et al., *JLab proposal PR-03-014* (2003).
14. Korotkov, V., Nowak, W.-D., and Oganessian, K., *Eur. Phys. J.*, **C18**, 639 (2001).
15. Boer, D., *Phys. Rev.*, **D60**, 014012 (1999).
16. Boer, D., Brodsky, S., and Hwang, D., *Phys. Rev.*, **D67**, 054003 (2003).
17. Jiang, X., et al., *JLab proposal PR-04-113* (2004).
18. Airapetian, A., et al., *Phys. Rev. Letts.*, **92**, 012005 (2004).
19. Zhu, L., et al., *JLab proposal PR-04-114* (2004).
20. Ackerstaff, K., et al., *Phys. Rev. Letts.*, **81**, 5519 (1998).
21. Garvey, G. T., and Peng, J. C., *Prog. Part. Nucl. Phys. Phys.*, **47**, 203 (2001).
22. Alberg, M., et al., *Phys. Lett.*, **B389**, 367 (2003).
23. Kniehl, B., Kramer, G., and Potter, B., *Nucl. Phys.*, **B582**, 514 (2000).

Semi-Inclusive DIS in NLO QCD

Marco Stratmann

Institute for Theoretical Physics, University of Regensburg, D-93040 Regensburg, Germany

Abstract. We review the theoretical framework for semi-inclusive deep-inelastic scattering in the current fragmentation region including next-to-leading order QCD corrections. Special emphasis is put on the large x behavior of the coefficient functions. Some applications are briefly discussed.

MOTIVATION FOR STUDYING SIDIS

Fully inclusive deep-inelastic scattering (DIS) measurements can only provide information on the sum of quark and antiquark parton distribution functions. Contributions from the various quark and antiquark flavors can be separated by identifying a hadron in the final state in coincidence with the scattered lepton. To the lowest order (LO) of QCD, cf. Fig. 1, the semi-inclusive deep-inelastic scattering (SIDIS) cross section $l(P_l)N(P_N) \to l'(P_l')H(P_H)X$ is given by

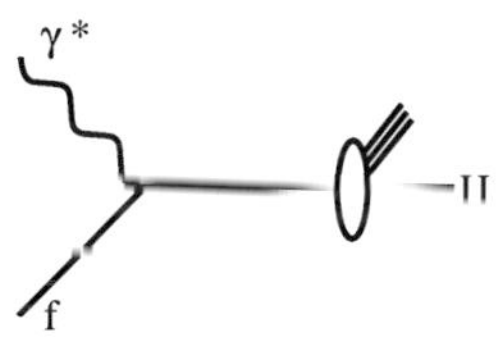

FIGURE 1. Lowest order contribution $q\gamma^* \to H$ to SIDIS.

$$\frac{d^3\sigma}{dxdydz} = \frac{2\pi\alpha^2}{Q^2}\,\frac{1+(1-y)^2}{y}\sum_{f=q,\bar{q}} e_f^2\, f(x,Q^2) D_f^H(z,Q^2) \quad , \tag{1}$$

and similarly for a longitudinally polarized beam and target (and/or observed hadron H), see, e.g., [1]. Most notably, the dependence on the momentum fraction of the struck quark f, $x = Q^2/(2P_N \cdot q)$, and $z = P_N \cdot P_H / P_N \cdot q$ (fractional energy of the observed hadron in the lab frame) "factorizes" in the LO expression *if* the hadronization is dominated by one fragmentation function as, e.g., for π^0 production where $D_u = D_{\bar{u}} = D_d = D_{\bar{d}}$. This property, however, turns out to be an artifact of the trivial LO kinematics and is no longer valid beyond the LO approximation of QCD as will be shown below.

The benefits from measurements of SIDIS are twofold: first and foremost it provides valuable insight into the nucleon (spin) structure. In particular for spin-dependent parton densities SIDIS is currently the only source of information for a separation of the various

CP747, *Structure of the Nucleon at Large Bjorken x*, edited by C. Bourrely, Z.-E. Meziani, and J. Soffer

quark and antiquark flavors [2]. Upcoming $W^{\pm}$ boson production data from BNL-RHIC will add to this [3]. Future SIDIS measurements have the potential to provide the best direct constraint on the polarized strange density Δs, which is difficult to access in pp collisions at RHIC. Current data are, unfortunately, inconclusive [2].

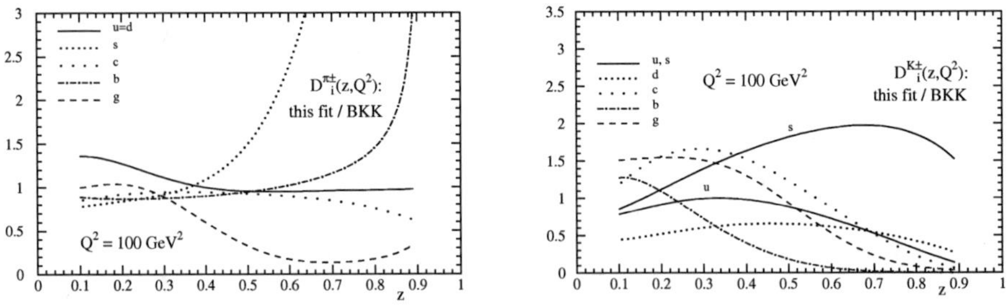

FIGURE 2. Comparison of two sets of parton-to-pion fragmentation functions (taken from [4]).

On the other hand, SIDIS data can considerably improve also our understanding of the hadronization process as modeled by parton-to-hadron fragmentation functions D_f^H in Eq. (1). Currently, the bulk of information comes from high-precision LEP data at the Z pole which mainly constrain, however, the flavor singlet fragmentation function [4]. For practical purposes we need fragmentation functions for *individual* flavors (including the gluon) at much lower scales than M_Z. Figure 2 shows pronounced differences between two sets of parton-to-pion fragmentation functions, see [4] for details, both fitted to LEP data. These two sets lead, for instance, to a considerable spread in the theoretical predictions for $pp \to \pi^0 X$ which was recently measured at RHIC [5]. Clearly, the task for the near future is a global QCD analysis of all available data constraining fragmentation functions to quantify the remaining uncertainties more precisely.

NLO QCD FRAMEWORK

As is well known, LO approximations give at best a rough qualitative picture for a given process and are clearly not sufficient for quantitative analyses. Next-to-leading order (NLO) QCD corrections often turn out to be sizable and, more importantly, significantly reduce the residual dependence on the choice of the completely arbitrary factorization scale(s) – the prime source of theoretical errors. In case of SIDIS, NLO corrections comprise of virtual and real gluon emission corrections to the LO reaction $q\gamma^* \to q$ as well as of two new genuine $\mathcal{O}(\alpha_s)$ subprocesses $q\gamma^* \to gX$ and $g\gamma^* \to qX$, cf. Fig. 3.

Along the calculation one encounters the usual bunch of divergences which are regulated in $n = 4 - 2\varepsilon$ dimensions. All singularities which still remain after adding real emission and virtual contributions are due to collinear parton-parton splittings and can be either factorized into the bare parton densities or into the bare fragmentation functions. These subtractions introduce two arbitrary factorization scales μ_f and μ_f' mentioned above. A third scale μ_r controls the running of the strong coupling α_s. Putting

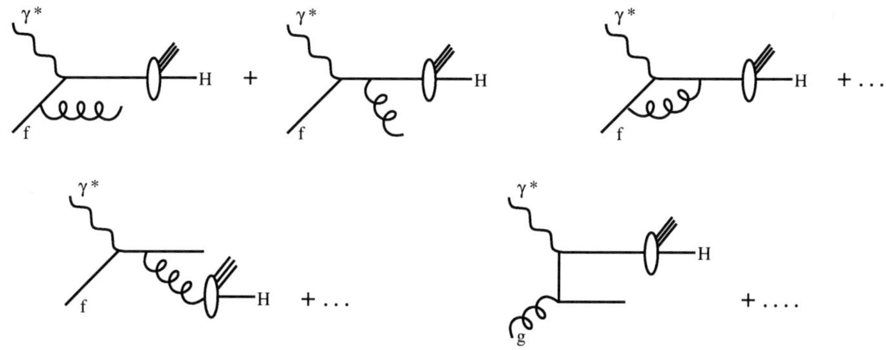

FIGURE 3. NLO real gluon emission and virtual QCD corrections to the LO process $q\gamma^* \to q$ (upper row); genuine $\mathcal{O}(\alpha_s)$ processes $q\gamma^* \to gX$ and $g\gamma^* \to qX$ (lower row)

everything together, the cross section for SIDIS has a more complicated structure than in LO, e.g., in the longitudinally polarized case $\vec{l}\vec{N} \to l'HX$ one has (up to trivial kinematical factors from the leptonic part)

$$\begin{aligned}\frac{d^3\Delta\sigma}{dxdydz} \simeq & \sum_{q=u,\bar{u},\ldots,\bar{s}} e_q^2 \left[\Delta q\left(x,\mu_f\right) D_q^H(z,\mu_f') + \int_x^1 \frac{d\hat{x}}{\hat{x}} \int_z^1 \frac{d\hat{z}}{\hat{z}} \Bigg\{ \right. \\ & \frac{\alpha_s(\mu_r)}{2\pi} \Delta q\left(\frac{x}{\hat{x}},\mu_f\right) \Delta C_{qq}^{(1)}(\hat{x},\hat{z},\frac{\mu_f}{Q},\frac{\mu_f'}{Q}) D_q^H\left(\frac{z}{\hat{z}},\mu_f'\right) + \\ & \frac{\alpha_s(\mu_r)}{2\pi} \Delta q\left(\frac{x}{\hat{x}},\mu_f\right) \Delta C_{gq}^{(1)}(\hat{x},\hat{z},\frac{\mu_f}{Q},\frac{\mu_f'}{Q}) D_g^H\left(\frac{z}{\hat{z}},\mu_f'\right) + \\ & \left. \frac{\alpha_s(\mu_r)}{2\pi} \Delta g\left(\frac{x}{\hat{x}},\mu_f\right) \Delta C_{qg}^{(1)}(\hat{x},\hat{z},\frac{\mu_f}{Q},\frac{\mu_f'}{Q}) D_q^H\left(\frac{z}{\hat{z}},\mu_f'\right) \Bigg\} \right] . \end{aligned} \tag{2}$$

The $\mathcal{O}(\alpha_s)$ coefficient functions $\Delta C_{ij}^{(1)}$ are now non-trivial functions of *both* x and z. The naive "factorization" of the x and z dependence in the LO expression (1) breaks down (not to be confused with a breakdown of the factorization procedure for collinear singularities which is, of course, still okay). To give one explicit example, the spin-dependent $\overline{\text{MS}}$ coefficient function $\Delta C_{qq}^{(1)}$ for the process $q\gamma^* \to qX$ is given by [1]

$$\begin{aligned}\Delta C_{qq}^{(1)} = C_F \Bigg[& -8\delta(1-x)\delta(1-z) + \\ \delta(1-x) & \left[\tilde{P}_{qq}(z)\ln\frac{Q^2}{\mu_f'^2} + (1+z^2)\left(\frac{\ln(1-z)}{1-z}\right)_+ + \frac{1+z^2}{1-z}\ln z + (1-z)\right] + \end{aligned}$$

$$\delta(1-z)\left[\tilde{P}_{qq}(x)\ln\frac{Q^2}{\mu_f^2}+(1+x^2)\left(\frac{\ln(1-x)}{1-x}\right)_+-\frac{1+x^2}{1-x}\ln x+(1-x)\right]+$$

$$2\frac{1}{(1-x)_+}\frac{1}{(1-z)_+}-\frac{1+z}{(1-x)_+}-\frac{1+x}{(1-z)_+}+2(1+xz)\Bigg]-2C_F(1-x)(1-z). \quad (3)$$

It should be noted that Eq. (3) contains mathematical distributions which are relevant for the $x \to 1$ and/or $z \to 1$ behavior. Also, the appearance of $\ln Q^2/\mu_f^2$ and $\ln Q^2/\mu_f'^2$ motivates usual choice of scales $\mu_f \simeq \mu_f' \simeq Q$ which avoids large logarithmic contributions in the hard coefficient functions. Variations around this standard choice of scale give a flavor of the theoretical uncertainties due to the truncation of the perturbative series at the NLO level.

A collection of all relevant NLO ($\overline{\text{MS}}$) coefficient functions for SIDIS can be found in [1] including the unpolarized ($lN \to l'HX$), longitudinally polarized ($\vec{l}\vec{N} \to l'HX$) as well as "helicity transfer" ($\vec{l}N \to l'\vec{H}X, l\vec{N} \to l'\vec{H}X$) cases. Finally, it should be noted that the framework outlined above only applies to the so-called "current fragmentation region" which is usually selected by a suitable cut on the longitudinal momentum fraction of the observed hadron x_F. Target fragmentation can be accounted for by a transformation to $z \to \tilde{z} \equiv \frac{E_H}{E_N(1-x)}$ (with $E_{H,N}$ in the $\gamma^* N$ c.m.s.). The factorization procedure in this case requires one additional non-perturbative function, the so-called "fracture function". The NLO corrections are also known for the target fragmentation region and can be found in [6] (and references therein).

SOME APPLICATIONS

Let us apply the NLO machinery as outlined above in an analysis of the recent HERMES data [2] for the SIDIS double spin asymmetry $A_{1,N}^H \equiv \frac{d\Delta\sigma/dxdydz}{d\sigma/dxdydz}$ for $N = p, d$ targets and $H = \pi^\pm$. Our goal is to extract some information on $\Delta\bar{u} - \Delta\bar{d}$, i.e., on the SU(2) flavor breaking of the light sea, for which several different model calculations exist, see [2] and references therein. We introduce two functions f, g in such a way that they leave existing fits to inclusive DIS data invariant [7] (recall that DIS data probe only $\Delta q + \Delta\bar{q}$):

$$\Delta\bar{u}' = \Delta\bar{u} + f \quad , \quad \Delta u' = \Delta u - f \quad (4)$$

$$\Delta\bar{d}' = \Delta\bar{d} + g \quad , \quad \Delta d' = \Delta d - g \ . \quad (5)$$

Unfortunately it turns out that present data do not sufficiently constrain f and g. All data for $A_{1,N}^H$ can be equally well described with an SU(3) symmetric sea (taken from the "standard scenario" of the GRSV fit [8]) or with a sizable $\Delta\bar{u} - \Delta\bar{d} \neq 0$, see Fig. 4.

The NLO formalism outlined in the previous Section should be also applied for studies of multiplicities, $(d^3\sigma^H/dxdydz)/(d^2\sigma^{\text{DIS}}/dxdy)$ which may provide information about fragmentation functions. For $H = \pi^+ + \pi^-$, where u, $\bar{u}$, d, and $\bar{d}$ all fragment with equal probability into the sum of charged pions, one expects that for fixed z the multiplicities should show no dependence on x at the LO level. NLO QCD corrections

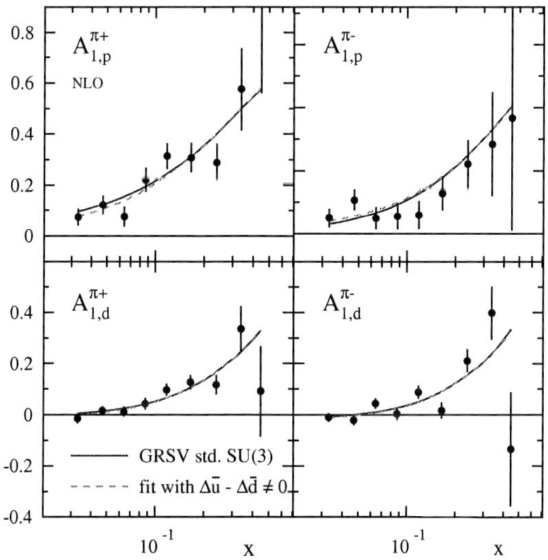

FIGURE 4. Comparison of two fits to the HERMES SIDIS data.

modify the expectation and predict a rather mild dependence on x due to the non-trivial dependence of the coefficient functions on x and z [9]. However, published HERMES data for charged pion multiplicites show a much more pronounced dependence towards smaller x [10]. Since smaller x also correspond to smaller Q^2 an explanation in terms of higher twist corrections seems to be tempting which would in turn imply that low Q^2 fixed target date cannot be used in standard global fits of parton densities and/or fragmentation functions. However, before arriving at such a conclusion one has to wait for a re-analysis of the available data which is currently underway.

SPOTLIGHT ON "LARGE X"

The emission of additional partons is strongly inhibited by kinematics as x and/or z approach 1. The "bremsstrahlung" spectrum of soft partons generates large logarithmic contributions $\alpha_s^N \ln^M(1-x)/(1-x)$, where $M \leq 2N-1$, in each order N of the perturbative series; for $N=1$ see Eq. (3). In this region of phase space the true expansion parameter is no longer α_s, and a fixed order perturbative QCD calculation is bound to fail. Clearly, a naive extraction of information about nucleon (spin) structure from data would be meaningless without first trying to resum these large logarithms to all orders.

The technical framework for soft gluon resummations is well-known [12] but can be quite tedious depending on the process under consideration. Resummations are performed in Mellin moment space, where complicated convolutions turn into simple multiplications and distributions into ordinary functions. In case of SIDIS we have to perform a double Mellin transformation [11, 7]

$$\Delta C_{ij}^{(1),nm}(\frac{\mu_f}{Q},\frac{\mu_f'}{Q}) \equiv \int_0^1 dx x^{n-1} \int_0^1 dz z^{m-1} \Delta C_{ij}^{(1)}(x,z,\frac{\mu_f}{Q},\frac{\mu_f'}{Q}) \tag{6}$$

which nevertheless can be done analytically. To give an explicit example, $\Delta C_{qq}^{(1)}(x,z)$ in Eq. (3) turns into [7]

$$\begin{aligned}\Delta C_{qq}^{(1),nm}(\frac{\mu_f}{Q},\frac{\mu_f'}{Q}) &= C_F\Bigg[-8-\frac{1}{m^2}+\frac{2}{(m+1)^2}+\frac{1}{n^2}+\frac{(1+m+n)^2+1}{m(m+1)n(n+1)}+3S_2(m)\\ &-S_2(n)+\left[S_1(m)+S_1(n)\right]\left\{S_1(m)+S_1(n)-\frac{1}{m(m+1)}-\frac{1}{n(n+1)}\right\}\\ &+\left[\frac{2}{n(n+1)}+3-4S_1(n)\right]\ln\left(\frac{Q}{\mu_f}\right)+\left[\frac{2}{m(m+1)}+3-4S_1(m)\right]\ln\left(\frac{Q}{\mu_f'}\right)\Bigg]\end{aligned} \tag{7}$$

In Mellin moment space $x \to 1$ corresponds to $n \to \infty$. The leading large n (or m) behavior is governed by $S_1(n) = \ln n + \gamma_E + \mathcal{O}(1/n)$ which can be resummed to all orders [12]. Resummations are expected to be of particular relevance for the interpretation of current and future SIDIS measurements at JLAB which focus at measurements at fairly large x (and z).

ACKNOWLEDGMENTS

I am grateful to the organizers of this interesting workshop for financial support.

REFERENCES

1. D. de Florian, M. Stratmann, and W. Vogelsang, Phys. Rev. **D57**, 5811 (1998).
2. A. Airapetian et al., HERMES Collab., Phys. Rev. Lett. **92**, 012005 (2004); hep-ex/0407032.
3. G. Bunce et al., Ann. Rev. Nucl. Part. Sci. **50**, 525 (2000).
4. S. Kretzer, Phys. Rev. **D62**, 054001 (2000).
5. S.S. Adler et al., PHENIX Collab., Phys. Rev. Lett. **91**, 241803 (2003).
6. D. de Florian et al., Nucl. Phys. **B470**, 195 (1996); Nucl. Phys. **B488**, 367 (1997).
7. M. Stratmann and W. Vogelsang, Phys. Rev. **D64**, 114007 (2001).
8. M. Glück, E. Reya, M. Stratmann, and W. Vogelsang, Phys. Rev. **D63**, 094005 (2001).
9. S. Kretzer, M. Stratmann, and W. Vogelsang, work in progress.
10. A. Airapetian et al., HERMES Collab., Eur. Phys. J. **C21**, 599 (2001).
11. G. Altarelli et al., Nucl. Phys. **B160**, 301 (1979).
12. W. Vogelsang, these proceedings (and references therein).

Helicity and Transversity Distributions at HERMES

Lara De Nardo
(for the HERMES Collaboration)

DESY, Notkestrasse 85, 22607 Hamburg, Germany

Abstract. Results obtained by the HERMES Collaboration for the polarized parton distributions Δu, Δd, $\Delta \bar{u}$, $\Delta \bar{d}$, Δs extracted from inclusive and semi-inclusive asymmetries on H and D in a LO analysis are presented. The *up* and *down* distributions are measured with good precision and are determined to be positive and negative respectively. The sea distributions have been extracted for the first time and the strange sea is consistent with zero within errors.
Single spin asymmetries on transversely polarized hydrogen target provide evidence of transversity distributions.

INTRODUCTION

A complete understanding of the structure of nucleons cannot leave aside the study of their spin. In the constituent quark model the spin of nucleons is carried only by quarks. Deep-inelastic scattering (DIS) experiments have shown that this view of the nucleon is too naive, being the total quark contribution to the nucleon spin only of the order of 30%. Therefore also the angular momentum and the gluon helicity have to play a role.

In deep inelastic scattering a lepton scatters off a quark inside a nucleon through the exchange of a virtual photon of squared four-momentum $q^2 = -Q^2$. Variables usually employed in the description of the interaction are the Bjorken variable x (the nucleon's momentum fraction carried by the struck quark), and z (the fractional energy of the outgoing hadron with respect to the virtual photon energy). The last variable is used in semi-inclusive DIS, where a hadron is detected in coincidence with the scattered lepton.

Semi-inclusive DIS can be used to isolate the different contributions to the nucleon spin by different quark flavors. Averaging over the quark intrinsic transverse momentum, three fundamental distributions in longitudinal quark momentum can be interpreted as probability densities: the momentum $q(x) = q^+(x) + q^-(x)$, the helicity $\Delta q(x) = q^+(x) - q^-(x)$, and the transversity $\delta q = q^\uparrow(x) - q^\downarrow(x)$, in a basis of transverse spin eigenstates. The size and the Q^2 dependence of the transversity and helicity distributions can be different because of relativistic motion of quarks and of their different coupling with gluons.

The HERMES experiment has measured the helicity distributions with unprecedented precision and has done the first measurement of transversity on a hydrogen target. Here both measurements will be reviewed.

CP747, *Structure of the Nucleon at Large Bjorken x*, edited by C. Bourrely, Z.-E. Meziani, and J. Soffer

THE HERMES EXPERIMENT

HERMES is a fixed-target experiment located at the east straight section of the HERA positron-proton storage ring at the DESY laboratory in Hamburg. The 27.6 GeV positrons are transversely self-polarized due to a small spin asymmetry in the emission of synchrotron radiation. In order to achieve longitudinal polarization, the electrons pass through spin rotators before entering the experimental area, and are rotated back afterwards to restore the initial conditions inside the ring.

One of the peculiarities of HERMES is the purity of the target, being the dilution factor (i.e. the ratio of polarizable nucleons divided by the total number of nucleons in the target) equal to unity for hydrogen and deuterium. The polarized gas enters into a windowless storage cell located coaxially inside the beam pipe. Gas escapes out of the ends of the cell and is pumped away by the high-speed differential pumping system.

HERMES [1] is a conventional forward spectrometer with a dipole magnet providing an integrated field of 1.3 Tm. The spectrometer is divided into two symmetric top-bottom halves by the HERA beam lines, which are shielded by a horizontal iron plate.

The acceptance in the scattering angle is limited at small angles by the iron plate to 0.04 rad, while the maximum angles in the horizontal and vertical directions are 0.17 and 0.14 rad, respectively. Each half of the detector has a series of drift-chamber planes for tracking. An angular resolution of $\delta\theta < 0.6$ mrad and a momentum resolution of $\delta p/p \sim 1\%$ are achieved. The trigger is formed by a combination of signals from three hodoscopes and a lead-glass calorimeter with an energy deposition greater than 1.4 GeV. Lepton-hadron discrimination is achieved by a probability analysis of the signals from the transition radiation detector, the preshower hodoscope and the calorimeter, while the RICH (Ring Imaging Cerenkov detector) provides p, K and π separation over almost all the momentum range.

HELICITY DISTRIBUTIONS AT HERMES

In polarized DIS, when both target and lepton are longitudinally polarized, the virtual photon γ^* is emitted with its spin in the same direction as the lepton that originated it. Because of conservation of angular momentum, it can only interact with a quark with spin orientation in the opposite direction of its own. By changing the orientation of the target polarization one is then sensitive to the quarks polarized in the same or in the opposite direction of the nucleon (q^+ and q^-). The quark helicity distributions are defined as the net quark polarization in the nucleon's polarization direction: $\Delta q = q^+ - q^-$. The measured asymmetry $A^h_{||}$, obtained from the number of hadrons of one given type detected in the parallel or anti-parallel spin-target configurations, can be related to the polarized quark distributions through the relation:

$$\frac{A^h_{||}(x,Q^2)}{D(1+\eta\gamma)} \simeq A^h_1(x,Q^2) \simeq \sum_q \frac{e_q^2\, q(x)\, \int dz\, D^h_q(z)}{\sum_{q'} e^2_{q'}(x)\, q'(x)\, \int dz\, D^h_{q'}(z)} \frac{\Delta q(x)}{q(x)} = \sum_q P^h_q(x,z) \frac{\Delta q(x)}{q(x)} \quad (1)$$

where D is the photon depolarization factor, γ and η are kinematic factors, D_q^h are the fragmentation functions of quark q into the hadron h, $q(x)$ are the unpolarized quark distributions, and $P_q^h(x,z)$ are the so-called purity factors, which describe the probability that a hadron h originates from an event in which the quark q was struck. By using measured inclusive and semi-inclusive asymmetries and models for the fragmentation functions and for the unpolarized quark distributions, it is then possible to extract the distributions $\Delta q/q$ from the system of equations (1).

Double spin asymmetries were measured on hydrogen and deuterium targets ([2],[3]). The number of inclusive DIS events was 1.7M on hydrogen and 6.7M on deuterium. Pion asymmetries on hydrogen and pion and kaon asymmetries on deuterium target were measured for the first time, taking advantage of the hadron discriminating power of the RICH detector. All asymmetries are large and positive, a fact attributed to the high probability of scattering off a u quark, except the K^- one, giving a hint that, being K^- composed by sea quarks only, the sea may result in being unpolarized.

The measured asymmetries were corrected for effects arising from the non-isotropic acceptance of the spectrometer, which resulted to be $\simeq$10% for $x < 0.1$ and negligible at higher x. Also, they were corrected for radiative and smearing effects. Such a correction accounts for the kinematic migration of events through the bins, and has the effect that the errors of the final Born asymmetries are correlated. The correction is mainly due to QED effects at low x and detector effects at high x, with an uncertainty inflation on the inclusive asymmetries up to 1.8 and up to 1.4 respectively. The effect is much smaller for SIDIS asymmetries.

The system of equations (1) was solved for ($\Delta u/u$, $\Delta d/d$, $\Delta\bar{u}/\bar{u}$, $\Delta\bar{d}/d$, $\Delta s/s$), with the hypothesis that $\Delta\bar{s}/\bar{s}$ is zero. This hypothesis was used because $\Delta\bar{s}/\bar{s}$ is not very well constrained by data. The purity matrix was multiplied by a nuclear mixing matrix, which takes into account the probability of scattering off a given nucleon in the deuterium target, and the nucleon's relative polarization.

The purities were obtained using the LUND string model implemented in the JETSET7.4 package [4] and the CTEQ5L parton distribution functions [5]. The string breaking LUND parameters were obtained after tuning the model to reproduce the HERMES hadron multiplicities. Systematic errors on the purities were assigned only from the fragmentation model, since the CTEQ5L PDF's are provided without uncertainties.

The main sources of systematic uncertainties on the measured asymmetries come from the polarization of beam (4.2% for H and 2.3% for D) and target (5.1% and 5.2%), acceptance correction (3.0% and 3.1%, for SIDIS), smearing (2.0%) and radiative corrections (2.0% for DIS and 1.0% for DIS), the model for R (1.1%) and g_2 (0.6%).

The resulting PDF's are shown in Figure 1, evolved to the common scale of 2.5GeV^2, and compared to recent fits performed on inclusive world data. The resulting Δu is found to be positive, while Δd is negative and smaller in magnitude. The sea distributions $\Delta\bar{u}$, $\Delta\bar{d}$ and Δs are extracted separately for the first time. Within the uncertainties the strange sea is in agreement with both fits as well as with the unpolarized hypothesis.

The breaking of the flavor symmetry in the unpolarized light sea quarks distributions is well established. In order to investigate the polarized case, a modified fit was used to extract the flavor asymmetry $\Delta\bar{u} - \Delta\bar{d}$.The results are shown in Figure 1, compared with predictions from the χQSM [6] and a meson cloud model [7]. The HERMES data favor

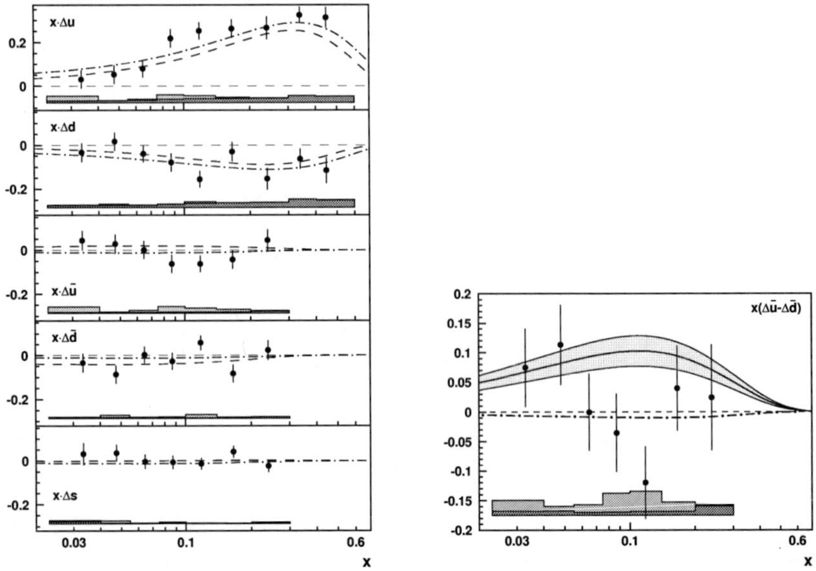

FIGURE 1. Left plot: polarized quark densities evolved at the common scale of 2.5 GeV^2 ([3]). The data points are compared with GRSV2000 ([8], dashed line) and BB ([9], dash-dotted line). The dark systematic bands come from the uncertainties on the asymmetries, while the light ones come form the uncertainties on the fragmentation model. Right plot: the light quark sea flavor asymmetry at $<Q^2>=2.5$ GeV^2, compared with the predictions [7] (dash-dotted line) and [6] (solid line with its $\pm 1\sigma$ error band).

a symmetric polarized light flavor sea and preclude the prediction of the χQSM at the 97% confidence level within the context of the LO analysis used.

TRANSVERSITY AT HERMES

Transversity has so far never been measured because it is a chiral odd quantity, and hard interactions conserve chirality. It can nevertheless be probed in semi-inclusive DIS, where the cross-section can be written as the sum of products of fragmentation functions (FF's) and distribution functions (DF's). The transversity distribution function δq may couple with the chiral-odd Collins fragmentation function, $H_1^{\perp}$, thus giving a finite result. Single spin asymmetries obtained from two transversely polarized target states, can arise from this 'Collins' mechanism as well as from the 'Sivers' mechanism, where the DF $f_{1T}^{\perp q}$, relating the quark transverse momentum with the transverse polarization of the nucleon, couples with the ordinary unpolarized FF. The Sivers function is odd under time reversal (T-odd), a fact that was believed for a long time to forbid its existence, until it was realized that final state interactions via a soft gluon could provide a mechanism for the existence of such a term in the cross section.

During the HERA shutdown period in 2001 a transverse target magnet was installed at HERMES. The 0.3T magnetic field is used to transversely polarize the gaseous target.

The field uniformity is better than $4.5 \cdot 10^{-5}$T along the beam direction in order to avoid beam-induced depolarization resonances. The results shown here were obtained from data taken during the years 2002-2003 with a transversely polarized hydrogen target and an unpolarized positron beam.

Angles used to define the scattering are ϕ, the azimuthal angle about the virtual photon direction relative to the lepton scattering plane, and ϕ_S, the angle of the target spin with respect to the lepton scattering plane (see Figure 2).

After applying standard SIDIS cuts, an asymmetry was obtained in each x and z bin as a function of ϕ and ϕ_S:

$$A^h_{UT}(\phi,\phi_S) = \frac{1}{|S_T|} \frac{\left(N_h^{\uparrow}(\phi,\phi_S) - N_h^{\downarrow}(\phi,\phi_S)\right)}{\left(N_h^{\uparrow}(\phi,\phi_S) + N_h^{\downarrow}(\phi,\phi_S)\right)}, \tag{2}$$

where $N_h^{\uparrow}$ and $N_h^{\downarrow}$ are the luminosity normalized yields in the two transversely polarized target states. Such an asymmetry was fitted with the functions $\sin(\phi \pm \phi_S)$ and the Collins ($\sin(\phi+\phi_S)$) and Sivers ($\sin(\phi-\phi_S)$) moments were thus extracted:

$$\frac{1}{2}A^h_{UT}(\phi,\phi_S) = <\sin(\phi+\phi_S)>^h_{UT} \frac{B(<y>)}{A(<x>,<y>)}\sin(\phi+\phi_S) + <\sin(\phi-\phi_S)>^h_{UT}\sin(\phi-\phi_S) \tag{3}$$

where $B(<y>) = (1-y)$ and $A(<x>,<y>) = \frac{1}{2}y^2 + (1-y)(1+R(x,y))/(1+\gamma^2)$, with R being the ratio of longitudinal to transverse DIS cross section, and $\gamma^2 = 2M_p x/(Ey)$. The azimuthal moments averaged over the experimental acceptance are shown in Figure 2, together with simulations from PYTHIA6 ([10]) of the fraction of pion yields coming from vector mesons in each x and z bin. The major source of systematic error comes from the target, which gives an overall scale uncertainty of 8%.

In the measured kinematic range the average Collins moment for π^+ is 0.021±0.007, while for π^- it is -0.038±0.008, where the errors are purely statistical ([11]). This result resembles that on helicity distributions, where Δu is positive and Δd is negative, giving a hint that this may apply also for transversity distributions. However the magnitude of these asymmetries hints to $|\delta d| > |\delta u|$ which has no correspondence to the helicity case, where $|\Delta d| < |\Delta u|$, and is thus unexpected. A possible explanation of the large size of the Collins π^- moment at high x is that the ratio of favored Collins fragmentation function (describing e.g. the fragmentation of u to π^+) to disfavored Collins fragmentation function (describing d into π^+) is of order -1. This opposite sign may be explained using a JETSET [10] simulation based on the string model of fragmentation. Here a favored pion is obtained from a first string break and a second, disfavored pion is created in the second break with transverse momentum in the opposite direction than that of the first pion, showing that any correlation of spin or kinematic observables should have opposite sign for favored and disfavored pions.

The total moments of the Sivers asymmetries are 0.017±0.004 for π^+ and 0.002±0.005 for π^-. The result on π^+ is the first evidence of a T-odd parton distribution in leptoproduction, and its sign implies a negative sign for the Sivers function of the u flavor, which dominates the π^+ production. Studies on pions deriving from exclusive ρ^0 events indicate that their asymmetry could be of the same order of magnitude of the semi-inclusive DIS π^+ asymmetry, thus showing that the value of the

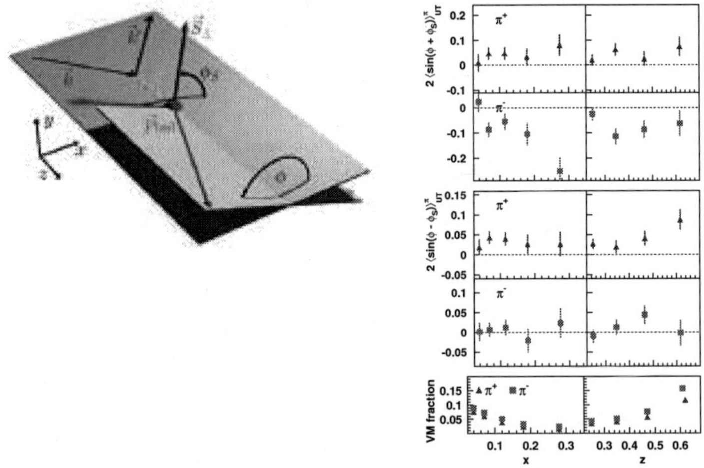

FIGURE 2. Left: a diagram showing the angles ϕ and ϕ_S. Right: Collins (upper panels) and Sivers (middle panels) azimuthal moments for positive and negative pions as a function of x and z ([11]). The bottom panels show the fraction of events coming from vector meson background, estimated from a Monte-Carlo simulation.

extracted Sivers asymmetry has to be interpreted with caution, as the exclusive meson contribution could be dominant.

CONCLUSIONS

Recent results concerning helicity and transversity distributions from the HERMES experiment have been shown. While the helicity distributions have been extracted with good precision, only hints of transversity have been obtained until now. The aim of the HERMES experiment is to use additional data with transverse target to extract transversity distributions with higher accuracy.

REFERENCES

1. The HERMES Collaboration, *Nuc. Instr. Meth.* **A417**, 230 (1998).
2. The HERMES Collaboration, *Phys. Rev. Lett.* **92**, (2004) 012005.
3. The HERMES Collaboration, hep-ex/0407032.
4. V. Barone et al., *Phys. Lett.* **B571**, 50 (2003).
5. H. Lai, et al., *Eur. Phys. J.* **C12**, 375 (2000).
6. B. Dressler et al., *Eur. Phys. J.* **C14**, 147 (2000).
7. F.-G. Cao and A.I.Signal, *Phys. Rev.* **D68**, (2003) 074002.
8. M. Glück et al., *Phys. Rev.* **D63**, 094005 (2001).
9. J. Blümlein and H. Böttcher, *Nucl. Phys.* **B636**, 225 (2002).
10. T. Sjöstrand et al., *Comput. Phys. Commun.* **135**, 238 (2001).
11. The HERMES Collaboration, hep-ph/0408013.

Novel Azimuthal Asymmetries in Drell Yan and Semi-inclusive Deep Inelastic Scattering

Leonard P. Gamberg*, Gary R. Goldstein† and Karo A. Oganessyan**

*Division of Science, Penn State-Berks Lehigh Valley College, Reading, PA 19610, USA
†Department of Physics and Astronomy, Tufts University, Medford, MA 02156, USA
** INFN-Laboratori Nazionali di Frascati, E. Fermi 40, 00044 Frascati, Italy

Abstract. We consider the leading and sub-leading twist T-odd and even contributions to the $\cos 2\phi$ azimuthal asymmetry in unpolarized dilepton production in Drell-Yan Scattering We estimate the contributions' effects at 500 GeV, 50 GeV, and 25 GeV energies in the framework of the parton model using a quark diquark-spectator model of the nucleon to approximate the soft contributions.

Keywords: T-odd effects, Transversity
PACS: 13.87.Fh 13.60.-r 13.88.+e 14.20.Dh

INTRODUCTION

One of the most interesting results in spin physics has been the discovery of a class of chirally odd quark distribution functions. Considerable attention has been focused on the transversity or covariant transverse spin distribution h_1 which provides information on the quark transverse spin distribution in a transversely polarized nucleon [1, 2, 3]. Chiral odd distribution functions require a quark helicity flip and are difficult to probe in inclusive deep inelastic scattering due to the helicity conserving property of quantum chromodynamics (QCD) interactions. However, when two hadrons participate in the scattering process, the nucleon's transversity can be accessed, for example, the double transverse spin asymmetry in Drell-Yan scattering [1, 4]. Alternatively, transversity can be probed in semi inclusive deep inelastic scattering (SIDIS) where outgoing hadrons are produced in the current fragmentation region [5, 6]. In this case the cross sections and distribution functions are sensitive to the transverse momentum of the quarks. Sensitivity to transverse momentum dependence in parton distribution functions leads to a class of leading twist spin dependent effects which are T-odd [7, 8, 9, 10]. The distribution functions exist by virtue of non-zero parton transverse momenta and would vanish at tree-level in any T-conserving model of hadrons and quarks. In this sense they are similar to the decay amplitudes for hadrons that involve single spin asymmetries which are non-zero due to final (and/or initial) state strong interactions (FSI) [5].

Such a T-odd distribution was proposed by Sivers in the context of inclusive pion production [7]. How such a distribution could actually arise without violating conservation laws remained an open question for some years [8]. Recently, however, a mechanism was introduced that could produce that Sivers effect without violating invariance principles. A reaction mechanism in terms of FSI (and using the quark-diquark model of the nucleon) was used to calculate this effect by Brodsky, Hwang and Schmidt [11] (BHS). That calculation was subsequently recast in terms of a color gauge invariant treatment of

CP747, *Structure of the Nucleon at Large Bjorken x*, edited by C. Bourrely, Z.-E. Meziani, and J. Soffer

transverse momentum dependent distribution functions [12, 13, 14]. Gluon loop corrections to tree-level calculations for the transversity of quarks and hadrons emerge from the gauge link insertion into the formal definitions of the transversity distributions. At first order in the strong coupling α_s the correction involves the single gluon approximation to FSI, a single loop diagram.

These FSI corrections, implemented in parton model inspired calculations, give rise to novel single spin and azimuthal asymmetries in SIDIS [11, 10, 15, 16, 17, 18, 19, 20] and Drell-Yan processes [21, 22] resulting in significant asymmetries. This is all the more interesting in light of the fact that in the perturbative QCD regime of quark and gluon dynamics such asymmetries are expected to be small; that is, of the order, $\alpha_s m_q/\sqrt{s}$ [23, 24]. One T-odd transversity distribution function (introduced by Boer and Mulders [9]) is $h_1^\perp$, which depends on variables x, Q^2 and $p_\perp$ of the struck quark; it measures the amount of quark transversity from an unpolarized nucleon. At moderate energies such T-odd effects arising from the non-perturbative $h_1^\perp$ may be the source of non-trivial $\cos 2\phi$ asymmetries in SIDIS [25, 15, 17] and in Drell-Yan scattering [26, 21, 22].

In this letter we will report our results on SIDIS [15, 16, 17] and present new results on the Drell-Yan process [27]. The latter process is interesting in light of the possibility of accessing this transvesity property of quarks in the proposed proton anti-proton experiments at Darmstat GSI [28], where an anti-proton beam is an ideal tool for studying transversity due to the dominance of valence quark effects. We demonstrate at proposed energies that both the T-odd and sub-leading twist T-even distributions in $\bar{p}\, p \to \ell\, \ell^+\, X$ provide contributions to the $\cos 2\phi$ asymmetry that are significant. We estimate that the sub-leading twist contribution is a non-trivial fraction of the large leading twist T-odd contribution using the parton model.

THE AZIMUTHAL ASYMMETRIES IN UNPOLARIZED SIDIS AND DRELL YAN SCATTERING

The angular asymmetries that can arise in *unpolarized* Drell-Yan scattering (*e.g.* $\bar{p}+p \to \mu^-\mu^+ + X$) and SIDIS ($e+p \to e'hX$) are obtained from the differential cross section expressions:

$$\frac{1}{\sigma}\frac{d\sigma}{d\Omega} = \frac{3}{4\pi}\frac{1}{\lambda+3}\left(1+\lambda\cos^2\theta+\mu\sin^2\theta\cos\phi+\frac{\nu}{2}\sin^2\theta\cos 2\phi\right) \quad (1)$$

and

$$\frac{d\sigma}{dxdydzdP_{h\perp}^2\, d\phi_h} = A+B+C\cos\phi+D\cos 2\phi, \quad (2)$$

respectively [29, 30, 31]. In the Drell-Yan process the angles refer to the lepton pair orientation in their rest frame relative to the boost direction and the initial hadron's plane, the Collins Soper frame [29]. The dependence on the other independent variables, $s, x, m_{\mu\mu}^2, q_T$, is suppressed, but the asymmetry parameters, λ, μ, ν, depend on those variables. In SIDIS, the azimuthal angle refers to the relative angle between the hadron production plane and the lepton scattering plane. A, B, C, and D are functions of $x, y, z, Q^2, |\mathbf{P}_{h\perp}|$. It is especially interesting that the $\cos 2\phi$ azimuthal asymmetry in Drell

Yan depends on the T-odd distribution $h_1^\perp$ and its anti-quark distribution, $\bar{h}_1^\perp$; whereas in SIDIS one essentially replaces the anti-quark distribution with the Collins function, $H_1^\perp$ [5].

$\cos 2\phi$ *Azimuthal Asymmetry in Drell Yan*

To leading order in Q^2, *i.e.* leading twist, the asymmetry ν is given by [21]

$$\nu_2 = \frac{\sum_a e_a^2 \mathscr{F}\left[(2\hat{\mathbf{h}}\cdot\mathbf{k}_\perp\cdot\hat{\mathbf{h}}\cdot\mathbf{p}_\perp - \mathbf{p}_\perp\cdot\mathbf{k}_\perp)h_1^\perp(x,\mathbf{k}_\perp)\bar{h}_1^\perp(\bar{x},\mathbf{p}_\perp)/(M_1M_2)\right]}{\sum_a e_a^2 \mathscr{F} f_1(x,\mathbf{k}_\perp)\bar{f}_1(\bar{x},\mathbf{p}_\perp)} \tag{3}$$

where the convolution is, $\mathscr{F} \equiv \int d^2\mathbf{p}_\perp d^2\mathbf{k}_\perp \delta^2(\mathbf{p}_\perp + \mathbf{k}_\perp - \mathbf{q}_\perp) f^a(x,\mathbf{k}_\perp)\bar{f}^a(\bar{x},\mathbf{p}_\perp)$. As pointed out by Collins and Soper [29], well before $h_1^\perp$ was identified, there is a higher twist T-even contribution to the $\cos 2\phi$ asymmetry which is not small at center of mass energies of 50 GeV^2

$$\nu_4 = \frac{\frac{1}{Q^2}\sum_a e_a^2 \mathscr{F}\left[2\left(\hat{\mathbf{h}}\cdot(\mathbf{k}_\perp - \mathbf{p}_\perp)\right)^2 - (\mathbf{k}_\perp - \mathbf{p}_\perp)^2 f_1(x,\mathbf{k}_\perp)\bar{f}_1(\bar{x},\mathbf{p}_\perp)\right]}{\sum_a e_a^2 \mathscr{F}\left[f_1(x,\mathbf{k}_\perp)\bar{f}_1(\bar{x},\mathbf{p}_\perp)\right]}. \tag{4}$$

Collins and Soper considered this effect as a possible source for the azimuthal asym-

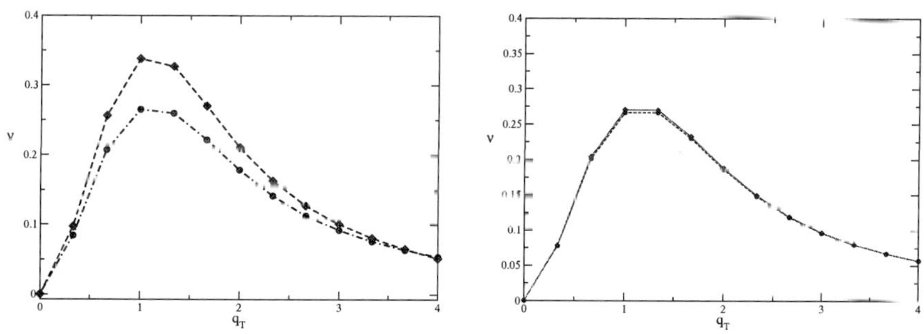

FIGURE 1. Left panel: ν plotted as a function of q_T for $s = 50$ GeV^2 and x in the range $0.2 - 1.0$, and q ranging from $2.5 - 5.0$ GeV: Right panel: $s = 500$ GeV^2 and q from $4.0 - 8.6$ GeV.

metry that had been measured. It arises from a dependence on the relative quark and anti-quark azimuthal orientation that enters in the convolution.

We estimate the leading twist 2 and twist 4 contributions using a parton model with the quark diquark spectator framework to estimate the T-odd and even distribution functions. Previously, we calculated these functions to predict single spin asymmetries and azimuthal dependences in SIDIS [10, 15, 16, 17]; $h_1^\perp$ is given by

$$h_1^\perp(x,k_\perp) = \frac{e_1 e_2 g^2}{2(2\pi)^4}\frac{(m+xM)(1-x)}{\Lambda(k_\perp^2)}\frac{1}{k_\perp^2}e^{-2b(k_\perp^2-\Lambda(0))}\left[\Gamma(0,2b\Lambda(0)) - \Gamma(0,2b\Lambda(k_\perp^2))\right], \tag{5}$$

and the unpolarized quark distribution function is

$$f(x) = \frac{g^2}{(2\pi)^2}(1-x)\cdot\left\{\frac{(m+xM)^2-\Lambda(0)}{\Lambda(0)} - \left[2b\left((m+xM)^2-\Lambda(0)\right)-1\right]\right.$$
$$\left.\times e^{2b\Lambda(0)}\Gamma(0,2b\Lambda(0))\right\}. \tag{6}$$

We have included a Gaussian damping of the quark transverse momenta in the nucleon. This models the known intrinsic $k_\perp$ distribution and also regularizes the convolutions that have to be done to obtain observable cross sections and asymmetries.

In Fig. 1, at center of mass energy of $s = 50$ GeV^2, the leading order T-odd contribution contributes about 28% with an additional 10% from sub-leading order T-even contributions. At center of mass energy of $s = 500$ GeV^2 the distinction between the leading order T-odd and additional sub-leading order T-even contributions are small. This reflects the diminution of the non-leading contribution with increasing s and Q^2. Note that we have not taken account of the evolution of the distribution $h_1^\perp$ with Q^2 scale. This evolution has not been worked out in general at this time. In Fig. 2, ν is plotted versus x at $s = 50$ GeV^2 where q_T ranges from 2 to 4 GeV. Again the higher twist contribution is significant.

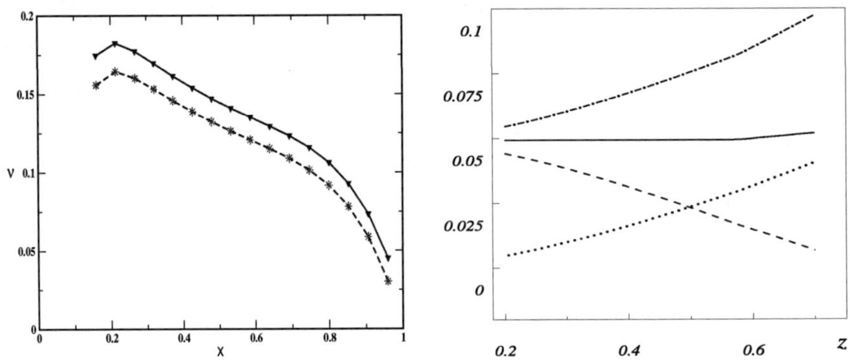

FIGURE 2. Left panel: ν plotted as a function of x for $s = 50$ GeV^2 q_T ranging from 2 to 4 GeV. Right panel: The z-dependence of the $\cos 2\phi$ asymmetry at HERMES [6] kinematics. The full and dotted curves correspond to the T-even and T-odd terms of asymmetry, respectively. The dot-dashed and dashed curves are the sum and the difference of those terms, respectively.

$\cos 2\phi$ *Azimuthal Asymmetry in SIDIS*

We use the conventions established in [9] for the asymmetries. Being T-odd, $h_1^\perp$ appears with the $H_1^\perp$, the T-odd fragmentation function in observable quantities. In particular, the following weighted SIDIS cross section projects out a leading double

T-odd $\cos 2\phi$ asymmetry,

$$\langle \cos 2\phi \rangle_{UU} = \frac{8(1-y)\sum_q e_q^2 h_1^{\perp(1)}(x) z^2 H_1^{\perp(1)}(z)}{(1+(1-y)^2)\sum_q e_q^2 f_1(x) D_1(z)}, \tag{7}$$

where the subscript UU indicates unpolarized beam and target. The Collins fragmentation function is given by [16],

$$H_1^{\perp}(z,k_\perp) = \frac{f_{qq\pi}^2 g^2}{(2\pi)^4} \frac{M\mu/z}{\Lambda(k_\perp^2)} \frac{(1-z)}{4z^2} \frac{1}{k_\perp^2} e^{-2c\left(k_\perp^2-\Lambda(0)\right)} \left[\Gamma(0,2c\Lambda(0)) - \Gamma(0,2c\Lambda(k_\perp^2))\right]. \tag{8}$$

The $\cos 2\phi$ asymmetry originating from T-even distribution and fragmentation function appears at order $1/Q^2$ [30, 31]. The $\langle \cos 2\phi \rangle$ from ordinary sub-sub-leading T-even and leading double T-odd (up to a sign) effects to order $1/Q^2$ is given by [17]

$$\langle \cos 2\phi \rangle_{UU} = \frac{2\frac{\langle k_\perp^2 \rangle}{Q^2}(1-y) f_1(x) D_1(z) \pm 8(1-y) h_1^{\perp(1)}(x) H_1^{\perp(1)}(z)}{\left[1+(1-y)^2 + 2\frac{\langle k_\perp^2 \rangle}{Q^2}(1-y)\right] f_1(x) D_1(z)}. \tag{9}$$

The z-dependence of this asymmetry at HERMES kinematics [6] are shown in the right panel of Fig. 2. The full and dotted curves correspond to the T-even and T-odd terms in the asymmetry, respectively. The dot-dashed and dashed curves are the sum and the difference of those terms, respectively. From the figure, one can see that the double T-odd asymmetry behaves like z^2, while the T-even asymmetry is flat in the whole range of z. Therefore, aside from the competing T-even $\cos 2\phi$ effect, the experimental observation of a strong z-dependence (especially at high z region) would indicate the presence of T-odd structures in *unpolarized* SIDIS implying that novel transversity properties of the nucleon can be accessed without involving spin polarization.

Conclusions

The interdependence of intrinsic transverse quark momentum and angular momentum conservation are intimately connected with studies of T-odd effects underlying the $\cos 2\phi$ azimuthal asymmetries in Drell-Yan and semi-inclusive deep inelastic scattering. Using re-scattering as a mechanism to generate T-odd distribution functions opens a new window into the theory and phenomenology of transversity in hard processes. We have also demonstrated that at moderate energies sub-leading twist contributions are non-trivial.

Acknowledgments

L.P.G. thanks J. Soffer, C. Bourrely and Z.E. Meziani and the organizers of the HiX2004 Workshop for the opportunity to present this work.

REFERENCES

1. J. Ralston and D. E. Soper, Nucl. Phys. **B152**, 109 (1979).
2. X. Artru and M. Mekhfi, Z. Phys. C **45**, 669 (1990); R. L. Jaffe and X. Ji, Phys. Rev. Lett. **67**, 552 (1991).
3. V. Barone, A, Drago, and P. G. Ratcliffe, Phys.Rept. **359**, 1 (2002).
4. G. Bunce, N. Saito, J. Soffer, and W. Vogelsang, Ann. Rev. Nucl. Part. Sci. **50**, 525 (2000).
5. J. C. Collins, Nucl. Phys. **B396**,161 (1993).
6. A. Airapetian *et al.*, Phys. Rev. Lett. **84**, 4047 (2000); Phys. Lett. B **562**, 182 (2003); arXiv: hep-ex/0408013.
7. D. Sivers, Phys. Rev D **41**, 83 (1990) ; Phys. Rev. D **43**, 261 (1991).
8. M. Anselmino, M. Boglione, and F. Murgia, Phys. Lett. B **362**, 164 (1995); M. Anselmino and F. Murgia, Phys. Lett. B **442**, 470 (1998).
9. D. Boer and P. J. Mulders, Phys. Rev. D **57**, 5780 (1998).
10. G. R. Goldstein and L. P. Gamberg, hep-ph/0209085, *Proceedings of* 31 st *International Conference on High Energy Physics (ICHEP 2002)*, Amsterdam, The Netherlands, 452 (2002).
11. S. J. Brodsky, D. S. Hwang, and I. Schmidt, Phys. Lett. B **530**, 99 (2002).
12. X. Ji and F. Yuan, Phys. Lett. B **543**, 66 (2002); A.V. Belitsky, X. Ji and F. Yuan, Nucl. Phys. **B656**, 165(2003) .
13. J. C. Collins, Phys. Lett. B **536**, 43 (2002).
14. D. Boer, P.J. Mulders, and F. Pijlman, Nucl. Phys. **B667**, 201 (2003).
15. L. P. Gamberg, G. R. Goldstein, and K. A. Oganessyan, Phys. Rev. D **67**, 071504 (2003).
16. L. P. Gamberg, G. R. Goldstein, and K. A. Oganessyan, Phys. Rev. D **68**, 051501 (2003).
17. L. P. Gamberg, G. R. Goldstein, and K. A. Oganessyan, arXiv: hep-ex/0309137; Nucleon Transversity Properties Through *ep* Scattering, *Proceedings* 11 th *International Workshop on Deep Inelastic Scattering (DIS 2003)*, St. Petersburg, Russia (2003).
18. A. Bacchetta, A. Metz, and J.-J. Yang, Phys. Lett. B **574**, 225 (2003).
19. L.P. Gamberg, D. S. Hwang, and K. A. Oganessyan, Phys. Lett. B **584**, 276 (2004).
20. A. Bacchetta, A. Schaefer, J.J. Yang, Phys. Lett. B **578**, 109 (2004).
21. D. Boer, Phys. Rev. D**60**, 014012 (1999)
22. D. Boer, S. J. Brodsky, and D. S. Hwang, Phys. Rev D **67**, 054003 (2003).
23. G. L. Kane, J. Pumplin, and K. Repko, Phys. Rev. Lett. **41**, 1689 (1978).
24. W. G. D Dharmarathna and G. R. Goldstein, Phys. Rev. D **41** , 1731 (1990); Phys. Rev. D **53**, 1073 (1996) .
25. D. M. Jansen, Ph.D. thesis, University of Washington, 1991.
26. J. S. Conway *et al.*, Phys. Rev. D **39**, 92 (1989).
27. L. P. Gamberg and G. R. Goldstein, *In preparation.*
28. QCD-PAC REPORT on PAX and ASSIA - Common report (2003); http://www.fz-juelich.de/ikp/pax
29. J. Collins and D. E. Soper, Phys. Rev. D**16**, 2219 (1977).
30. R.N. Cahn, Phys. Lett. B **78**, 269 (1978); Phys. Rev. D **40**, 3107 (1989).
31. J. Levelt and P.J. Mulders, Phys. Rev. D **49**, 96 (1994).

EXCLUSIVE REACTIONS

GPDs from small to large x

A. Schäfer

Institute for Theoretical Physics, University of Regensburg, D-93040 Regensburg, Germany

Abstract. My talk was devoted to various aspects of GPDs, ranging from small to large x. This article covers only a small fraction of these, namely the exclusive photoproduction of dijets and heavy quarkonia. Both examples demonstrate that a quantitative theoretical controle of some of the many processes which can be parametrized by GPDs is highly non-trivial.

GPDS IN A NUTSHELL

It is obvious that many aspects of hadron physics which traditionally are described with seperate formulations in QCD are ultimately related. Often, however, this connection could only be explored within the framework of specific models. (An example are e.g. form factors and distribution functions.) The basic idea of GPDs is to develop a stringent QCD description which is broad enough to cover most of the relevant hadronic processes. For a very nice review see [1] It turns out that the ambiguities involved in model building reappear in the GPD formalism as well defined QCD problems concerning e.g. the size of factorization-breaking terms, the convergence of expansions in twist and α_s, the analytic properties of the parametrizing functions etc. This rather remarkable fact opens the possibility for tedious but steady progress in making the various relations more and more stringent. Not surprisingly these problems become especially demanding both for large and very small values of x.

GPDs are basically response functions of hadrons: you extract some quark or gluon with longitudinal momentum fraction $x+\xi$ from a hadron and put it back with longitudinal momentum fraction $x-\xi$ plus some transverse momentum.

Meanwhile a pletora of reactions has been analysed in terms of GPDs and many

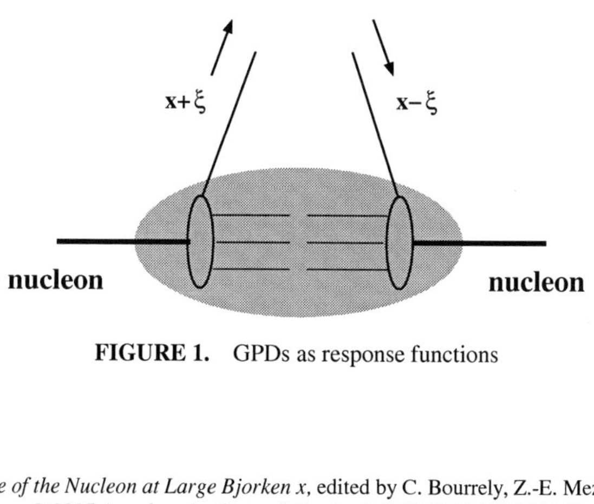

FIGURE 1. GPDs as response functions

CP747, *Structure of the Nucleon at Large Bjorken x,* edited by C. Bourrely, Z.-E. Meziani, and J. Soffer

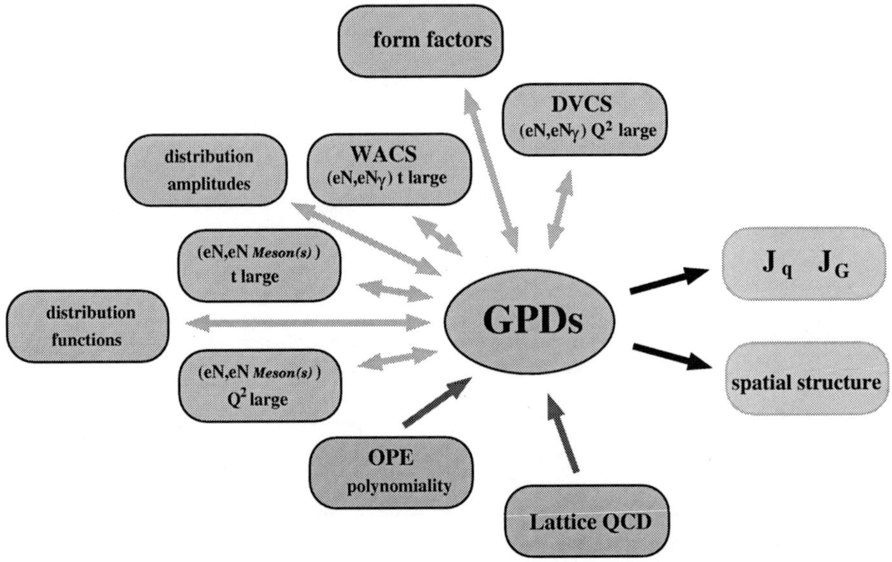

FIGURE 2. GPDs allow to parametrize a large number of processes. They are in addition constraint by polynomiality properties and results from lattice QCD. To the extent that GPDs are better and better determined they provide ever more precise results for the total parton angular momentum and the transverse structure of hadrons, informations which cannot be obtained otherwise.

interesting relations have been established, see Fig. 2. We use the notation of [2] and asign moment acording to (for DVCS)

$$h(p_\mu) + \gamma^*(q_\mu) \to h(p'_\mu) + \gamma(q'_\mu)$$

for any hadron h and define $\Delta_\mu = p'_\mu - p_\mu$, $t = \Delta^2$, $\bar{p}_\mu = \frac{1}{2}(p'_\mu + p_\mu)$; $n_\mu = (1,0,0,-1)/\sqrt{2}\bar{p}^+ \to n \cdot \bar{p} = 1$ and $\xi = -n \cdot \Delta$. Then e.g. the quark GPDs H_q and E_q are defined as:

$$\int \frac{dz^-}{2\pi} e^{ix\bar{p}^+ z^-} \langle p'|\, \bar{q}(-\frac{1}{2}z)\, \gamma^+ q(\frac{1}{2}z)\, |p\rangle \Big|_{z^+=0,\, z_\perp=0}$$
$$= \frac{1}{P^+} \left[H_q(x,\xi,t)\, \bar{N}(p')\gamma^+ N(p) + E_q(x,\xi,t)\, \bar{N}(p') \frac{i\sigma^{+\alpha}\Delta_\alpha}{2M} N(p) \right]$$

DIFFRACTIVE DIJET PRODUCTION

In principle this process (see Fig. 3) allows to extract the gluon GPD H_g which certainly contains some of the most interesting and novel information on the internal structure and dynamics of the nucleon. The cross section has a chirallity conserving (CC) and a

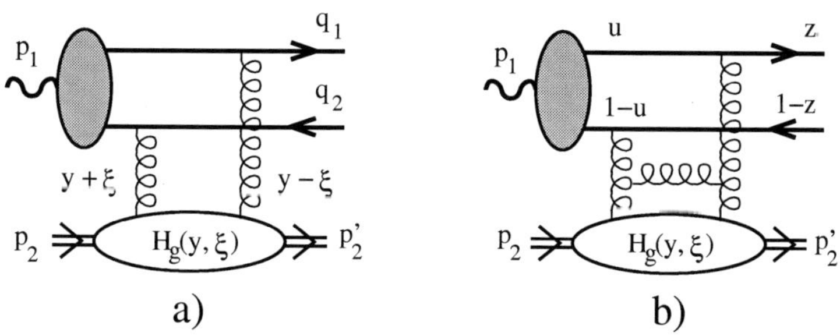

FIGURE 3. Two contributing amplitudes, while a) is the leading term in the collinear expansion, b) is an example for a contribution which is not suppressed by energy logarithms and thus has to be taken into account.

chirality violating (CV) part [3].

$$2\pi \frac{d\sigma_{\gamma\to 2\text{jets}}}{d\phi\, dq_\perp^2\, dt\, dz}\bigg|_{t=0} = \sum_q e_q^2 \frac{\alpha_{EM}\,\alpha_s^2\pi^2}{2N_c q_\perp^6} \times \left[(1-4z\bar{z}\cos^2\phi)|\mathscr{I}_{CC}|^2 + \frac{\pi^2\alpha_s^2\chi^2\langle\bar{q}q\rangle^2}{N_c^2 q_\perp^2}|\mathscr{I}_{CV}|^2 \right] \tag{1}$$

The chirality conserving amplitude contains the matrix element ($e^{(\lambda)}$ is the photon polarization vector)

$$\langle 0|\bar{q}(0)\gamma_+\frac{1\pm\gamma_5}{2}q(x_-,\mathbf{r})|\gamma^{(\lambda)}(q)\rangle = \frac{iN_c e_q}{4\pi^2\mathbf{r}^2}q_+\int_0^1 du\, e^{-iuq_+x_-}\left[(e^{(\lambda)}\cdot\mathbf{r})(2u-1)\pm i\varepsilon_{ik}\mathbf{r}_i e_k^{(\lambda)}\right] \tag{2}$$

and the chirality violating amplitude the following matrix element

$$\langle 0|\bar{q}(0)\sigma_{\alpha\beta}q(x)|\gamma^{(\lambda)}(q)\rangle = i e_q\chi\langle\bar{q}q\rangle\left(e_\alpha^{(\lambda)}q_\beta - e_\beta^{(\lambda)}q_\alpha\right)\int_0^1 du\, e^{-iu(qx)}\,\phi_\gamma(u,\mu) \tag{3}$$

We calculated the CC cross section to NLO and corrrspondingly the CV one to LO. Let us not that the CV amplitude is especially interesting because it depends both on the magnetic susceptibility of the vacuum $\chi\langle\bar{q}q\rangle$ [4], defined by

$$\langle 0|\bar{q}\sigma_{\alpha\beta}q|0\rangle_F = e_q\chi\langle\bar{q}q\rangle F_{\alpha\beta} \tag{4}$$

and on the photon distribution amplitude $\phi_\gamma(u,\mu)$, which are both phenomenologically important QCD properties. Typically CV processes are suppressed by the hard scale, e.g.

$$\frac{\text{chirality breaking}}{\text{chirality conserving}} \sim \frac{4\pi^2\chi\langle\bar{q}q\rangle}{N_c Q} \sim \frac{650\text{MeV}}{Q} \tag{5}$$

but in this case the explicite calculation gives

$$\frac{d\sigma_{CV}/dq_\perp^2}{d\sigma_{CC}/dq_\perp^2} \simeq (7\pm 2\text{ GeV})^2 \cdot \frac{\alpha_s(q_\perp)^2}{q_\perp^2}\left(\frac{\chi\langle\bar{q}q\rangle}{50\text{ MeV}}\right)^2. \tag{6}$$

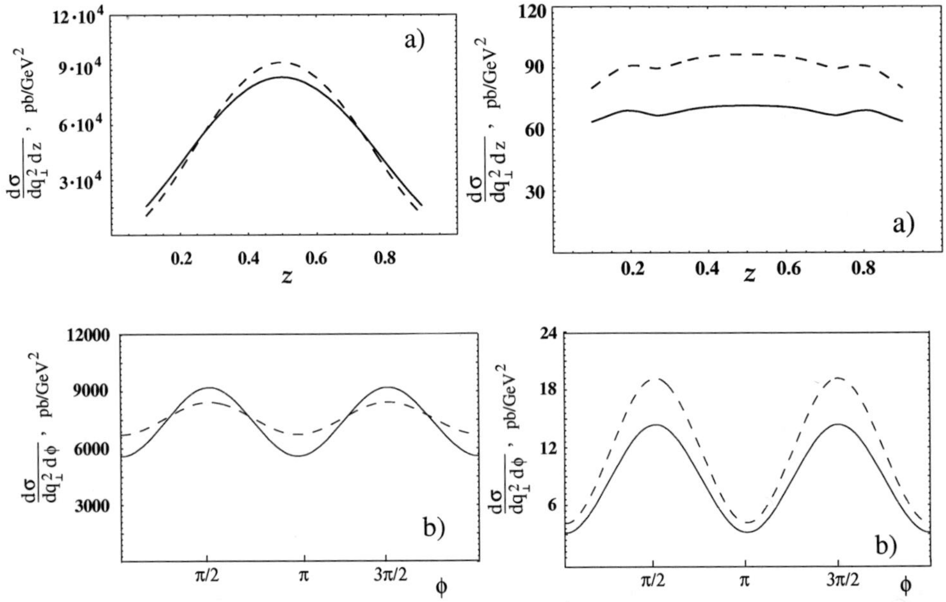

FIGURE 4. Cross sections for $q_\perp = 2$ GeV (left) and $q_\perp = 5$ GeV (rigth). For $q_\perp = 2$ GeV the CV contribution dominates, for $q_\perp = 5$ GeV the CC contribution.

implying that for typical kinematics the CV part is sizeable. Furthermore the CC contribution is $\sim 1 - \cos^2\phi$ where ϕ is the angle between the photon polarization vector and $\vec{q}$, which induces a marked change in the ϕ-dependence as function of $q_\perp$. The observation of this systematic dependece would clearly allow to identify both contributions. Let us note in passing that similar results are obtained for diffractive pion dissociation into two jets, see [5].

To get these results we had, however, to introduce an infrared cut-off to treat a factorization-breaking end-point divergence in the CV part. This is compatible with the usual factorization proof, because the CV amplitude is by itself power-suppressed, but it demonstrates that this suppression is for some processes only effective at really hard scales. It also demonstrates that to access some of the most interesting physics the theoretical analysis of exclusive reactions has to be pushed beyond leading twist and LO, or even NLO. Let us also remark that the factorization-violating contribution has the same functional form as the asymptotic photon wave function, $\sim z(1-z)$. This illustrates that if data favour a photon or hadron wave function close to the asymptotic one this does not necessarily imply a dominance of the leading contribution.

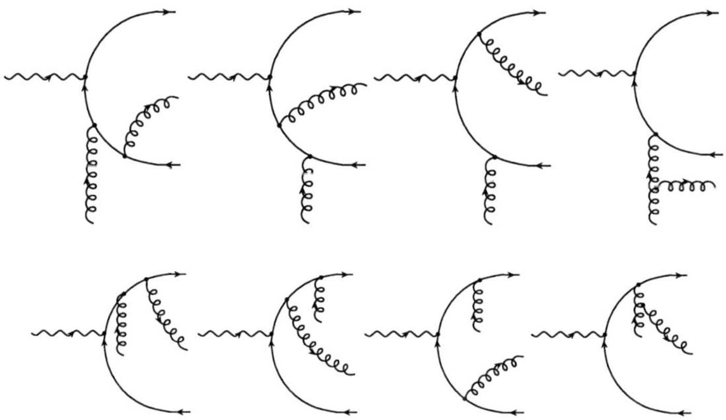

FIGURE 5. NLO contributions to the photon-$\bar{Q}Q$ vertex only. These have to be contracted with all possible outgoing amplitudes.

DIFFRACTIVE PRODUCTION OF HEAVY QUARKONIA

An alternative process usually favoured to obtain information on the gluon structure of e.g. the nucleon is heavy quarkonium production. This holds also true for diffractive charmonium and bottonium production. However, especially after our experience with jet production, one might wonder how heavy 'Heavy' would have to be such that the 'leading' terms really dominate. To find this out requires a full NLO calculation which we did [6]. This involves a very large number of amplitudes, see Fig. 5 and some highly non-trivial decissions on how to treat the wave function renormalization.

The result is pretty sobering: The scale dependence is so strong that far more work will be needed, before this process can serve as an efficent tool to extract the gluon GPD H_g. Fig. 6 shows the cross sections for Υ photoproduction. The theoretical predictions are at LO (left) and NLO (right). The factorization and renormalization scales where chosen identical to be $\mu_F = \mu_R = 1.3$ GeV and $\mu_F = \mu_R = 7$ GeV. The data are from ZEUS and H1. Note that even the ordering of the two curves change between LO and NLO. Let us stress that we were forced to make some approximations in the course of our calculations, which certainly can be improved, but it is hard to believe that the extremely large effects could be caused purely by these approximations.

SUMMARY

We have discussed two examples for exclusive processes which are sensitive to the gluon GPD H_g. Our results were meant to illustrate the following facts: The physics of the processes, which can be parametrized by GPDs is very rich but also highly non-trivial. All the different processes alluded to in Fig.2 have to be analysed in great detail in at least NLO and taking power-corrections into account before one can judge how far the truely comprehensive, unifying description of hadron physics which is so strongly hoped

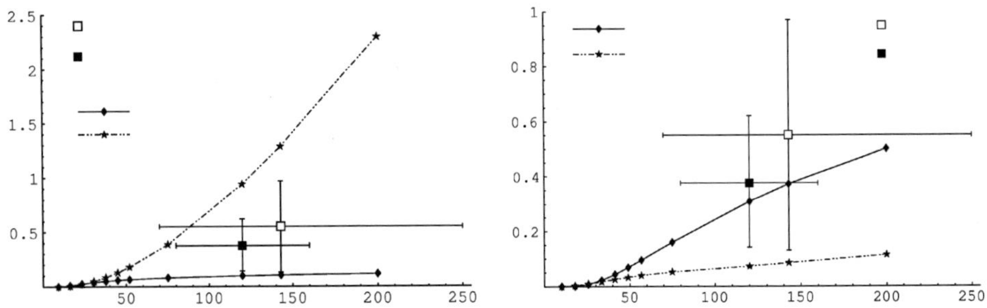

FIGURE 6. Results for Υ production in LO (left) and NLO (right)

for will become reality.

ACKNOWLEDGMENTS

This work was supported by BMBF and HGF. Let me stress that for the two projects presented the most important contributions came D. Ivanov and G. Krasnikov.

REFERENCES

1. Diehl, M., *Phys. Rep.*, **388**, 41 (2003).
2. Ji, X., *J.Phys.*, **G24**, 1181 (1998).
3. Braun, V. e. a., *Phys. Rev. Lett.*, **89**, 172001 (2002).
4. Ioffe, B., and Smilga, A., *Nucl. Phys.*, **B232**, 109 (1984).
5. Braun, V., *Phys. Lett.*, **B509**, 43 (2001).
6. Braun, V. e. a., *Phys. Rev. Lett.*, **89**, 172001 (2002).

Generalized Parton Distributions and Deep Exclusive Reactions: Present Program at JLab

Michel Garçon

DAPNIA/SPhN, CEA-Saclay, 91191 Gif-sur-Yvette, France

Abstract. We review briefly the physical concept of generalized parton distributions and the experimental challenges associated with the corresponding measurements of deep exclusive reactions. The first data obtained at Jefferson Lab for exclusive photon (DVCS) and vector meson (DVMP) electroproduction above the resonance-excitation region are described. Two upcoming dedicated DVCS experiments are presented in some detail.

GENERALIZED PARTON DISTRIBUTIONS

The physical concepts and underlying theory of generalized parton distributions (GPD) were to be reviewed at this Workshop by A. Schäfer, who could not attend. We certainly missed his insight and rigor on the subject. In a much more heuristic way than he would have done, I recall some of the main features of GPDs.

In the Bjorken limit, exclusive processes of the type $\gamma^* p \rightarrow p\gamma, \pi, \rho, \omega \cdots$ are described in term of a handbag diagram. The corresponding amplitude factorizes [1, 2] into a hard scattering process involving one participant parton (e.g. $\gamma^* q \rightarrow \gamma q$) and a soft rearrangement in the nucleon parameterized by four GPDs, denoted by H, E ("unpolarized" or more exactly independent of the quark helicity) and $\tilde{H}, \tilde{E}$ ("polarized"). In addition to the scale variable Q^2, these are functions of three variables: x, the average longitudinal parton momentum fraction before and after the hard scattering; ξ, half the difference between these two momentum fractions, and a four-momentum transfer squared t.

An oversimplified physical image can be given in terms of ordinary quantum mechanics. Schematically, a nucleon reduced to a 3-quark configuration can be described by a wave function $\psi(x_1, \vec{k}_1, x_2, \vec{k}_2, x_3, \vec{k}_3)$ where x is again a parton longitudinal momentum fraction and $\vec{k}$ its transverse momentum. An "ordinary" parton distribution is given by the probability integral

$$q(x) \sim \int |\psi(x, \vec{k}_1, x_2, \vec{k}_2, x_3, \vec{k}_3)|^2 [dX] \quad (1)$$

CP747, *Structure of the Nucleon at Large Bjorken x,* edited by C. Bourrely, Z.-E. Meziani, and J. Soffer

with $[dX] = \delta(x+x_2+x_3-1)\delta^{(2)}(\vec{k}_1+\vec{k}_2+\vec{k}_3)dx_2dx_3d\vec{k}_1d\vec{k}_2d\vec{k}_3$. In this representation, a GPD can be thought of as an overlap integral

$$H(x,\xi,t) \sim \int \psi^*(x-\xi,\vec{k}_1+\vec{\Delta}_\perp,x_2,\vec{k}_2,x_3,\vec{k}_3)\cdot\psi(x+\xi,\vec{k}_1,x_2,\vec{k}_2,x_3,\vec{k}_3)[dX] \quad (2)$$

The off-diagonal character of the GPD appears in the change of longitudinal momentum: a GPD thus expresses the *coherence* between states of different longitudinal momenta. Kinematically, this rearrangement cannot occur without a non-zero squared momentum transfer $t = f(\xi,\vec{\Delta}_\perp)$. Through a Fourier transform with respect to $\vec{\Delta}_\perp$ at $\xi = 0$, GPDs can be related to spatial transverse distributions of partons. Thus, through the x-t correlation, the GPDs contain information about the transverse distribution of partons of given x. This opens the way for a femto-photography of the nucleon. Qualitatively, this correlation between the transverse position $\vec{b}$ and the longitudinal momentum $\vec{p}$ gives also access to information about the orbital angular momentum of the partons ($\vec{b}\wedge\vec{p}$).

This picture can be extended to include quark-antiquark configurations in the initial and/or final states.

EXPERIMENTAL CHALLENGES

At high virtuality Q^2 of the exchanged photon, the measurements of exclusive production reactions are very challenging. Precision measurements require high luminosity $\times$ acceptance since the cross sections are small. Exclusivity is necessary for a quantitative comparison with theory and for a detailed study of scaling laws. This puts high demands on resolution and/or complete coverage of reaction products. For example, in the $ep \to ep\gamma$ process, it is in principle sufficient to identify two of the three particles in the final state and measure their momentum vector (this is the case in the first two graphs of figure 1). In practice, as beam and secondary particles energies increase, it becomes increasingly difficult to separate DVCS from exclusive π^0 production when measuring only the scattered electron and proton (third graph of fig. 1). It is then necessary to detect all three particles in the final state.

DEEPLY VIRTUAL COMPTON SCATTERING

The first measurements of beam spin asymmetry in the $\vec{e}p \to ep\gamma$ exhibit an azimuthal dependence characteristic of the interference of a helicity conserving process such as DVCS with the Bethe-Heitler process [4, 5]. Moreover the magnitude of this asymmetry is agreement with models based on GPDs, including small higher-twist contributions and next-to-leading order α_S corrections. Note however that the same results may also be described within a Regge model [6].

Since then, other CLAS data taken at beam energies of 4.8 and 5.75 GeV are being analyzed as well. This is still based on the study of $ep \to epX$ configurations. At 4.8 GeV, the resolution still allows to perform a shape analysis of the missing mass spectra, similar to the procedure in Ref. [4], thus separating the γ and π^0 contributions [7]. An

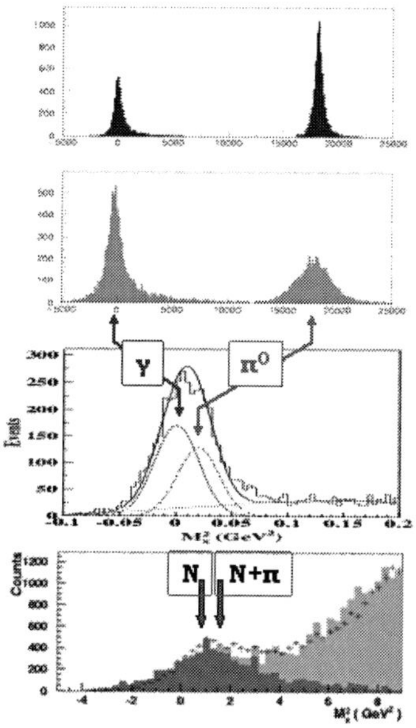

FIGURE 1. Squared missing mass for performed VCS or DVCS experiments. From top to bottom: $ep \rightarrow epX$ from MAMI (850 MeV beam energy) [3], from JLab/Hall A (4 GeV) [3], and from JLab/CLAS (4.2 GeV) [4], $ep \rightarrow e\gamma X$ from HERMES (27 GeV) [5].

exemple of such a separation is given in the third graph of fig. 1, where each contribution is independently calibrated using subsets of unambiguously identified events [4]. At 5.75 GeV, the missing mass resolution worsens and precludes such an analysis. The pion contribution is then minimized (but not yet quantified) through an appropriate choice of kinematical cuts, guided by a Monte-Carlo simulation of both processes [8]. In both cases, earlier results are confirmed, and the higher statistics allow for a finer binning, in Q^2, x_B and/or t. The measured beam spin asymmetry, at fixed x_B and t, is consistent with a $1/Q$ behaviour, as expected from the leading-twist (handbag) contribution to DVCS.

Target spin asymmetries, using a polarized NH_3 target, were investigated as well [9]. Here again the contribution of the $ep \rightarrow ep\pi^0$ process remains to be quantified.

Dedicated experiments now aim at detecting the three particles in the final state, in order to ensure a complete exclusivity in the measurements. See figure 2.

In CLAS [10], the photons emitted in a forward cone (3-14°) will be detected in a new inner calorimeter, consisting of 424 small lead-tungstate crystals read by avalanche photodiodes. In addition, a specifically designed two-coil solenoid will focus the background low energy electrons along the beam axis. A 100-crystal prototype was tested with success in December 2003. The whole experiment is to be mounted in February

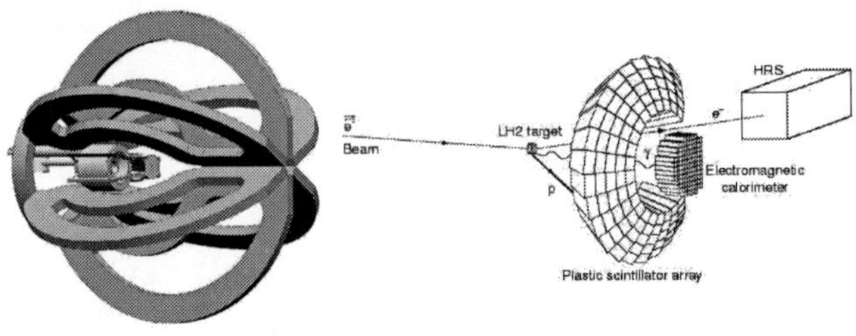

FIGURE 2. Shematic views of dedicated CLAS (left) and Hall A (right) DVCS experiments. The standard CLAS detection is not shown; the new solenoid and calorimeter are positioned at the center of the CLAS toroidal magnet.

2005.

In Hall A [11], a lead-fluoride calorimeter is mounted in direct view of the target, along the direction of the virtual photon, for each of three spectrometer settings for the scattered electron. The recoil protons are detected in a concentric annular plastic scintillator array. The experiment is currently running at the designed luminosity of 10^{37} cm^{-2}s^{-1}. A follow-up experiment will study DVCS on the neutron [12]. The quasi-free $e(n) \to en\gamma$ reaction will be measured using a liquid deuterium target and complementing the proton array with thin scintillator for proton/neutron discrimination. The DVCS beam spin asymmetry on the neutron has been shown to be more sensitive to the quark total angular momentum in the nucleon.

Double DVCS ($ep \to epe^+e^-$ proceeding through $\gamma^* q \to \gamma^* q$) has the potential to access GPDs at a tunable kinematical point $x = f(\xi, q'^2)$, where q'^2 is the virtuality of the outgoing photon (or equivalently the invariant mass of the corresponding lepton pair) [17]. Positrons were cleanly selected in CLAS and $ep \to epe^+X$ configurations analyzed [18]. Although $ep \to epe^+e^-$ events could be clearly identified, the analysis faces a theoretical and experimental challenge: there are two electrons in the final state. Antisymmetrisation was not taken into account in the calculations. Experimentally, the kinematical variables depend on the identification of an electron either as scattered or coming from the lepton pair. One of the two possible choices result in the observation of clear peaks associated with very low Q^2 vector meson production. Our preliminary conclusion is then that the observed $ep \to epe^+e^-$ events cannot be interpreted in the framework of DDVCS and GPDs.

DEEPLY VIRTUAL MESON PRODUCTION

The hard electroproduction of mesons, or DVMP, should also proceed through the same handbag diagram. In this case, factorization has been demonstrated for longitudinal virtual photons [2]. According to the nature of the emitted meson, one probes different GPDs (H and E for vector mesons, $\tilde{H}$ and $\tilde{E}$ for pseudoscalar mesons) and different flavor combinations (e.g. ρ vs ω, or π^0 vs η). The difficulty here is in isolating the contribution of longitudinal photons and reaching high enough Q^2, since the additional gluon exchange necessary to produce the meson moves the asymptotic regime to higher values of Q^2.

Longitudinal cross sections have been extracted for ρ electroproduction at 4.2 GeV [13]. A similar analysis for higher energy (5.75 GeV), with higher statistics data, is underway. In this channel, both the CLAS and HERMES data are in semi-quantitative agreement with a model based on GPDs, and including an effective description of higher-twist effects.

The ϕ [14] and ω [15] channels have been analyzed as well. In the latter case, a determination of the emitted ω spin density parameters led to the conclusion that helicity is not conserved between the virtual photon and the emitted meson. Thus the handbag diagram does not dominate the reaction $ep \to ep\omega$ and the longitudinal cross sections could not be extracted. The ω channel thus appears the most challenging one to access the GPDs. Within a Regge approach, it was shown that this behaviour is due to the dominantly transverse t-channel π^0 exchange, or rather the exchange of the corresponding saturating Regge trajectory [16]. Only for the ω channel is this specific exchange contribution dominant. This calculation gives a new insight in the high-Q^2 and high-t behaviour of such exclusive reactions [16].

CONCLUSIONS

Generalized Parton Distributions have merged as a powerful, attractive and unifying concept for the nucleon structure. The relation between GPDs and deep exclusive reactions still needs experimental validation. There are experimental indications that the handbag diagram is at work in DVCS and this is being studied with increased statistics data sets at CLAS. In DVMP, extensive data sets on vector meson production, also from CLAS, will be coming shortly. Dedicated DVCS experiments, in Hall A (2004) and CLAS (2005), will measure beam spin asymmetries, detecting all three particles in the final state to ensure full exclusivity. They should establish on firm grounds the validity of the approach. Detailed tests of scaling will be performed. If scaling is indeed observed, or deviations thereof understood, these experiments, in conjunction with theoretical input, will provide the first significant measurements of GPDs.

REFERENCES

1. X. Ji, Phys. Rev. D **55** (1997) 7114; A.V. Radyushkin, Phys. Rev. D **56** (1997) 5524.
2. J.C. Collins, L. Frankfurt, and M. Strikman, Phys. Rev. D **56** (1997) 2982-3006.

3. H. Fonvieille, private communication.
4. S. Stepanyan *et al.* (CLAS collaboration), Phys. Rev. Lett. **87** (2001) 182002.
5. A. Airapetian *et al.* (HERMES collaboration), Phys. Rev. Lett. **87** (2001) 182001.
6. F. Cano and J.-M. Laget, Phys. Lett. **B**551 (2003) 317.
7. G. Gavalian *et al.* (CLAS collaboration), preliminary results.
8. H. Avakian *et al.* (CLAS collaboration), preliminary results.
9. S. Cheng *et al.* (CLAS collaboration), preliminary results.
10. V. Burkert, L. Elouadrhiri, M. Garçon, S. Stepanyan *et al.*, CEBAF experiment 01-113.
11. P. Bertin, C. Hyde-Wright, R. Ransome, F. Sabatié *et al.*, CEBAF experiment 00-110.
12. P. Bertin, C. Hyde-Wright, F. Sabatié, E. Voutier *et al.*, CEBAF experiment 03-106.
13. C. Hadjidakis *et al.* (CLAS collaboration), hep-ex/0408005.
14. J. Santoro *et al.* (CLAS collaboration), preliminary results.
15. L. Morand *et al.* (CLAS collaboration), in preparation. M. Garçon, contr. to the *10th Int. Symp. on Meson-Nucleon Physics and the Nucleon Structure*, Beijing, Aug. 2004.
16. J.-M. Laget, Phys. Rev. D **70** (2004) 054023.
17. M. Guidal and M. Vanderhaeghen, Phys. Rev. Lett. **90** (2003) 012001; A.V. Belitsky and D. Müller, Phys. Rev. Lett. **90** (2003) 022001.
18. S. Morrow, M. Garçon *et al.* (CLAS collaboration), preliminary results.

Pion photoproduction on the nucleon

Haiyan Gao[1], Lingyan Zhu[2]

[1] *Triangle Universities Nuclear Laboratory and the Department of Physics, Duke University, Durham, North Carolina, U.S.A.*
[2] *University of Illinois, Urbana-Champaign, Urbana, IL, U.S.A.*
for the Jefferson Lab E94-104 and E02-010 Collaboration

Abstract. The $\gamma n \to \pi^- p$ and $\gamma p \to \pi^+ n$ reactions are essential probes of the transition from meson-nucleon degrees of freedom to quark-gluon degrees of freedom in exclusive processes. The cross sections of these processes are also advantageous, for the investigation of oscillatory behavior around the quark counting prediction, since they decrease relatively slower with energy compared with other photon-induced processes. In this talk, we discuss recent results on the $\gamma p \to \pi^+ n$ and $\gamma n \to \pi^- p$ processes from Jefferson Lab experiment E94-104. We also discuss a new experiment in which singles $\gamma p \to \pi^+ n$ measurement from hydrogen, and coincidence $\gamma n \to \pi^- p$ measurements at the quasifree kinematics from deuterium for center-of-mass energies between 2.3 GeV to 3.4 GeV in fine steps at a center-of-mass angle of 90° are planned. The proposed measurement will allow a detailed investigation of the oscillatory scaling behavior in photopion production processes.

INTRODUCTION

Exclusive processes are essential to studies of transitions from the non-perturbative to the perturbative regime of Quantum Chromodynamics (QCD). The differential cross-section for many exclusive reactions [1] at high energy and large momentum transfer appear to obey the quark counting rule [2]. The constituent counting rule predicts the energy dependence of the differential cross section at fixed center-of-mass angle for an exclusive two-body reaction at high energy and large momentum transfer as follows:

$$d\sigma/dt = h(\theta_{cm})/s^{n-2}, \tag{1}$$

where s and t are the Mandelstam variables, s is the square of the total energy in the center-of-mass frame and t is the momentum transfer squared in the s channel. The quantity n is the total number of elementary fields in the initial and final states, while $h(\theta_{cm})$ depends on details of the dynamics of the process. The quark counting rule was originally obtained based on dimensional analysis of typical renormalizable theories. The same rule was later obtained in a short-distance perturbative QCD approach by Brodsky and Lepage[3]. Despite many successes, a model-independent test of the approach, called the hadron helicity conservation rule, tends not to agree with data in the similar energy and momentum region. The presence of helicity-violating amplitudes indicates that the short-distance expansion cannot be the whole story. In addition some of the cross-section data can also be explained in terms of non-perturbative calculations [4].

In recent years, a renewed trend has been observed in deuteron photo-disintegration experiments at SLAC and JLab [5, 6, 7]. Onset of the scaling behavior has been observed

CP747, *Structure of the Nucleon at Large Bjorken x*, edited by C. Bourrely, Z.-E. Meziani, and J. Soffer

in deuteron photo-disintegration [5, 6, 7] at a surprisingly low momentum transfer of 1.0 (GeV/c)^2 to the nucleon involved. However, a recent polarization measurement on deuteron photo-disintegration [8] shows disagreement with hadron helicity conservation in the same kinematic region where the quark counting behavior is apparently observed. These paradoxes make it essential to understand the exact mechanism governing the early onset of scaling behavior.

The elastic proton-proton (pp) scattering data at high energy and large momentum transfer have shown very interesting characteristics: the scaled 90° center-of-mass angle data, $s^{10}\frac{d\sigma}{dt}$ show substantial oscillations about the power law behavior. One theoretical interpretation for such oscillatory behavior and the striking spin-correlation [9] in pp scattering involves interference between hard pQCD short-distance and long-distance (Landshoff) [10] amplitudes [11] along with an energy dependent phase arising from gluonic radiative corrections. This effect is believed to be analogous to the Coulomb-nuclear interference that is observed in low-energy charged-particle scattering. Brodsky and de Teramond [12] suggest that the structure seen in $s^{10}\frac{d\sigma}{dt}(pp \rightarrow pp)$ and the large spin correlations can be attributed to $c\bar{c}uuduud$ resonant states. The opening of this channel gives rise to an amplitude with a phase shift similar to that predicted for gluonic radiative corrections. Deviations (oscillations) from the pQCD counting rule above the resonance region has recently been shown in a model of a composite system with two spinless charged constituents [13], employing the so-called concept of "restricted locality" of quark-hadron duality [14].

Ji, Ma and Yuan [15] derived a generalized counting rule for exclusive processes at fixed angles involving parton orbital angular momentum and hadron helicity flip. This generalized counting rule opens a new window for probing the quark orbital angular momentum inside the nucleon. Recently, a non-perturbative derivation of generalized counting rules including the orbital angular momentum was obtained by Brodsky and de Teramond [16]. Therefore, it is very important to investigate the scaling behavior in details and to search for QCD oscillations. The photo-pion production reactions are well suited for these purposes because the cross section of these processes decrease relatively slower with energy ($\frac{ds}{dt} \propto s^{-7}$) compared with other photon-induced processes. Rough power-law dependence of meson photoproduction seems to agree with the constituent quark counting rule prediction [17] within experimental uncertainties, for example in the case of the $\gamma p \rightarrow \pi^+ n$ process at a center-of-mass angle of 90°. Yet it is not clear whether the counting rule scaling behavior has been observed in the $\gamma\, p \rightarrow \pi^0\, p$ process because discrepancies exist between different measurements. For the $\gamma\, n \rightarrow \pi^-\, p$ process, no cross section data exist above a photon energy of 2.0 GeV prior to the recent Jefferson Lab experiment E94-104 [18].

JEFFERSON LAB EXPERIMENT E94-104

Experiment E94-104 was carried out in Hall A [19] at the Thomas Jefferson National Accelerator Facility (JLab). The continuous electron beam, at a current around 30 μA and energies from 1.1 to 5.5 GeV, impinged on a 6% copper radiator and generated an untagged bremsstrahlung photon beam. The production data were taken with the 15 cm cryogenic liquid hydrogen (LH2) target for singles $p(\gamma,\pi^+)n$ measurement, or with the liquid deuterium (LD2) target for coincidence $d(\gamma,\pi^- p)p$ measurement. The two High Resolution Spectrometers (HRS) in Hall A were used to detect the outgoing pions and recoil protons. Two new aerogel Čerenkov detectors in the left spectrometer were constructed for this experiment to provide particle identification for positive particles, mainly pions and protons, since the time-of-flight technique fails at high momentum. Details of the Hall A spectrometers can be found [19].

Based on two-body kinematics, the incident photon energy was reconstructed from final states, i.e. the momentum and angle of the π^+ in the singles measurement, momenta and angles of the π^- and p in the coincidence measurement. A 100 MeV bin with the center of the bin 75 MeV from the beam energy, was chosen for the data analysis, where the multi-pion contribution was negligible. The data after background subtraction, with cuts on trigger type, coincidence timing, PID (particle identification), acceptance and photon energy, were compared to a modified Monte Carlo simulation code for this experiment based on MCEEP [20] with the same cuts on acceptance and photon energy. The raw cross section was extracted by comparing data and simulation. The distributions of acceptance, reconstructed momentum and photon energy were in good agreement with results obtained from simulations. Details on the simulation and the bremsstrahlung photon flux calculation can be found [21].

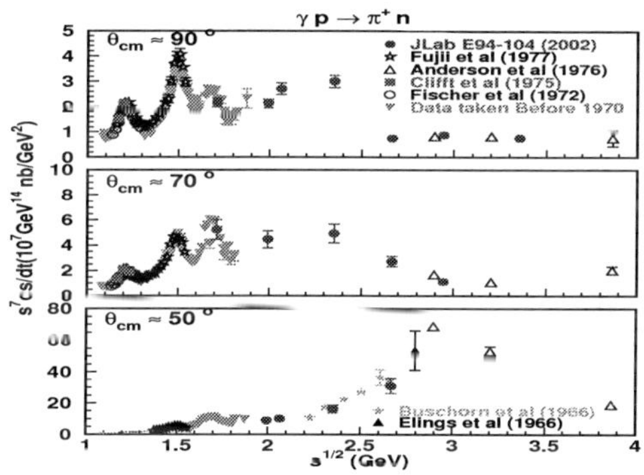

FIGURE 1. *The scaled differential cross section $s^7 \frac{d\sigma}{dt}$ versus center-of-mass energy for the $\gamma p \rightarrow \pi^+ n$ at $\theta_{cm} = 90°, 70°, 50°$.*

Important correction factors were applied to deduce the final cross section such as the nuclear transparency factor in the deuteron due to the final state interaction, material absorption of pions and protons, pion decay loss, detection efficiency, etc.. Analysis details can be found in [21, 22]. The total errors were dominated by systematic errors and were estimated to be 10% in cross section, while point-to-point uncertainties for the three kinematics at 3.3, 4.2 and 5.5 GeV to be 5%. The statistical errors were approximately 2%. The results [23], as shown in Figure 1 and Figure 2, agree with the world data within uncertainties in the overlapping region. The data at $\theta_{cm} = 70°, 90°$ exhibit a global scaling behavior predicted by the constituent counting rule in π^- channel, similar to what was observed in the π^+ channel at similar center-of-mass angles. The data at $\theta_{cm} = 50°$ do not display scaling behavior and may require higher photon energies for the observation of the onset of the scaling behavior. The data suggest that a transverse

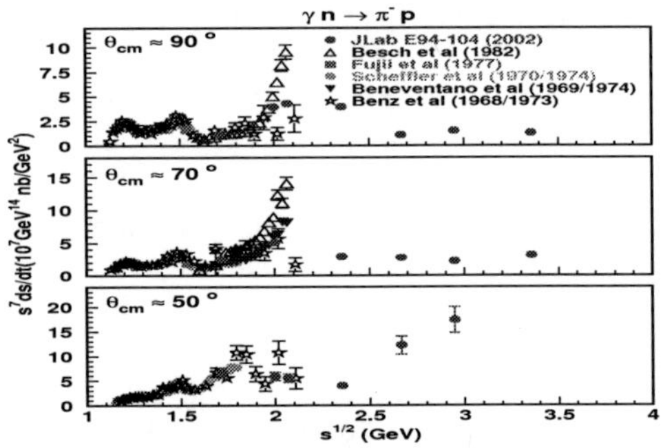

FIGURE 2. *The scaled differential cross section $s^7\frac{d\sigma}{dt}$ versus center-of-mass energy for the $\gamma n \rightarrow \pi^- p$ at $\theta_{cm} = 90°, 70°, 50°$.*

momentum of around 1.2 GeV/c might be the scale governing the onset of scaling for the photo-pion production, which is consistent with what has been observed in deuteron photodisintegration [6]. Data in these two channels at 90° show possible oscillations around the scaling behavior in similar ways as suggested by the insets in Fig. 3 and Fig. 4. Note that this possible oscillatory behavior occurs above the known baryon resonance region. Unfortunately, the coarse and few photon energy settings of this experiment do not allow us to claim the observation of oscillations. Measurements with much finer binning, planned at JLab [29], are essential for the confirmation of such oscillatory scaling behavior.

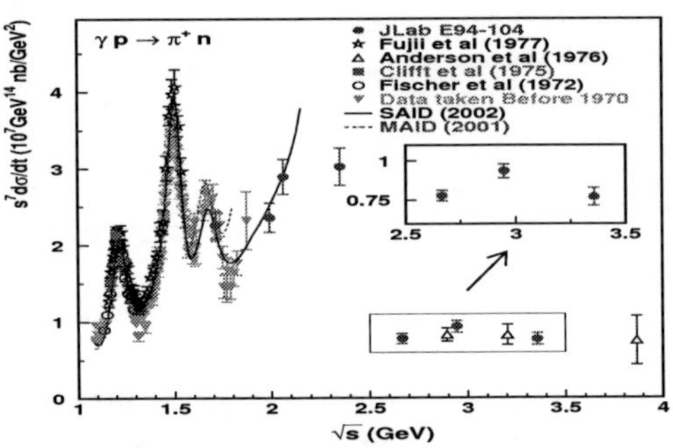

FIGURE 3. *The scaled differential cross section $s^7\frac{d\sigma}{dt}$ versus center-of-mass energy for the $\gamma p \rightarrow \pi^+ n$ at $\theta_{cm} = 90°$. The data from JLab E94-104 are shown as solid circles. The error bars for the new data include statistical and systematic uncertainties, except those in the inset in which only point-to-point systematic uncertainties are included to highlight the possible oscillatory scaling behavior.*

JEFFERSON LAB EXPERIMENT E02-010

Recently a new experiment [29] was approved to carry out a measurement of the photo-pion production cross-section for the fundamental $\gamma n \rightarrow \pi^- p$ process from a ^2H target

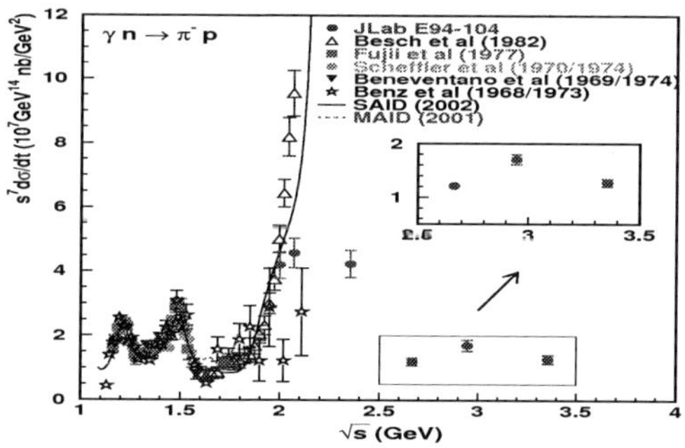

FIGURE 4. *The scaled differential cross section $s^7 \frac{d\sigma}{dt}$ versus center-of-mass energy for the $\gamma n \rightarrow \pi^- p$ at $\theta_{cm} = 90°$. Only point-to-point systematic uncertainties are shown in the inset as in Fig.3.*

and for the $\gamma p \rightarrow \pi^+ n$ process from a hydrogen target at a center-of-mass angle of 90°, at $\sqrt{s} \sim 2.25$ GeV to 3.41 GeV in steps of approximately 0.07 GeV. The new experiment will make individual cross-section measurement with a 2% statistical uncertainty and a point-to-point systematic uncertainty of < 3%. Such precision will allow the test of the oscillatory behavior in the scaled fundamental cross section measurement. The proposed experiment will only be possible with the unique JLab capability of high luminosity and such an experiment will be carried out in Hall A at JLab. The projected measurements of E02-010 on the $\gamma n \rightarrow \pi^- p$ process at $\theta_{cm} = 90°$ together with future energy upgraded 12 GeV CEBAF are shown in Fig. 5. Also shown is a fit of the E94-104 data based on a two-component model of Jain and Ralston [30].

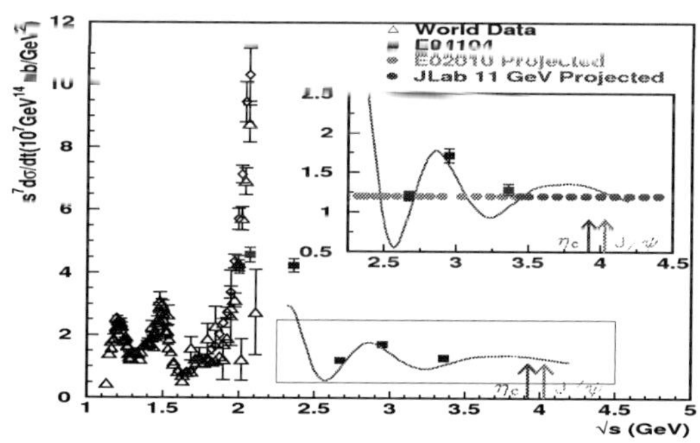

FIGURE 5. *The projected measurement from E02-010 and future 12 GeV energy upgrade at JLab on the scaled differential cross section $s^7 \frac{d\sigma}{dt}$ versus center-of-mass energy for the $\gamma n \rightarrow \pi^- p$ at $\theta_{cm} = 90°$.*

ACKNOWLEDGMENTS

We thank John Arrington and Roy Holt for careful reading of this manuscript, and Dipangkar Dutta in preparing Figure 5. This work is supported by the U.S. Department of Energy under contract number DE-FC02-94ER40818 and DE-FG02-03ER41231.

REFERENCES

1. G. White *et al.*, Phys. Rev. **D49**, 58 (1994).
2. S.J. Brodsky and G.R. Farrar, Phys. Rev. Lett.**31**, 1153 (1973); Phys. Rev. D **11**, 1309 (1975); V. Matveev *et al.*, Nuovo Cimento Lett. **7**, 719 (1973);
3. G.P. Lepage, and S.J. Brodsky, Phys. Rev. D **22**, 2157 (1980).
4. N. Isgur and C. Llewelyn-Smith, Phys. Rev. Lett. **52**, 1080 (1984).
5. J. Napolitano *et al.*, Phys. Rev. Lett. **61**, 2530 (1988); S.J. Freedman *et al.*, Phys. Rev. C **48**, 1864 (1993); J.E. Belz *et al.*, Phys. Rev. Lett. **74**, 646 (1995); C. Bochna *et al.*, Phys. Rev. Lett. **81**, 4576 (1998).
6. E.C. Schulte, *et al.*, Phys. Rev. Lett. **87**, 102302 (2001);
7. P. Rossi *et al.*, hep-ph/0405207, submitted to Phys. Rev. Letts.; M. Mirazita *et al.*, Phys. Rev. C **70**, 014005 (2004).
8. K. Wijesooriya, *et al.*, Phys. Rev. Lett. **86**, , (2)975 (2001).
9. D. G. Crabb *et al.*, Phys. Rev. Lett. **41**, 1257 (1978).
10. P. V. Landshoff, Phys. Rev. D **10**, 1024 (1974).
11. S.J. Brodsky, C.E. Carlson, and H. Lipkin, Phys. Rev. D **20**, 2278 (1979); J.P. Ralston and B. Pire, Phys. Rev. Lett. **61**, 1823 (1988); C.E. Carlson, M. Chachkhunashvili, and F. Myhrer, Phys. Rev. D **46**, 2891 (1992).
12. S. J. Brodsky, and G. F. de Teramond, Phys. Rev. Lett. **60**, 1924 (1988).
13. N. Isgur, S. Jeschonnek, W. Melnitchouk, and J.W. Van Orden, Phys. Rev. D **64**, 054005 (2001); F.E. Close and N. Isgur, Phys. Lett. **B 509**, 81 (2001); S. Jeschonnek and J.W. Van Orden, Phys. Rev. D **65**, 094038 (2002); F.E. Close and Q. Zhao, Phys. Rev. D **66**, 054001 (2002).
14. Q. Zhao and F. Close, Phys. Rev. Lett. **91**, 022004 (2003).
15. X. Ji, J.-P. Ma, F. Yuan, Phys. Rev. Letts. **90**, 241601 (2003).
16. S.J. Brodsky, G.F. de Teramond, Phys.Lett.B **582**, 211 (2004), also hep-th/0310227.
17. R.L. Anderson *et al.*, Phys. Rev. **D14**, 679 (1976).
18. Jefferson Lab Experiment E94-104, Spokespersons: H. Gao, R.J. Holt.
19. J. Alcorn *et al.*, Nucl. Instrum. Methods A522, 294 (2004).
20. P. Ulmer *et al.*, http://www.physics.odu.edu/~ulmer/mceep/mceep.html.
21. L.Y. Zhu, Ph.D. thesis, February, 2004, Massachusetts Institute of Technology (unpublished).
22. L.Y. Zhu *et al.*, Phys. Rev. Lett. **91**, 022003 (2003).
23. L.Y. Zhu *et al.*, nucl-ex/0409018, submitted to Phys. Rev. C.
24. K.-H. Hellwege, Numerical Data and Functional Relationships in Science and Technology, Group I, Volume 8: H. Genzel, P. Joos and W. Pfeil, Photoproduction of Elementary Particles, (1973).
25. HEPDATA: REACTION DATA Database, http://durpdg.dur.ac.uk/hepdata/reac.html.
26. H.-J. Besch *et al.*, Z. Phys. C **16**, 1 (1982).
27. I. I. Strakovsky (private communication); R. A. Arndt *et al.*, nucl-th/0205067, (2002), accepted by Phys. Rev. C; http://gwdac.phys.gwu.edu/.
28. I. I. Strakovsky (private communication); S. S. Kamalov, *et al.*, Phys. Rev. C **64**, 032201 (2001)(A dynamical model); D. Drechsel, *et al.*, Nucl. Phys. A **645**, 145 (1999)(a unitary isobar model); http://www.kph.uni-mainz.de/MAID/. MAID2001 refers to the Nov. 2001 version of the MAID solution from S. S. Kamalov.
29. Jefferson Lab Proposal PR02-010, Spokespersons: D. Dutta, H. Gao and R.J. Holt.
30. J.P. Ralston and B. Pire, Phys. Rev. Lett. **61**, 1823 (1988); J.P. Ralston and B. Pire, Phys. Rev. Lett. **65**, 2343 (1990); P. Jain, private communications.

Generalized Parton Distributions and Constituent Quark Models

M. Traini*, B. Pasquini† and S. Boffi†

Dipartimento di Fisica, Università degli Studi di Trento, Povo (Trento), and INFN, Gruppo Collegato di Trento, Trento, and ECT, Villazzano, Trento, Italy
†Dipartimento di Fisica Nucleare e Teorica, Università degli Studi di Pavia and INFN, Sezione di Pavia, Pavia, Italy

Abstract. Generalized parton distributions (GPDs) are studied at the hadronic (nonperturbative) scale within different assumptions based on a relativistic constituent quark model. In particular, by means of a meson-cloud model we investigate the role of nonperturbative antiquark degrees of freedom and the valence quark contribution. A QCD evolution of the obtained GPDs is used to add perturbative effects and to investigate the GPDs' sensitivity to the nonperturbative ingredients of the calculation at larger (experimental) scale.

INTRODUCTION

We study the nucleon's GPDs within specific hadron models and address the problem of evolving the input distributions to the experimental scale investigating the effects of different dynamical assumptions. In particular, we want to investigate both the quark core structure of the nucleon and its chiral properties [1]. In fact, the new aspects of the GPDs with respect to the better known parton distributions are related to the socalled ERBL region where the presence of dynamical $q\bar{q}$ pairs, both in the nonperturbative and perturbative regimes, plays a crucial role [2].

MODELING GPDS WITH DOUBLE DISTRIBUTIONS

We will assume, following ref. [3], that GPDs can be written in terms of double distributions involving a given profile function and the forward parton distribution derived in some model. We shall concentrate our attention on the chiral even (helicity conserving) distribution $H^q(\bar{x},\xi,Q^2,t)$ for partons of flavor q at the hadronic scale where the models we are going to discuss are assumed to be valid to evaluate the twist-two amplitude.

For sake of simplicity we follow the common notation which explicitly indicates three variables only $(\bar{x},\xi,t)$ instead of four $(\bar{x},\xi,Q^2,t)$. We shall come back to the definition of Q^2 when we discuss the values of the hadronic scale Q_0^2 and the QCD evolution of H^q in Q^2 (in a symmetric frame of reference where q^μ and the average nucleon momentum $\overline{P}^\mu = (P^\mu + P'^\mu)/2$ are collinear along the z axis and in opposite directions, the quark light-cone momentum is $\bar{k}^+ = \bar{x}\overline{P}^+$, the invariant momentum square is $t = \Delta^2 = (P'^\mu - P^\mu)^2$, and the skewedness ξ describes the longitudinal change of

CP747, *Structure of the Nucleon at Large Bjorken x*, edited by C. Bourrely, Z.-E. Meziani, and J. Soffer

the nucleon momentum, $2\xi = -\Delta^+/\overline{P}^+$). We also introduce non-singlet (valence) and singlet quark distributions,

$$H^{NS}(\bar{x},\xi,t) \equiv \sum_q [H^q(\bar{x},\xi,t) + H^q(-\bar{x},\xi,t)] = H^{NS}(-\bar{x},\xi,t), \quad (1)$$

$$H^S(\bar{x},\xi,t) \equiv \sum_q [H^q(\bar{x},\xi,t) - H^q(-\bar{x},\xi,t)] = -H^S(-\bar{x},\xi,t), \quad (2)$$

respectively. Besides being symmetric or antisymmetric in $\bar{x}$, they are also symmetric under $\xi \to -\xi$ due to the polynomiality property [2].

The analogous GPD for gluons is symmetric in $\bar{x}$, i.e. $H^g(\bar{x},\xi,t) = H^g(-\bar{x},\xi,t)$, and reduces to the gluon density $g(x)$ in the forward limit ($\bar{x} \to x$) $H^g(x,0,0) = xg(x), \quad x > 0$.

There are two distinct regions: the Dokshitzer-Gribov-Lipatov-Altarelli-Parisi (DGLAP) region, $|\bar{x}| > \xi$, and the Efremov-Radyushkin-Brodsky-Lepage (ERBL) region, $|\bar{x}| < \xi$. The naming derives from the fact that the GPD evolution equations in the region $|\bar{x}| > \xi$ ($|\bar{x}| < \xi$) reduce to the familiar DGLAP (ERBL) equations in the limit $\xi = 0$ ($\xi = 1$).

The singlet and gluon distributions mix under evolution, while the non-singlet distribution does not.

In the two following subsections we recall the derivation of $q(x)$.

Parton distributions at low resolution scale

In two recent papers [4] we discussed a method to evaluate GPDs within light-front constituent quark models (CQMs) at the scale dominated by valence (constituent) degrees of freedom. Therefore, the calculation is restricted to the DGLAP region $|\bar{x}| \geq \xi$ and the generation of $q\bar{q}$ contributions can have a perturbative origin only. In contrast, within the DD-based model both ERBL and DGLAP regions of quark GPDS are populated with any input quark distribution $q(x)$ (with or without sea contribution). In this subsection we only consider valence quarks leaving for the next subsection the case including the sea.

The quark distribution $q(x)$ is obtained as the forward limit of GPDs where $\Delta^\mu \to 0$ and $\bar{x}$ goes over to the parton momentum fraction x. Consequently, $H^q(\bar{x},\xi,Q^2,t)$ reduces to the ordinary DIS parton distribution, i.e. $H^q(x,0,\mu_0^2,0) = q(x,\mu_0^2)$, where $q(x,\mu_0^2)$ is the (unpolarized) quark distribution of flavor q at the hadronic scale $Q^2 = \mu_0^2$ consistent with the assumed CQM.

The distribution automatically fulfills the support condition and satisfies the (particle) baryon number and momentum sum rules at the hadronic scale μ_0^2 where the valence contribution dominates the twist-two response.

The meson cloud model introduces the sea by incorporating $q\bar{q}$ contributions into the valence-quark model of the parton distribution discussed in the previous section.

The basic hypothesis of the meson-cloud model is that the physical nucleon state can be expanded (in the infinite momentum frame (IMF) and in the one-meson approximation) in a series involving bare nucleons and two-particle, meson-baryon states [5]. Its wave function is written as the sum of meson-baryon Fock states: $|p\rangle = \sqrt{Z}|p\rangle_{bare} + \Sigma_{BM} \int dy\, d^2\vec{k}_\perp\, \phi_{BM}(y, k_\perp^2)\, |B(y, \vec{k}_\perp); M(1-y, -\vec{k}_\perp)\rangle$. Here $\phi_{BM}(y, k_\perp^2)$ is the probability amplitude for the proton to fluctuate into a virtual baryon-meson BM system with the baryon and meson having longitudinal momentum fractions y and $1-y$ and transverse momenta $\vec{k}_\perp$ and $-\vec{k}_\perp$, respectively. Z is the wave function renormalization constant and is equal to the probability to find the bare proton in the physical proton. The lowest lying fluctuations for the proton which we include in our calculation involve p, n, Δ^{++}, Δ^+, Δ^0, π^+, π^-, π^0. More precisely:

The nucleon. forthe partonic content at the scale μ_0^2 for the nucleon, the Δ and the pion, we make use of the light-front approach discussed in sect. 2.1 and calculate the diagonal component of the generalized parton distributions, i.e. the inclusive parton distributions by means of $q_p^{bare}(x) \equiv q(x, \mu_0^2)$, where $q(x, \mu_0^2)$ is given by the relativistic quark model;

The Δ. we use the results of the relativistic model for the nucleon and the isospin symmetries;

The pion. the wave function of the pion is taken from ref. [6]. The choice of the model is consistent with the hypercentral CQM we adopt for the nucleon.

In order to extract the parton distributions of the nucleon including the sea (cloud) contributions we need to match sea, valence and gluons within the radiative parton model and to identify the matching scale Q_0^2 consistent with QCD evolution equation [7, 8].

We assume that the gluon distribution has a valence-like form and reads $G(x, Q_0^2) = \frac{N_g}{3}\left[u_V(x, Q_0^2) + d_V(x, Q_0^2)\right]$, where N_g represents the number of gluons. Since $\int dx \left[u_V(x, Q_0^2) + d_V(x, Q_0^2)\right] = 2 + 1$ because of baryon number conservation, we have $\int dx\, G(x, Q_0^2) = \frac{N_g}{3} \int dx \left[u_V(x, Q_0^2) + d_V(x, Q_0^2)\right] = N_g$. Following Glück *et al.* [7], we fix $N_g = 2$, the minimum number of gluons one needs to build a color singlet state. The total momentum carried at the scale Q_0^2 (the scale of the *physical* proton, which will be larger than the scale of the *bare* proton μ_0^2 made up of three valence only) must fulfill the requirement $\int dx\, x \left[G(x, Q_0^2) + u_V(x, Q_0^2) + d_V(x, Q_0^2) + 2\left[\bar{u}(x, Q_0^2) + \bar{d}(x, Q_0^2)\right]\right] = 1$, and consequently $\int dx\, x \left[u_V(x, Q_0^2) + d_V(x, Q_0^2)\right] = \left[1 - \int dx\, 2x \left[\bar{u}(x, Q_0^2) + \bar{d}(x, Q_0^2)\right]\right] / (1 + N_g/3) = 0.52$. In conclusion, by using the notation $\langle f \rangle_n = \int dx\, x^{n-1} f(x)$ for the moments, we have $\langle u_V + d_V \rangle_2 = 0.52$, $\langle G \rangle_2 = 0.35$, $\langle 2\left(\bar{u} + \bar{d}\right) \rangle_2 = 0.13$, consistent with a next-to-leading order (NLO) evolution of the moments (in the DIS renormalization scheme) with $Q_0^2 = 0.27$ GeV^2 and $\Lambda_{\rm NLO} = 0.248$ GeV [7, 8].

The actual values of Q_0^2 are fixed [8] evolving back unpolarized data, until the valence distribution matches the required momentum $\langle u_V + d_V \rangle_2 = 0.52$.

The value of $\Lambda_{\rm NLO}$ is suggested by the analysis of Glück *et al.* [7], the coupling $\alpha_s(Q_0^2)|_{\rm NLO}$ is obtained evolving back the valence distribution as previously discussed, and Q_0^2 is found by solving numerically the NLO transcendental equation $\ln \frac{Q_0^2}{\Lambda_{\rm NLO}^2} - \frac{4\pi}{\beta_0 \alpha_s} + \frac{\beta_1}{\beta_0^2} \ln \left[\frac{4\pi}{\beta_0 \alpha_s} + \frac{\beta_1}{\beta_0^2}\right] = 0$,

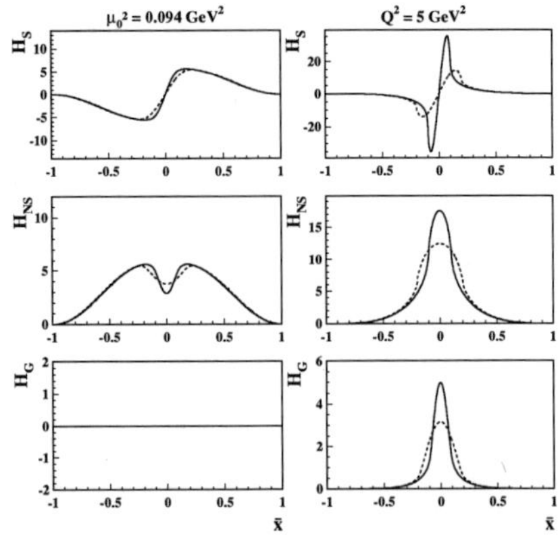

FIGURE 1. Singlet quark (upper panels), non-singlet quark (middle panels), and gluon (lower panels) GPDs for the DD-based model using input parton distributions of the hypercentral CQM. Full lines for $\xi = 0.1$, dashed lines for $\xi = 0.2$. Left panels refer to the input hadronic scale μ_0^2, right panels to the corresponding results obtained from NLO evolution up to $Q^2 = 5$ GeV^2.

The hadronic scale, μ_0^2, consistent with the presence of valence degrees of freedom only, is much lower and consistent with the constituent quark mass, its actual value being $\mu_0^2 = 0.094$ GeV^2. Since, in our case, the starting points for the evolution (μ_0^2, Q_0^2) are rather low, the form of the equations must guarantee complete symmetry for the evolution from μ_0^2 to $Q^2 \gg \mu_0^2$ and *back* avoiding additional approximations associated with Taylor expansions and not with the genuine perturbative QCD expansion [8, 9].

The QCD evolution was numerically performed to NLO accuracy according to a modified version of the code of ref. [10] (see also [11]).

RESULTS AND DISCUSSION

In fig. 1 results for the singlet quark, non-singlet quark and gluon GPDs obtained with no initial gluons and an input $q(\beta)$ simply given by the bare (valence) parton distribution $q_p^{bare}(x)$, derived from an hypercentral CQM. The model already gives a nonvanishing contribution to quark GPDs in the ERBL region at the hadronic scale μ_0^2 without introducing discontinuities at $\bar{x} = \xi$ and with a weak ξ dependence. In particular, the absence of the sea contribution gives $H^S = H^{NS}$ at $\bar{x} > \xi$. After evolution up to $Q^2 = 5$ GeV^2 GPDs are almost confined into the ERBL region with a significant ξ dependence.

The same GPDs obtained with the model based on DDs when the input is implemented with the sea according to the meson-cloud model and results are shown in fig. 2. Now, the gluon GPD does no more vanish at the initial hadronic scale, that has to be redefined at $Q_0^2 = 0.27$ GeV^2 according to the previous conclusions. After evolution the qualitative result is similar to the case without the sea in spite of a more evident ξ dependence of the singlet quark GPD at the input hadronic scale. However, the shape of

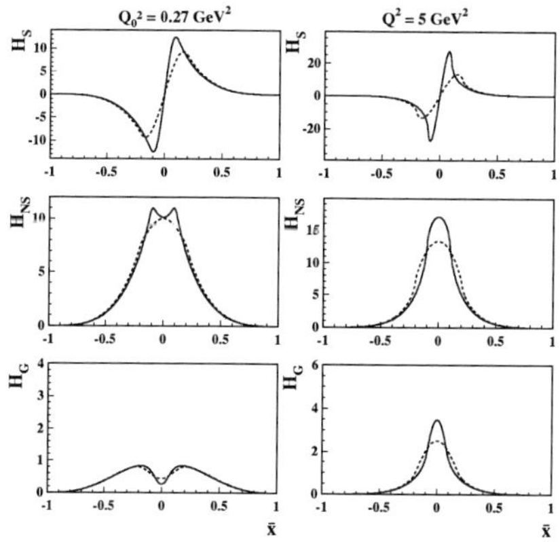

FIGURE 2. Same as in fig. 1, but with input parton distributions including the sea contribution at the initial hadronic scale Q_0^2.

the non-singlet quark GPD around $\bar{x} = 0$ is sensitive to the input, and the ξ dependence of the GPDs size is in general less pronounced.

As a final remark let us discuss the validity of our evolution procedure. As already mentioned the evolution has been performed by using a code due to Freund and McDermott and modified to restore the correct NLO coupling constant thorough the transcendental equation (see discussion in the previous section). The modification plays a crucial role when the evolution starts from low hadronic input as in the present investigation. Fig. 3 shows the results of LO and NLO evolution for singlet, nonsinglet and gluon distributions up to $Q^2 = 5$ GeV^2 starting from a hadronic scale as low as $Q_0^2 = 0.27$ GeV^2 (where sea contributions are included). The behaviour and the results in the DGLAP region in the singlet and nonsinglet sectors are consistent with the results obtained within analogous radiative parton model approaches for the standard (diagonal) distribution functions (e.g. ref. [8]). The Leading and Next-to-Leading predictions are quite close showing that the evolution is clearly under control and converging.

Larger differences are seen in the ERBL sector (which, however, remain within $\approx 30\%$) for singlet and nonsinglet GPDs. Only the gluon distribution shows larger discrepancy. This is not a drawback of our model as can be seen from an analogous calculation [1] which makes use of the standard (GRV98) [12] parametrization for the diagonal partonic input at $Q_0^2 = 0.36$ GeV^2, comparable with our input scale. A result already discussed by Freund and McDermott [10] for the GPDs calculated within the GRV98 input model. It is well known that also in the case of parton distribution functions generated from low scale parametrizations, NLO gluons are affected by renormalization scale dependence, a problem not yet addressed for GPDs and deserving further investigation.

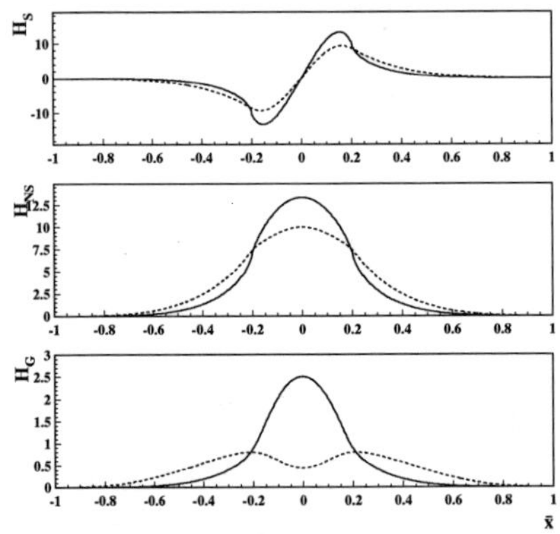

FIGURE 3. Singlet quark (upper panels), non-singlet quark (middle panels), and gluon (lower panels) GPDs for the DD-based model using input parton distributions of the hypercentral CQM with the sea contribution in the meson-cloud model, evolved at $Q^2 = 5$ GeV^2 and at $\xi = 0.2$. The full lines are the results from NLO evolution, and the dashed lines corresponds to LO evolution.

ACKNOWLEDGMENTS

We are grateful to A. Freund and M. McDermott for providing us with their evolution code. This research was partially supported by the Italian MIUR through the PRIN Theoretical Physics of the Nucleus and the Many-Body Systems.

REFERENCES

1. B. Pasquini, M. Traini and S. Boffi, hep-ph//0407228.
2. M. Diehl, Phys. Rep. 388, 41 (2003).
3. A.V. Radyushkin, Phys. Rev. D 59, 014030 (1999).
4. S. Boffi, B. Pasquini and M. Traini, Nucl. Phys. B 649, 243 (2003); 680, 147 (2004).
5. W. Melnitchouk, J. Speth and A.W. Thomas, Phys. Rev. D 59, 014033 (1999); S. Kumano, Phys. Rep. 303, 183 (1998); A.W. Thomas, W. Melnitchouk and F.M. Steffens, Phys. Rev. Lett. 85, 2892 (2000).
6. Ho-Meoyng Choi, Chueng-Ryong Ji, Phys. Rev. D 59, 074015 (1999).
7. M. Glück, E. Reya and A. Vogt, Z. Phys. C 53, 127 (1992); M. Glück, E. Reya and A. Vogt, Z. Phys. C 67, 433 (1995).
8. M. Traini, V. Vento, A. Mair and A. Zambarda, Nucl. Phys. A 614, 472 (1997); A. Mair and M. Traini, Nucl. Phys. A 628, 296 (1998).
9. T. Weigl and W. Melnitchouk, Nucl. Phys. B 465, 267 (1996).
10. A. Freund and M. McDermott, Phys. Rev. D 65, 056012 (2002); (E) 66, 079903 (2002) and http://durpdg.dur.ac.uk/hepdata/dvcs.html.
11. A. V. Belitsky, D. Müller, L. Niedermeier and A. Schäfer, Nucl. Phys. B 546, 279 (1999); Phys. Lett. B 437, 160 (1998); A. V. Belitsky, B. Geyer, D. Müller and A. Schäfer, Phys. Lett. B 421, 312 (1998); A. V. Belitsky, A. Freund and D. Müller, Nucl. Phys. B 574, 347 (2000). A. Freund and M. McDermott, Eur. Phys. J. C 23, 651 (2002); A. Freund, M. McDermott and M. Strikman, Phys. Rev. D 67, 036001 (2003);
12. M. Glück, E. Reya and A. Vogt, Eur. Phys. J. C 5, 461 (1998).

Nucleon DIS structure functions from semi-inclusive $ed \to enX$ reactions

Misak Sargsian

Florida International University, Miami, FL 33199, USA

Abstract. We discuss semiinclusive deep inelastic $d(e,e'N)X$ reaction in which one of the nucleons is detected in the target fragmentation region. Two main issues are addressed. One is the extraction of the "free" neutron DIS structure function from semiinclusive scattering off the deuteron. Other is the possibility to study the DIS mechanism of $eN \to e'X$ scattering at large x through the study of reinteraction of recoil nucleon with the products of the DIS scattering off bound nucleon target. We demonstrate that the extrapolation of the kinetic energy of recoil nucleon to the negative pole value of the target nucleon propagator allows us to extract the on-shell DIS structure function of the bound nucleon. We demonstrate also that in this extrapolation the final state interaction amplitudes are smooth functions and have negligible contribution in the cross section of the reaction.

Keywords: Semiinclusive, DIS
PACS: 25.30.-c, 25.30.Fj

GENERAL FRAMEWORK

We consider semiinclusive deep inelastic scattering (DIS) off the deuteron:

$$e+d \to e'+N+X \tag{1}$$

where nucleon N is detected in the target fragmentation region. In general the cross section of such reactions can be represented as follows:

$$\frac{d\sigma}{dxdQ^2d^3p_s/E_s} = \frac{4\pi\alpha_{em}^2}{xQ^4}(1-y-\frac{x^2y^2m_n^2}{Q^2})\times$$
$$\left[F_L^D+(\frac{Q^2}{2q^2}+tan^2(\frac{\theta}{2})\frac{\nu}{m_N}F_T^D+(\frac{Q^2}{q^2}+tan^2(\frac{\theta}{2}))^{\frac{1}{2}}cos(\phi)F_{TL}^D+cos(2\phi)F_{TT}^D\right]. \tag{2}$$

where four independent nuclear structure functions, $F^D_{L,T,TL,TT}$ depend on Q^2, x, α_s, p_{st}, with $\alpha_s = 2\frac{E_s-p_s cos(\theta_s)}{m_D}$ being light cone momentum fraction of the deuteron carried out by recoil nucleon. It is normalized in such way that the sum of $\alpha_s+\alpha=2$, where α is the similar quantity for the target nucleon. p_{st} is the transverse momentum of the recoil nucleon with the z axis aligned in the direction of $\vec{q}$. In many practical considerations one integrates over the azimuthal angles of the recoil nucleon, which yields:

$$\frac{d\sigma}{dxdQ^2d^3p_s/E_s} = \frac{4\pi\alpha_{em}^2}{xQ^4}(1-y-\frac{x^2y^2m_n^2}{Q^2})\left[F_{2D}^{SI}+2tan^2(\frac{\theta}{2}))\frac{\nu}{m_N}F_{1D}^{SI}\right], \tag{3}$$

where: $F^{SI}_{2D}(x,Q^2,\alpha_s,p_t) = F^D_L + \frac{Q^2}{2q^2}\frac{\nu}{m_N}F^D_T$ and $F^{SI}_{1D}(x,Q^2,\alpha_s,p_t) = \frac{F^D_T}{2}$

CP747, *Structure of the Nucleon at Large Bjorken x*, edited by C. Bourrely, Z.-E. Meziani, and J. Soffer

(a) (b)

FIGURE 1. IA (a) and FSI (b) amplitudes

Theoretical Description

The theoretical description is based on the assumption that the reaction of Eq.(1) proceeds through the interaction of virtual photon off one of the bound nucleons in the deuteron with subsequent recoiling of the spectator nucleon in the final state of the reaction. Two main diagrams will contribute to the reaction (1): impulse approximation (IA) (Fig.1a) and final state interaction (FSI) (Fig.1b) diagrams.

Impulse Approximation

In IA the recoil nucleon is a spectator of $\gamma^* N$ scattering off bound nucleon. One can apply Feynman diagram rules to write down the IA amplitude in the formal form:

$$A_{IA} = < X|J^{em}(Q^2,x)|n > \bar{u}(p_d - p_s)\bar{u}(p_s)\frac{\Gamma_d}{m_n^2 - (p_d - p_s)^2}, \tag{4}$$

where Γ_d is the covariant $d \rightarrow pn$ transition vertex, and $< X|J^{em}(Q^2,x)|n >$ represents the DIS amplitude of electron scattering off bound nucleon.

The nuclear DIS structure functions in this case can be expresses through the convolution of bound nucleon DIS structure functions and the nuclear spectral function, S which describes the probability of finding an interacting nucleon in the target with momenta α and p_t and a recoil nucleon in the final state of the reaction with momenta α_s and p_{st}. Note that in IA $\alpha_s + \alpha = 2$ and $\mathbf{p_t} = -\mathbf{p_{st}}$. Using these relations for the nuclear DIS structure functions one obtains[1, 2]:

$$\begin{aligned}
F_{2D}^{SI}(x,Q^2,\alpha_s,p_t) &= \frac{S(\alpha_s,p_t)}{n}\frac{m_N \nu}{pq} \\
&\times \left[(1+cos\delta)^2(\alpha + \frac{pq}{Q^2}\alpha_q)^2 + \frac{1}{2}sin^2\delta\frac{p_t^2}{m_N^2}\right] F_{2N}^{eff}(\tilde{x},Q^2,\alpha,p_t), \\
F_{1D}^{SI}(x,Q^2,\alpha_s,p_t) &= \frac{S(\alpha_s,p_t)}{n}\left[F_{1N}^{eff}(\tilde{x},Q^2,\alpha,p_t) + \frac{p_t^2}{2pq}F_{2N}^{eff}(\tilde{x},Q^2,\alpha,p_t)\right].
\end{aligned} \tag{5}$$

The nuclear spectral function, S is model dependent quantity whose form depends on the framework/formalism used to describe the interaction. We consider two theoretical approaches:
In *virtual nucleon formalism* scattering is described in the LAB frame of the nucleus

and electrons scatter off the virtual nucleon whose virtuality is defined by the kinematic parameters of spectator nucleon. In this case the form of the spectral function is defined through the evaluation of the IA amplitude at the one-mass shell pole of the spectator nucleon propagator in the Lab frame. Doing this one obtains:

$$S^{VN}(\alpha_s, p_{st}) = E_s \frac{M_d}{2(M_d - E_S)} \Psi_d^2(p_s). \tag{6}$$

Note that within virtual nucleon approximation, $n = 1$ in Eq.(5). It can be shown [2] also that S^{VN} has a proper normalization that conserves number of baryons involved in the scattering. On the other hand the momentum sum rule is not satisfied. This reflects the fact that due to virtuality of interacting nucleon part of the momentum is shared by the non-nucleonic degrees of freedom in nuclei.

In the approximation of *light cone* (LC) dynamics the scattering is described in the light cone reference frame. In this case the spectral function is evaluated at the pole of the spectator nucleon in the LC reference frame. This yields a spectral function [3]:

$$S^{LC}(\alpha_s, p_{st}) = E_k \Psi_d^2(k), \quad \alpha_s = \frac{E_k + k_z}{E_k}, \quad p_{st} = -k_t. \tag{7}$$

The remarkable property of the LC spectral function is that it fulfills both baryon number conservation and momentum sum rules ([3]) (for recent discussion see also [2]). Note also that within LC approximation $n = (2-\alpha_s)^2$ in Eq.(5).

Final State Interaction

Calculation of FSI amplitude of Fig.1b is in general a very complex task. It requires a detailed understanding of the dynamics of DIS as well as its hadronic structure at the final state of the reaction. One simplification is achieved if the energies involved in the scattering are large enough that the phase spaces of current and target fragmentations are isolated. In this case one can formally represent the FSI diagram of Fig.1b as follows (see e.g.[4]):

$$A_{FSI} = \int \frac{d^4 p_{s'}}{i(2\pi)^4} \frac{\langle X, s|A_{FSI}|X's'\rangle\langle X'|J^{em}(Q^2,x)|N\rangle \bar{u}(p_d - p_{s'})\bar{u}(p_{s'})}{p_{s'}^2 - m_N^2 + i\varepsilon} \frac{\Gamma_d}{(p_d - p_{s'})^2 - m_N^2}, \tag{8}$$

where $\langle X'|J^{em}(Q^2,x)|N\rangle$ is the DIS transition amplitude and $\langle X, s|A_{FSI}|X's'\rangle$ describes the FSI of the DIS product $|X's'\rangle$ off the spectator nucleon.

EXTRACTING "FREE" NEUTRON DIS STRUCTURE FUNCTION

The processes of Eq.(1) with recoil proton can be used to extract the DIS structure function of the neutron. Since the neutron is bound, it requires a careful treatment of off-shell effects in maximally model independent way. One such approach is of Chew and Low [5] who studied the issues of extracting $\pi^+\pi^0$ and $n+n \to n+n$ cross sections from $\pi^+ + p \to p + X$ and $n + d \to p + n + n$ reactions respectively. In their analysis they observed that the analytical structure of the IA amplitude (Eq.(4)) is such that it has a pole in negative (nonphysical) region of kinetic energies of spectator nucleon. Indeed straightforward manipulation of the amplitude of Eq.(4) yields:

$$A_{IA} = < X|J^{em}(Q^2,x)|n > \bar{u}(p_d - p_s)\bar{u}(p_s)\frac{\Gamma_d}{|\varepsilon_B|(M_d + m_n - m_p) + 2M_d T_s} \tag{9}$$

which shows that it has a pole at

$$T_s^{pole} = -\frac{|\varepsilon_B|}{2}(1 + \frac{m_n - m_p}{M_d}) \approx -\frac{|\varepsilon_B|}{2} \tag{10}$$

Thus the extrapolation of the measured semiinclusive cross section to the negative kinetic energy region of spectator nucleon will allow us to isolate the "free" cross section of electron - neutron scattering.

Loop Theorem

Procedure of pole extrapolation however is useful if one can prove in the model independent way that the contributions of all other amplitudes of semiinclusive DIS scattering (i.e. Fig.2b) does not contain singularities at the on-shell pole of the bound nucleon propagator. To prove this one observes from Fig.2b that all FSI type amplitudes contain at least one loop which involves the spectator nucleon. Therefore it is sufficient to prove that no singularities exist for the amplitude that contains at least one loop involving the spectator nucleon to prove that all higher order diagrams will have a smooth behavior at the negative kinetic energy pole. For this one considers the amplitude of the form of Eq.(8) which contains explicitly a loop involving a rescattering of the "X" state off the spectator nucleon. Taking the integral over $d^3p_{s'}$ at the pole of spectator nucleon: $\int dp^0_{s'} \frac{1}{p^2_{s'} - m^2_N + i\varepsilon} = -\frac{2\pi}{2E_{s'}}$ and introducing $k = p_s - p_{s'}$ one obtains:

$$A_{FSI} = \int \frac{d^3k}{(2\pi)^3} \frac{\langle X,s|A_FSI|X's'\rangle\langle X'|J^{em}(Q^2,x)|N\rangle\bar{u}(p_d - p_s - k)\bar{u}(p_s - k)}{2(m_N + T_s - k_0)} \frac{\Gamma_d}{-M^2_d + 2M_d(m + T_s - k_0)}. \tag{11}$$

To estimate the contribution of this amplitude at the pole of struck nucleon it is enough to estimates the integral

$$\int \frac{d^3k}{M_d|\varepsilon_b| + 2M_d T_s - 2M_d k_0}. \tag{12}$$

Using the relation: $k_0 = \sqrt{m^2_N + p^2_s} - \sqrt{m^2_N + (p_s - k)^2} = \frac{\vec{p}_s \vec{k}}{m_N} - \frac{k^2}{2m_N}$ it can be shown that the analytic continuation of $T_s \to -\frac{|\varepsilon_b|}{2}$ yields a finite result for Eq.(12), namely

$$\frac{k^2 dk}{2M_d(\frac{k^2}{2m_N})} \to \frac{dk}{2}. \tag{13}$$

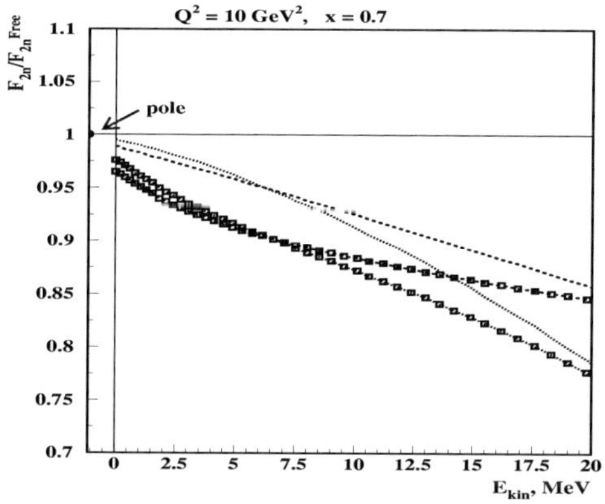

FIGURE 2. Ratio of bound nucleon structure function to that of free nucleon as a function of kinetic energy of spectator nucleon. Dashed lines IA, solid lines IA+FSI predictions. For unmarked lines $p_{st} = 0$ and marked lines correspond to the production of recoil nucleon at 90^0 with respect to $\mathbf{q}$.

Now one can see that the integral in Eq.(11) is finite since at large k limit integrand is suppressed due to the amplitude of rescattering of system X off the spectator nucleon. This result demonstrates that all contributions that contain at least one rescattering with spectator nucleon have smooth behavior at the pole of the struck nucleon propagator as compared to the singular behavior of IA amplitude in Eq.(9). In Fig.2 we quantified this result considering different rescattering scenarios for kinematics maximally favorable for FSI. This result demonstrates that the extrapolation to the negative kinetic energy pole leaves FSI diagrams finite and constraints them on the level of 1-2%.

FSI AND THE MECHANISM OF DIS SCATTERING

The last issue we discuss is the possibility of using the spectator nucleon as a probe to study the mechanism of DIS scattering at large x. As it can be seen in Fig.3 the spectator nucleon will probe a very different final states depending on whether mechanisms of DIS off the bound nucleon is dominated by Feynman mechanism (Fig.3(a)) or mechanism in which DIS is dominated by photon scattering off the minimal fock component of quark-gluon wave function of the bound nucleon.

Our initial estimates demonstrate that these two mechanisms may yield markedly different reinteraction for the produced system X and the spectator nucleon in the reaction.

Another situation that allows meaningful modeling of the FSI amplitude is the semi-inclusive scattering at relatively moderate values of final invariant masses produced off the bound nucleon in the deuteron ($W \leq 2\ GeV$). In this case the rescattering can be modeled through the sum of FSI amplitudes in which resonances R is produced at the

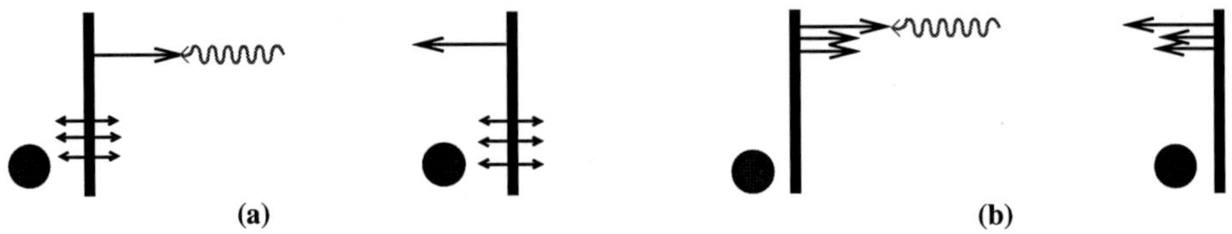

FIGURE 3. Semiinclusive DIS scattering in Feynman (a) and minimal Fock component approximation mechanism

intermediate state of the reaction:

$$F_{FSI} = -\frac{\sqrt{(2\pi)^3 E_s}}{2}\sum_R \int \frac{d^3 p'_s}{(2\pi)^3} \frac{f^{RN}(p_{st}-p'_{st}) < R|J(Q^2)|\gamma^*, N >}{p'_{sz} - p_{sz} + \Delta_R + i\varepsilon} \psi_D(p'_s) \quad (14)$$

where

$$\Delta_R = \frac{M_d + q_0}{q} T_s + \frac{W_N^2 - M_R^2}{2q}. \quad (15)$$

Our preliminary results of comparison of calculations based on Eq.(14) with the preliminary data of experiment[6] indicate a reasonable agreement for the kinematics of $Q^2 = 1.8$ GeV^2, $x = 0.45$ and $W = 1.7$ GeV.

CONCLUSION

We considered the theoretical framework of investigation of semiinclusive DIS scattering off the deuteron target. Two applicataions of these reactions are discussed. One is the use of the pole extrapolation procedure to extract the "free" DIS structure function of the neutron. In this case it is demonstrated that the only IA term has a singular behaviour. As a result the FSI effects are negligible in this extrapolation. This is a model independent observation, based on the theorem that any additional loop in the scattering amplitude involving the spectator nucleon has a smooth behaviour at the pole of the bound nucleon propagator.

The second application of seminclusive DIS scatterings is the possibility to use the spectator nucleon as a probe of DIS mechanism at large x and large values of the spectator nucleon transverse momenta. We demonstrate that at moderate values of W the FSI can be described through the rescaattering of produced resonances off the spectator nucleon. The first calculations yield a reasonable agreement with the preliminary data from d(e,e'p)X experiment at Jefferson Lab.

REFERENCES

1. W. Melnitchouk, M. Sargsian and M. I. Strikman, Z. Phys. A **359**, 99 (1997).
2. M. M. Sargsian, S. Simula and M. I. Strikman, Phys. Rev. C **66**, 024001 (2002).
3. L. L. Frankfurt and M. I. Strikman, Phys. Rept. **76**, 215 (1981).
4. M. M. Sargsian, Int. J. Mod. Phys. E **10**, 405 (2001).
5. G. F. Chew and F. E. Low, Phys. Rev. **113**, 1640 (1959).
6. S.E. Kuhn and K.A. Griffioen (spokespersons), JLab Proposal, CEBAF-PR-102-94, 39pp. (1994).

PARITY VIOLATING EXPERIMENTS AND DRELL-YAN REACTIONS

Parity at High x

P. A. Souder

Department of Physics, Syracuse University, Syracuse, NY 13244

Abstract. We discuss some interesting physics goals for measurements of parity violation in deep inelastic scattering with the proposed new 11 GeV polarized electron beam at JLab. Topics discussed include higher twist effects, charge symmetry violation, and the ratio d/u of up quarks to down quarks in the proton structure functions. In addition, we sketch a new type of apparatus that will provide ample statistics at the highest possible values of Q^2.

INTRODUCTION

The upgrade of JLab to an 11 GeV facility will open the door for precision measurements of parity violation in the deep inelastic (DIS) region, especially at higher values of Bjorken x. Historically, the observation of parity violation in DIS played a key role in establishing the validity of the Standard Model [1]. More recently, the NuTeV [2] neutrino experiment at Fermilab has measured the Paschos-Wolfenstein ratio R^{PW} of neutral current DIS cross sections normalized to charged current DIS cross sections with a precision $\delta R^{PW}/R^{PW} \sim 0.7\%$. The result is about 3σ higher than naive expectations. One explanation is that the NuTeV result has seen interesting QCD effects such as charge symmetry violations in the structure functions [3]. Parity experiments at JLab will also be sensitive unexplored QCD effects. Moreover, the experiments at JLab will not have to integrate over a large range of kinematics as was done for NuTev, and thus may isolate kinematic regions where the novel effects may be most significant. We describe below a program to shed light on these issues.

KINEMATICS

The relevant variables in DIS are $Q^2 = 4EE'\sin^2[\theta/2]$, $\nu = E - E'$, $x = Q^2/2M\nu$, and $y = [E-E']/E$, where M is the nucleon mass. The cross section is given by

$$\frac{d\sigma}{d\Omega dE'} = \frac{\alpha^2}{4E^2}\frac{\cos^2(\theta/2)}{\sin^2(\theta/2)}[W_2 + 2W_1\tan^2(\theta/2)].$$

Commonly used structure functions are $F_2(x,Q^2)$ and $R(x,Q^2)$ where $\nu W_2 = F_2(x,Q^2)$, $R(x,Q^2) = \sigma_L/\sigma_T = \left[1+Q^2/\nu^2\right]F_2(x,Q^2)/2xF_1(x,q^2) - 1$ and $MW_1 = F_1(x,Q^2)$.

CP747, *Structure of the Nucleon at Large Bjorken x*, edited by C. Bourrely, Z.-E. Meziani, and J. Soffer

For parity violation in polarized electron scattering, we define the asymmetry $A = (\sigma_R - \sigma_L)/(\sigma_R + \sigma_L)$. Then:

$$A^{PV} = Q^2 \frac{G_F}{2\sqrt{2}\pi\alpha}\left[a(x) + \frac{1-(1-y)^2}{1+(1-y)^2} b(x)\right],$$

with $a(x) = \sum_i f_i(x) C_{1i} Q_i / \sum_i f_i(x) Q_i^2$, and $b(x) = \sum_i f_i(x) C_{2i} Q_i / \sum_i f_i(x) Q_i^2$. Here $f_i(x)$ is the probability of a parton with flavor i having a fraction x of the momentum of the nucleon. Also Q_i is the electromagnetic charge and $C_{1i}(C_2 i)$ are the weak vector (axial vector) charges for the ith flavor. For an isoscalar target such as deuterium, we have

$$a(x) = \frac{6}{5}\left[(C_{1u} - \frac{1}{2}C_{1d}) + \frac{2}{15}\frac{s(x)+\bar{s}(x)-c(x)-\bar{c}(x)}{D}\right]$$

$$b(x) = \frac{6}{5}\left[(C_{2u} - \frac{1}{2}C_{2d})\frac{q(x)-\bar{q}(x)}{D} + \text{heavy quark terms}\right] \quad (1)$$

where $D(u) = u(x) + d(x) + \bar{u}(x) + \bar{d}(x) + \frac{2}{5}[s(x)+\bar{s}(x)] + \frac{8}{5}[c(x)+\bar{c}(x)]$ and $q(x) = u(x) + d(x) + s(x) + c(x)$. The $a(x)$ term is largest since $C_{1u} - C_{1d}/2 \sim -0.36$ whereas $C_{2u} - C_{2d}/2 \sim 0.06$.

SURVEY OF THE PROGRAM

To be useful, experiments must measure the asymmetries with errors on the order of 1-2%. It is unlikely that interesting physics can be established with less precise results. Since the cross sections and asymmetries are reasonable large, achieving the required errors in the raw asymmetries is expected to be straightforward. The challenges more likely will be beam polarimetry, measuring the average value of Q^2, and evaluating backgrounds.

Most of the measured asymmetry arises from the large $a(x)$ term. To a good approximation, it is a known constant: $a(x) = 1.15$. Any deviation from this value might be a signature for new QCD effects. The advantage is that no fits or GLAP evolution is required that might complicate the interpretation.

At JLab kinematics, A^{PV} is much more sensitivity to the small $b(x)$ terms than in any previous measurement. Thus the results are also is sensitivity to physics beyond the Standard Model [4].

We expect that the program at JLab should cover as broad a kinematic range as possible. This might include points at $x = 0.4$, 0.5, 0.6 and 0.7. In addition, varying Q^2 is important to evaluate the possibility of higher twist (HT) terms discussed below. The sensitivity of an experiment usually maximizes at the lowest value of Q^2. However, one important criterion for DIS is that the missing mass squared $W^2 > 4.0$ GeV^2. Then the relation

$$Q^2 = (W^2 - M^2)/(1/x - 1) \quad (2)$$

TABLE 1. Higher twist coefficients D_i from Ref [5]. Values are given for LO and NNNLO. Also given are Q^2_{min} and the percent effects that might be seen at Q^2_{min}.

x	D_i (LO)	D_i (NNNLO)	Q^2_{min}	$D_i/Q^2_{min}(\%)$ (LO)	$D_i/Q^2_{min}(\%)$ (NNNLO)
0.1-0.2	-0.07	0.01	0.5	-14	2
0.2-0.3	-0.11	0.0(3)	1.0	-11	0
0.3-0.4	-0.06	-0.01	1.7	-3.5	-0.5
0.4-0.5	0.22	.11	2.6	8	4
0.5-0.6	0.85	0.39	3.8	22	10
0.6-0.7	2.6	1.4	5.8	45	24
0.7-0.8	7.3	4.4	9.4	78	47

with $W^2 = 4.0$ defines the smallest possible Q^2, denoted Q^2_{min}. We would anticipate that the x-points could be made at Q^2_{min} and also at $2Q^2_{min}$ except at $x = 0.7$, where even a single Q^2 point is ambitious.

HIGHER TWIST CORRECTIONS

One interesting candidate for hadronic corrections are the higher twist (HT) effects which might add Q^2 dependence to the asymmetry:

$$A^{PV}(x,Q^2) = A^{PV}(x)(1 + C(x)/Q^2).$$

The values for $C(x)$ are not known. However, HT effects have been extracted from GLAP fits to structure function data $F_2(x,Q^2)$ [5] using

$$F_2(x,Q^2) = F_2(x)(1 + D(x)/Q^2).$$

Values of the $D(x)$ are given in Table I. For our purposes here, we will make the very naive assumption that

$$C(x) \sim D(x).$$

It turns out that the values of the $D(x)$ depend upon how many orders of α_s is taken in the GLAP evolution. At leading order (LO), the higher twist contributions are significant and similar to the results of older analyses [6, 7]. However, as higher orders are taken, NLO, NNLO, and NNNLO, $D(x)$ becomes quite small for $x < 0.4$.

A summary of the $D(x)$ is given in Table I. To design an experiment, one needs to know the maximum fractional effect. This occurs at the lowest possible Q^2, which we have denote Q^2_{min} and defined in Eq. 2. Fractional HT effect at Q^2_{min} are also given in Table I. If the HT effects in A_{PV} follow the pattern of Table I, we might see easily observed effects at the highest x-range and no effect at lower x values.

Some contributions to $D(x)$ may very well cancel in the ratio that comprises A^{PV} One interesting effect that probably does not cancel can be described as scattering by diquarks. Indeed, diquarks might be more important at High x. The theory for this specific term in parity violation has not been well developed. Nevertheless, there is interest in such a term [8]. We note that there are few experiments that provide clean data for large HT effects, leaving room for an important contribution from parity violation.

TRANSVERSE BEAM ASYMMETRIES

There has been very recent interest in measuring asymmetries from the scattering of transversely polarized electrons from fixed targets [9, 10]. The interesting feature about this process is that hadronic effects that make small contributions to many processes become the dominant contribution to transverse asymmetries. It is possible, for example, that higher twist effects could be studied in this fashion [11]. The experimental techniques for measuring transverse beam asymmetries are very similar to those of parity experiments, so it is appropriate to consider both kinds of physics in the same proposals.

CHARGE SYMMETRY VIOLATION

One of the assumptions in deriving Eq. 1 is that $u^p = d^n$ and $d^p = u^n$. Since this is violated at some level, we define charge symmetry violation (CSV) parameters:

$$\delta u(x) = u^p(x) - d^n(x); \;\; \delta d(x) = d^p(x) - u^n(x).$$

The effect of the CVS parameters on R^{PW} is[12]:

$$\frac{\delta R^{PW}}{R^{PW}} \sim 0.85 \frac{\delta u - \delta d}{u+d}.$$

Of course, this must be averaged over the kinematics of the experiment. For parity, we have

$$\frac{\delta A^{PV}}{A^{PV}} = 0.28 \frac{\delta u - \delta d}{u+d}. \tag{3}$$

Knowledge about $(\delta u - \delta d)/(u+d)$ is limited. A bag model calculation suggests $(\delta u - \delta d)/(u+d) \sim 1\%$ at $x = 0.4$ and a factor of 2 or so higher at $x = 0.6$. To explain NuTeV, $(\delta u - \delta d)/(u+d)$ would have to be a factor of 3 higher than this. Empirical limits from the MRST group are also a factor of 3 higher. Thus CSV is a possible explanation for the NuTev result.

The asymmetry A^{PV} is less sensitive to CSV than is R^{PW}. However, the x-dependence of CSV is not known. If the CSV ratio in Eq. 3 is larger at high x, perhaps because of the falling values of the structure functions, measurements of APV at JLab might have sufficient sensitivity to observe an effect. Additional studies as to how large the x-dependent CSV effects might be and what limits might be set from existing data high x would be very useful. Any experimental demonstration of an x-dependent CSV would be extremely important.

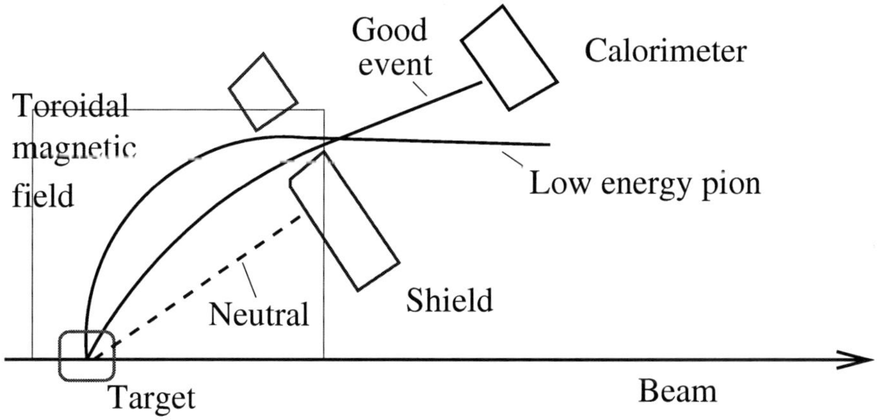

FIGURE 1. Sketch of a calorimetric detector for high statistics studies at high x.

MEASURING D/U AT HIGH X

A topic that has received much attention is the determination of the ratio d/u for the proton at high x [13, 14, 15, 16]. For the proton,

$$a(x) \approx \frac{3}{4}\left[\frac{6C_{1u}u(x) - 3C_{1d}d(x)}{u(x)+\frac{1}{4}d(x)}\right] \sim \left[\frac{u(x)+0.912d(x)}{u(x)+0.25d(x)}\right]$$

This quantity is sensitive to the ratio d/u. The advantage of parity violation is that there is no need for nuclear targets, and controversial corrections for nuclear effects are totally absent. If issues of higher twist can be resolved, parity violation can make an important contribution to this field.

If the higher twist effects can be controlled, kinematics at $x = 0.6-0.7$ would be most appropriate. For the proton, twice as much beam time is required as for the deuteron for the same fractional error $\delta A/A$.

REQUIRED APPARATUS

The lowest x and Q^2 point can be done with existing facilities at JLab. With the advent of the 11 GeV upgrade and the MAD spectrometer, additional points can be reach. Also, useful data with $W < 2$ can be obtained. However, to reach the highest Q^2 points with sufficient statistics, a spectrometer with at least 50% acceptance in aizmuth is required. We do not know how to do this economically with a conventional magnetic spectrometer. We suggest that calorimeter such as the one used for A4 at Mainz might be more suitable. At the higher beam energies, however, the calorimeter must be shielded from photons and low energy pions originating in the target. Thus a sweeping magnet, such as the G_0 toroid, would be required. A sketch of the concept is given in Fig. 1. Such an apparatus would also be useful for many other physics topics. One obvious example is measurements of $A_1(x)$

CONCLUSIONS

We suggest a program of measurements of APV with both proton and deuterium targets covering the range $0.4 < x < 0.7$ with $W > 2$ GeV and with a Q^2 range of a factor of two. At the larger values of x, we might observe higher twist effects and charge symmetry violation. Information about physics beyond the Standard Model can be extracted at lower values of x, especially if the higher twist effects are small. In that case, data from the proton will provide important data on the value for d/u for the proton. We believe that a new calorimetric spectrometer that will operate with large solid angle at large scattering angles will be required to provide ample statistics at some of the more exciting kinematic points.

ACKNOWLEDGMENTS

I would like to thank A. Afanesev, D. Beck, P. Bosted, J. T. Londergan, K. Kumar, K. McFarland, F. Maas, Z. Meziani, W. Melnitchouk, M. J. Ramsey-Musolf, P. Reimer, A. Thomas, and M. Vanderhaeghen for useful discussions. This work is supported by the U. S. Department of Energy under Grant No. DE-FG020-84ER40146.

REFERENCES

1. C. Y. Prescott, *et al., Phys. Lett.* B 84, 524 (1979)) .
2. G. P. Zeller *et al.* [NuTeV Collaboration], Phys. Rev. Lett. **88**, 091802 (2002) [Erratum-ibid. **90**, 239902 (2003)].
3. J. T. Londergan, arXiv:hep-ph/0408243.
4. A. Kurylov, M. J. Ramsey-Musolf and S. Su, Phys. Lett. B **582**, 222 (2004)
5. A. D. Martin, R. G. Roberts, W. J. Stirling and R. S. Thorne, Eur. Phys. J. C **35**, 325 (2004)
6. M. Vircheau *et al., Phys. Lett.* B 274, 221 (1992) .
7. P. Amaudrez *et al., Nucl. Phys.* B 371, 3 (1992) .
8. S. Brodsky, in *"Proceedings from the JLab/Temple University HiX2000 Workshop,"*, 2000.
9. A. V. Afanasev and N. P. Merenkov, arXiv:hep-ph/0406127.
10. M. Gorchtein, P. A. M. Guichon and M. Vanderhaeghen, Nucl. Phys. A **741**, 234 (2004)
11. A. V. Afanasev, private communication.
12. J. T. Londergan and A. W. Thomas, Phys. Rev. D **67**, 111901 (2003)
13. W. Melnitchouk, I. R. Afnan, F. Bissey and A. W. Thomas, Phys. Rev. Lett. **84**, 5455 (2000)
14. W. Melnitchouk and A. W. Thomas, Phys. Lett. B **377**, 11 (1996)
15. S. I. Alekhin, Phys. Rev. D **63**, 094022 (2001)
16. S. Kuhlmann *et al., Phys. Lett.* B 476, 297 (2000) .

Current Status of the NuTeV Anomaly

J.T. Londergan

Dept. of Physics, Indiana University, Bloomington, IN, 47404, USA [1]

Abstract. The NuTeV neutrino measurement of the Weinberg angle differs by 3 standard deviations from measurements at the Z pole. We review the current status of attempts to explain this result, both within and outside the Standard Model.

The NuTeV experiment [1] produced separate beams of neutrinos and antineutrinos that were scattered from an iron target. Charged and neutral-current interactions were differentiated according to the length of the subsequent track and the amount of missing energy. Events were selected using the criterion that the visible energy in the calorimeter satisfied the relation $20 < E_{vis} < 180$ GeV, and that the event vertex was contained within the fiducial volume. This resulted in a data sample of 1.62 million ν's, and 0.351 million $\bar{\nu}$'s. These cross sections were used to extract a value for the Weinberg angle, through a relation originally derived by Paschos and Wolfenstein [2]

$$R^{PW} \equiv \frac{\langle \sigma_{NC}^{\nu N_0} \rangle - \langle \sigma_{NC}^{\bar{\nu} N_0} \rangle}{\rho_0^2 \left(\langle \sigma_{CC}^{\nu N_0} \rangle - \langle \sigma_{CC}^{\bar{\nu} N_0} \rangle \right)} = \frac{1}{2} - \sin^2 \theta_W = \frac{R^\nu - R^{\bar{\nu}}}{1 - r R^\nu} . \qquad (1)$$

In Eq. 1, $\langle \sigma_{NC}^{\nu N_0} \rangle$ is the total neutral-current inclusive cross section for neutrinos on an isoscalar target. The quantity $\rho_0 \equiv M_W/(M_Z \cos \theta_W)$ is one in the Standard Model. Alternatively, the Paschos-Wolfenstein (PW) ratio is obtained by combining ratios of neutral current to charged current cross sections for neutrinos and antineutrinos; R^ν is the ratio of neutral-current to charged-current total cross sections induced by neutrinos on an isoscalar target, and r is the ratio of $\bar{\nu}$ to ν charged-current total cross sections.

The NuTeV group obtained the neutral/charged ratios $R^\nu = 0.3916 \pm 0.0007$ and $R^{\bar{\nu}} = 0.4050 \pm 0.0016$, from which they extracted $\sin^2 \theta_W = 0.2277 \pm 0.0013\ (stat) \pm 0.0009\ (syst)$. This value is three standard deviations above the measured value for the Weinberg angle obtained from electroweak (EW) processes near the Z pole, $\sin^2 \theta_W = 0.2227 \pm 0.00037$ [3]. This unexpected and now-famous result corresponds to a 1.2% decrease in the left-handed coupling g_L^2 of hadrons to the EW current. Because cuts and acceptance corrections are different for neutrinos and antineutrinos, the resulting ratios R^ν and $R^{\bar{\nu}}$ cannot be combined simply, as in Eq. (1), to obtain the Weinberg angle. The NuTeV group extracted the neutral to charged-current ratios through a detailed Monte Carlo simulation of the reaction. As a result, corrections to the Weinberg angle calculated

---

[1] Research sponsored in part by US NSF under grant NSF PHY0244822

CP747, *Structure of the Nucleon at Large Bjorken x*, edited by C. Bourrely, Z.-E. Meziani, and J. Soffer

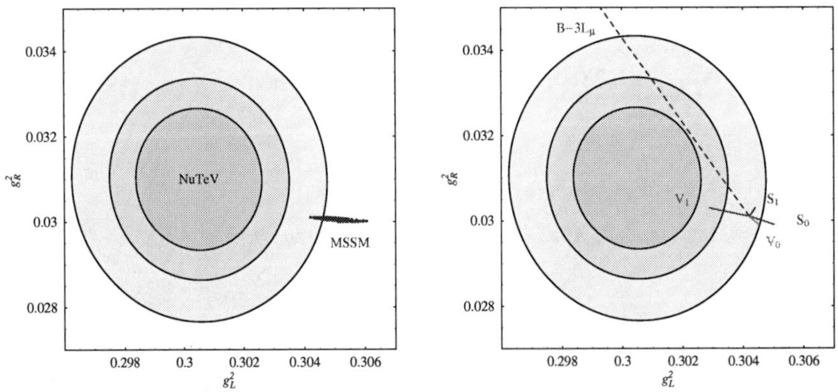

FIGURE 1. (Color online). Deviations from Standard Model for various types of new particles, without conflicting by more than 1σ with existing EW bounds, from Ref. [5], plotted vs. couplings of hadrons to EW currents g_L^2 and g_R^2. Circles represent standard deviations from NuTeV result. Left: minimal supersymmetry loops, from light sleptinos and gauginos. Right: solid lines, various leptoquark couplings; dashed line, extra $B-3L_\mu$ gauge boson.

from the PW ratio will not accurately predict the corresponding changes for the NuTeV experiment. The NuTeV group [4] has published functionals that provide the sensitivity of their observables to a given effect.

The Paschos-Wolfenstein ratio depends on a large number of assumptions. In this review we will examine a number of these assumptions and estimate the uncertainty in the Weinberg angle resulting from each effect. One obvious possibility is contributions from physics beyond the Standard Model. Davidson *et al.* have provided a comprehensive review of possible contributions from "new physics" [5]. Several observables measured at the Z pole are known to very great precision, in some cases at the 0.1% level; this makes it very difficult to find new particles that resolve the NuTeV neutrino measurement (at the 1% level), but leave all other measured quantities within their experimental limits. It is necessary to choose new particles whose number, masses and couplings are delicately adjusted to fit all existing data.

We briefly review a few of these attempts. An obvious candidate would be loop effects from the lightest supersymmetric particles. In general, SUSY loops in MSSM (minimal supersymmetric models) tend to have the wrong sign, and increase the size of the anomaly. Other SUSY candidates violate existing constraints. High-mass extra Z' vector bosons, unmixed with the known Z boson, might solve the NuTeV anomaly by tuning the mass(es) and couplings. Davidson *et al.* found that the most likely possibility was a Z' satisfying the gauge symmetry $B-3L_\mu$, where B is the baryon number and L_μ the lepton number. However, the couplings are tightly constrained by muon $g-2$ measurements. Leptoquarks are scalar or vector bosons that couple to both leptons and quarks. Assuming that the couplings conserve baryon and lepton number, it was found that only triplet $SU(2)_L$ leptoquarks, with carefully chosen mass splittings, could explain the NuTeV effect. Future LHC experiments could test the existence of leptoquarks.

Oblique corrections (from high-mass particles that couple only to vector bosons) are tightly constrained by EW data, and Davidson *et al.* conclude they are probably unable to resolve the NuTeV anomaly. Fig. 1 plots contributions from new particles. The left graph shows minimal SUSY loops from light sleptinos and gauginos. The right graph shows contributions from an extra Z' with gauge symmetry $B - 3L_\mu$ (dashed curve), and from minimal leptoquarks (solid curves). These are plotted vs. the left and right-handed couplings g_L^2 and g_R^2. Of the various $SU(2)_L$ symmetries for leptoquarks - scalar singlet S_0 and and triplet S_1, and vector singlet V_0 and triplet V_1 - only V_1 could remove the NuTeV anomaly.

New physics outside the Standard Model is very tightly constrained by the very precise results at the Z pole, and seems unlikely to resolve the NuTeV measurement. As a result, recent efforts have focused on "QCD Corrections" within the current Standard Model. We review the status of the most promising effects.

First, the Paschos-Wolfenstein ratio requires an isoscalar target; this produces a correction to the Weinberg angle due to the excess neutrons in the iron target. The NuTeV collaboration have taken this correction into account. The correction is large (of order -.008), but should be known to within a couple percent. Radiative corrections, which involve coupling of soft photons to the final muon line, occur only for the charged-current events, and constitute a substantial correction. It would be surprising if radiative corrections were not calculated accurately; however, two different groups are currently re-calculating radiative corrections [6].

Since NuTeV measurements were carried out on an iron target, there are nuclear corrections for the structure functions. Reactions induced by charged leptons have measured shadowing corrections at small x, EMC-type corrections at intermediate x, and Fermi motion corrections at large x. Miller and Thomas [7] suggested that shadowing effects for neutrinos (virtual W or Z) could be significantly different from shadowing for muons [8], and that shadowing would affect charged-current and neutral-current events quite differently. The NuTeV group countered that shadowing occurs at values of Q^2 significantly lower than their experiment. They also claim that the shadowing corrections should cancel when one subtracts neutrino cross sections from antineutrino reactions. Kovalenko *et al.* [9] calculated significant nuclear effects, combining shadowing with EMC-type effects calculated in an extended x-rescaling model. Their model predicted greater changes to $R^{\overline{\nu}}$ than to R^{ν}, whereas NuTeV found $R^{\overline{\nu}}$ almost identical to the Standard Model prediction, while R^{ν} was 3σ below the Standard Model value. Kumano [10] calculated a "modified Paschos-Wolfenstein relation" for nuclei, and Hirai *et al.* [11] report preliminary results in attempting to calculate nuclear effects on the NuTeV result. They assume that nuclear corrections for neutrinos are identical to those from charged lepton DIS.

In parton distributions, charge symmetry corresponds to interchange of protons and neutrons (simultaneously interchanging up and down quarks). In nuclear physics, charge symmetry is respected to a high degree of precision, generally at less than the 1% level [12, 13]. There is no direct evidence for charge symmetry violation (CSV) in high energy experiments [14], and until recently all phenomenological analyses assumed charge symmetric PDFs. CSV PDFs are defined by $\delta d_{\rm v}(x) \equiv d_{\rm v}^p(x) - u_{\rm v}^n(x)$ and $\delta u_{\rm v}(x) \equiv$

$u_v^p(x) - d_v^n(x)$. CSV and strange quark corrections to the PW ratio have the form

$$\Delta R^{PW} \approx \left[1 - \frac{7}{3}\sin^2\theta_W\right] \frac{\delta U_v - \delta D_v - S_v}{2(U_v + D_v)} \tag{2}$$

In Eq. (2), $Q_v = \langle x(q - \bar{q})\rangle$ represents the total momentum carried by valence PDFs, and we have for simplicity omitted a small QCD radiative correction proportional to α_s [5]. The first two terms in Eq. (2) are CSV corrections, while the last term represents a strange quark contribution. Only valence PDFs contribute to the PW ratio; sea quark contributions to the NuTeV result are substantially smaller than valence effects.

Recently, the MRST group [15] have included isospin violation in a phenomenological evaluation of PDFs. They carried out a global fit of PDFs using a wide variety of high energy data. MRST chose a specific form for valence quark charge symmetry violating PDFs. They assumed that δu_v and δd_v were equal and opposite, and given by

$$\delta u_v(x) = -\delta d_v(x) = \kappa(1-x)^4 x^{-0.5}(x - .0909). \tag{3}$$

At both small and large x the MRST CSV PDFs agree with their valence quark distributions, and the first moment is zero, a necessary condition to preserve valence quark normalization. The coefficient κ was varied in a global fit to high energy data. MRST obtained a shallow χ^2 minimum about the value $\kappa = -0.2$, with the 90% confidence level for the range $-0.8 \leq \kappa \leq +0.65$.

The MRST group also found evidence for sea quark CSV, although the sea quark correction to the Weinberg angle is much smaller than valence CSV. Using the best-fit CSV values obtained by MRST, would remove roughly 1/3 of the NuTeV anomaly. Note that the form chosen by MRST, Eq. (3), insures that δu_v and δd_v add coherently in the correction to the PW ratio, Eq. (2). The value $\kappa = -0.6$, within the 90% confidence limit found by MRST, would completely remove the NuTeV anomaly, while $\kappa = +0.6$ would double the discrepancy. The MRST results show that isospin violating PDFs could completely remove the NuTeV anomaly in the Weinberg angle, or make it twice as large, without serious disagreement with any of the data used to extract quark and gluon PDFs.

Earlier theoretical estimates of valence parton CSV were carried out by Sather [16] and Rodionov *et al.* [17]. Sather derived an analytic approximation to valence quark PDFs, while Rodionov *et al.* calculated valence CSV expressions from quark models. In both cases, the resulting CSV corrections tend to decrease the size of the NuTeV anomaly, and they remove roughly 30%, or 1σ of the discrepancy in the Weinberg angle. In Fig. 2 we show the valence quark CSV PDFs for the MRST best fit value $\kappa = -0.2$. This agrees very well with the theoretical estimates of both Sather and Rodionov [18]. Theoretical calculations were also carried out by the NuTeV collaboration [4], and by Gao and Signal [19], who estimated CSV effects in a meson-cloud model. Both groups obtained significantly smaller CSV contributions, although all theoretical results were well within the phenomenological limits established by MRST.

The contribution from strange quarks in Eq. (2) depends on the momentum asymmetry between strange quarks and antiquarks, $S_v \equiv \int_0^1 x[s(x) - \bar{s}(x)]dx$. The most precise measurements of strange quark PDFs are obtained from opposite sign dimuons produced from reactions induced by ν's or $\bar{\nu}$'s; the former (latter) is sensitive to the strange

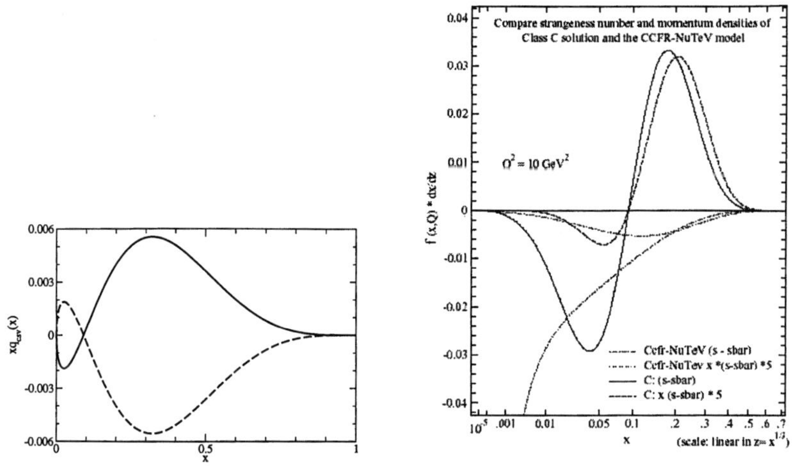

FIGURE 2. (Color online). Left: MRST valence CSV PDF from Ref. [15], corresponding to best fit value $\kappa = -0.2$ in Eq. (3). Solid curve: $x\delta d_{\rm v}(x)$; dashed curve: $x\delta u_{\rm v}(x)$. Right: Positive values at large x: $s - \bar{s}$ and $5x(s - \bar{s})$, from CTEQ global fit, Ref. [22]. Negative values: the same quantities calculated in earlier CCFR/NuTeV analysis of strange quark distributions, Ref. [21].

(antistrange) quark distributions. Both CCFR [20] and NuTeV [21] collaborations have measured these cross sections. The CTEQ group [22] analyzed this in a global fit that allowed the possibility of $s \neq \bar{s}$. They found a positive value for $S_{\rm v}$, which would decrease the Weinberg anomaly; the most likely value gave roughly a 30% decrease in the NuTeV anomaly for the Weinberg angle; in their analysis, the strange contribution alone is unable to remove the NuTeV anomaly. The NuTeV group initially reported a best fit to dimuon cross sections with a slightly negative value $S_{\rm v} = -.0027 \pm 0.0013$ [4]; this would increase the NuTeV anomaly. In Fig. 2 we plot the CTEQ and the original NuTeV analyses of strange quark distributions; we show both $s - \bar{s}$, and the second moment $x(s - \bar{s})$ (multiplied by 5).

The NuTeV and CTEQ groups are currently collaborating to resolve the differences in their analyses. At present, a few differences still persist: CTEQ uses a somewhat lower charm mass (necessitated by HERA data in their global fit); and the two groups use slightly different acceptance corrections and fragmentation functions. In their global fit, the CTEQ group calculates dimuon production using an LO fit, while NuTeV uses NLO calculations throughout. The CTEQ group enforced the condition that the proton have zero net strangeness $\langle s - \bar{s} \rangle = 0$, while at first NuTeV did not (this is apparent in Fig. 2). The values obtained by both groups have changed, but they are still not in agreement; currently, CTEQ find $-0.001 < S_{\rm v} < +0.004$ [23] while the NuTeV best fits still give $S_{\rm v}$ zero or slightly negative, but not positive. It is not clear why this is the case; although the CTEQ group carries out a global fit while NuTeV specifically fits dimuon production, the dimuon data dominate the determination of strange quark PDFs in the global fit. Catani *et al.* [24] have shown that, at NNLO, it is possible to perturbatively generate nonzero values for $S_{\rm v}$ even if it is zero at the starting scale (this is not possible in LO or NLO).

They generate a small negative value for S_V from perturbative three-loop contributions.

In conclusion, it is difficult to find physics outside the Standard Model that can reconcile the NuTeV measurement of the Weinberg angle, while agreeing with the extremely precise EW data at the Z pole. Recent attention has focused on QCD corrections. Of these effects, the most viable candidate is corrections from isospin violating PDFs. The MRST global fit shows that CSV effects are capable of removing 100% of the NuTeV anomaly, without serious disagreement with any high energy data. Partonic CSV effects large enough to explain the NuTeV anomaly would be visible at the few percent level in future experiments. Another possible explanation would be a strange quark momentum asymmetry. At present, the CTEQ group estimates that strange quark effects could remove roughly $1/3$ of the NuTeV anomaly, while NuTeV analysis of the same data suggests a small contribution that most likely increases the anomaly. Another correction arises from nuclear effects in neutrino scattering; currently both the magnitude and sign of nuclear effects are uncertain.

Theoretical work cited here was carried out with A.W. Thomas. The author thanks G. Zeller, K. McFarland, S. Kretzer, F. Olness, W-K Tung and R. Thorne for useful discussions.

REFERENCES

1. NuTeV Collaboration, G.P. Zeller *et al.* , Phys. Rev. Lett. **88**, 091802 (2002).
2. E. A. Paschos and L. Wolfenstein, Phys. Rev. **D7**, 91, (1973).
3. D. Abbaneo *et al.* , arXiv:hep-ex/0112021.
4. NuTeV Collaboration, G.P. Zeller *et al.* , Phys. Rev. **D65**, 111103 (2002).
5. S. Davidson, S. Forte, P. Gambino, N. Rius, and A. Strumia, JHEP **202**, 037 (2002).
6. K. McFarland, private communication.
7. G.A. Miller and A.W. Thomas, arXiv:hep-ex/0204007.
8. C. Boros, J. T. Londergan and A. W. Thomas, Phys. Rev. d**58**, 114030 (1998).
9. S. Kovalenko *et al.* , P. Lett. B**546**, 68 (2002).
10. S. Kumano, Phys. Rev. D**66**, 111301 (2002).
11. M. Hirai, S. Kumano and T-H. Nagai, arXiv:hep-ph/0408023.
12. G. A. Miller, B. M. K. Nefkens and I. Slaus, Phys. Rep. **194**,1 (1990).
13. E. M. Henley and G. A. Miller in *Mesons in Nuclei*, eds M. Rho and D. H. Wilkinson (North-Holland, Amsterdam 1979).
14. J. T. Londergan and A. W. Thomas, in *Progress in Particle and Nuclear Physics*, Volume 41, p. 49, ed. A. Faessler (Elsevier Science, Amsterdam, 1998).
15. MRST Collaboration, A.D. Martin *et al.* , arXiv:hep-ph/0308087.
16. E. Sather, Phys. Lett. **B274**, 433 (1992).
17. E. Rodionov, A. W. Thomas and J. T. Londergan, Mod. Phys. Lett. **A9**, 1799 (1994).
18. J. T. Londergan and A. W. Thomas, arXiv:hep-ph/0407247.
19. F.G. Cao and A.I. Signal, P. Lett. B**559**, 229 (2003).
20. CCFR Collaboration, A.O. Bazarko *et al.* , Z. Phys. C**65**, 189 (1995).
21. NuTeV Collaboration, M. Goncharov *et al.* , Phys. Rev. D**64**, 112006 (2001).
22. S. Kretzer, arXiv:hep-ph/0405221; S. Kretzer *et al.* arXiv:hep-ph/0312322.
23. F. Olness *et al.* arXiv:hep-ph/0312323.
24. S. Catani *et al.* , arXiv:hep-ph/0404240.

The Past and Future of Drell-Yan Physics at Fermilab

R.S. Towell for the FNAL E866/NuSea and E906 Collaborations

Abilene Christian University, Abilene, TX 79699

Abstract. The Fermilab E866/NuSea Collaboration has measured the Drell-Yan dimuon cross sections in 800 GeV/c pp and pd collisions. The results indicate that recent global parton distribution fits provide a good description of the light antiquark sea in the nucleon over the Bjorken-x range $0.03 < x < 0.15$, but overestimate the valence quark distributions as $x \to 1$. Future measurements by FNAL E906 will extend these data to higher x values.

Drell-Yan dimuon production provides valuable information about the partonic structure of hadrons that is complementary to deep-inelastic scattering (DIS) studies because it distinguishes between quarks and antiquarks. The Drell-Yan cross section for $pA \to \mu^+\mu^- X$ may be written in terms of the parton distribution functions (PDFs) of the colliding hadrons as

$$M^3 \frac{d^2\sigma}{dM\,dx_F} = \frac{8\pi\alpha^2}{9} \frac{x_1 x_2}{x_1 + x_2} \sum_q e_q^2 \left[q_1(x_1)\bar{q}_2(x_2) + \bar{q}_1(x_1)q_2(x_2)\right] \quad (1)$$

where the subscript 1 (2) denotes the beam (target) hadron. Measurement of the invariant mass, $M^2 = x_1 x_2 s$, and Feynman-x, $x_F = 2p_l/\sqrt{s} = x_1 - x_2$, of the muon pair (where p_L is the longitudinal momentum of the muon pair, and $\sqrt{s}$ the center-of-mass energy of the hadrons) determines the momentum fraction $x_{1(2)}$ of the beam (target) parton. Equation 1 is exact to leading order, but its general features are also preserved by next to leading order (NLO) calculations.

The Fermilab E866/NuSea Collaboration has measured the Drell-Yan cross sections in 800 GeV/c pp and pd collisions. This represents the first study of the Drell-Yan cross section in pp collisions over a broad kinematic range and the most comprehensive measurement in pd collisions. Previous publications [1, 2] have described the E866 measurement of the $\bar{d}/\bar{u}$ ratio in the proton as a function of x, based on the x_2 dependence of the Drell-Yan cross section ratio $\sigma^{pd}/2\sigma^{pp}$, and those results have been included in recent global PDF fits [3, 4, 5, 6, 7].

We present measurements of the pp and pd Drell-Yan dimuon cross sections for pair mass in the ranges $4.2 < M < 8.7$ GeV or $10.85 < M < 16.85$ GeV and $-0.05 < x_F < 0.8$. At the large values of x_F appropriate for much of the E866 data, the first term in the sum in Eq. 1 dominates and the cross sections are primarily sensitive to the valence distributions in the proton beam and the antiquarks at small x in the proton and deuteron targets. Thus, these measurements probe the partonic structure of the nucleon

CP747, *Structure of the Nucleon at Large Bjorken x*, edited by C. Bourrely, Z.-E. Meziani, and J. Soffer

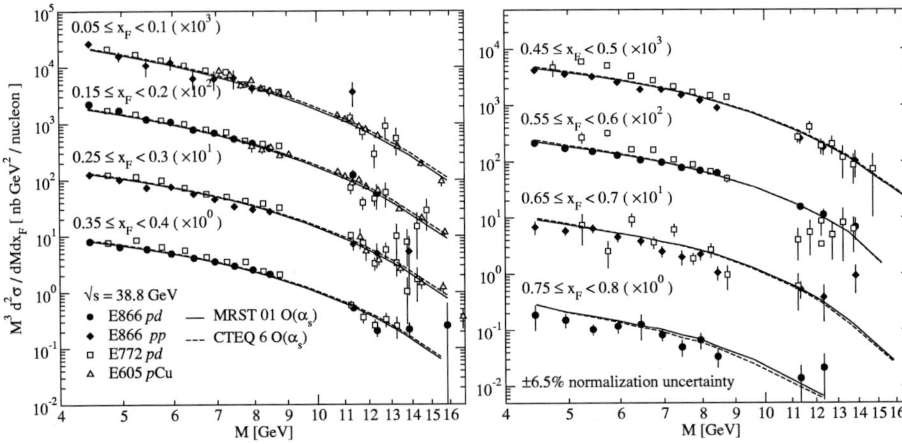

FIGURE 1. FNAL E866 Drell-Yan cross sections per nucleon for selected x_F bins. The E866 *pd* (solid circles) and *pp* (solid diamonds) cross sections are shown in alternate decades, compared with previous *p*Cu results from E605 [8] (open triangles) and *pd* results from E772 [12] (open squares). NLO cross section calculations based on the CTEQ6 [4] (dashed curves) and MRST2001 [6] (solid curves) PDF fits are also shown. The error bars on the E605 and E772 data points are statistical only. Those on the E866 data points are the sum in quadrature of the statistical and point-to-point systematic uncertainties. An additional $\pm 6.5\%$ normalization uncertainty is common to all E866 data points.

in two important kinematic regions and provide new constraints on the magnitudes of the antiquark sea for $0.03 < x < 0.15$ and of the valence quarks for $x \to 1$.

E866 used a 3-dipole magnetic spectrometer employed previously in E605 [8], E772 and E789, modified by the addition of new detectors at the first tracking station. Drell-Yan events were recorded using three different spectrometer magnet settings, chosen to focus low-, intermediate- and high-mass muon pairs and provide acceptance from below the J/ψ to above 15 GeV. A detailed description may be found in Ref. [2]. The present analysis used 55,000 *pp* and 121,000 *pd* Drell-Yan events.

The spectrometer acceptance was calculated separately for each spectrometer setting as a function of mass, x_F, and transverse momentum (p_T) using a detailed Monte Carlo simulation of the spectrometer. Triply-differential cross sections, $d^3\sigma/dMdx_Fdp_T$, were calculated for each kinematic bin, then integrated over p_T to obtain the invariant cross sections, $M^3d^2\sigma/dMdx_F$. The measured cross sections were corrected for radiative effects based on the formalism given by Baur *et al.* [11]. The magnitude of these corrections ranged from -2% at small x_F to +5% at large x_F.

A detailed study was performed of the point-to-point systematic uncertainties in the measured cross sections. The total point-to-point systematic uncertainties within any (M, x_F) bin were strongly correlated between the *pp* and *pd* cross sections. In addition to the point-to-point systematic uncertainties, there was a $\pm 6.5\%$ overall normalization uncertainty, associated with the calibration of the beam intensity. See Ref. [10] for additional details.

Figure 1 shows the Drell-Yan *pp* and *pd* invariant cross sections per nucleon for

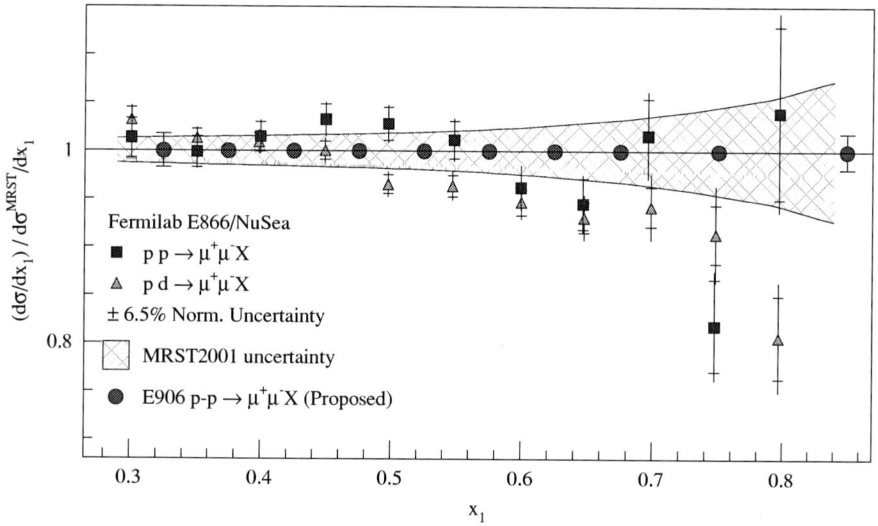

FIGURE 2. Ratios of the measured Drell-Yan *pp* (squares) and *pd* (triangles) cross sections to NLO calculations based on the MRST2001 [6] PDF fit plotted *vs.* x_1, averaged over x_2. The inner error bars represent the statistical uncertainty in the ratio, the outer error bar the sum in quadrature of the statistical and point-to-point systematic uncertainties. Solid lines represent the experimental uncertainty ranges [9] on $4u + d$ in the MRST2001 PDF fit. The circle data points show the precision of the expected results of the FNAL E906 experiment.

selected x_F bins. The results agree with previous 800 GeV/*c* Drell-Yan cross section measurements in *p*Cu collisions by E605 [8]. They also agree with previous measurements in *pd* collisions by E772 [12] for $x_F < 0.3$. Figure 1 also shows the results of next-to-leading-order calculations of the Drell-Yan cross sections based on the CTEQ6 [4] and MRST2001 [6] global PDF fits. The agreement with the global fits is very good over the entire kinematic region.

While the overall normalization is well reproduced, there are systematic deviations between the measurements and the predictions. To elucidate these deviations, it is useful to examine the experimental cross sections separately as functions of x_1 and x_2. Most of the events have $x_1 \gg x_2$, which implies that the x_1 dependence is primarily sensitive to the valence quarks in the proton beam and the x_2 dependence to the antiquarks in the target. The ratios of the experimental cross sections to NLO calculations using the MRST2001 global PDF fit as a function of x_2 averaged over x_1 shows the MRST2001 PDFs provide a very good description of the x_2 dependence of the *pp* cross sections over the full range, and a good description of the *pd* cross sections for $x_2 < 0.15$. The CTEQ6 fits describe the x_2 dependence equally well. This indicates that the current PDFs interpolate the $\bar{u}$ and $\bar{d}$ distributions successfully between the HERA measurements at small x and the E605 measurements for $x > 0.14$, but these new results provide tighter constraints on the magnitude of the antiquark distributions for $0.03 < x < 0.15$ than have existed to date [9].

Figure 2 shows the ratios of the cross sections to NLO calculations using the MRST2001 PDF fits as a function of x_1. The qualitative behavior of the CTEQ6 distributions is quite similar: both PDF fits overestimate the valence quarks by ≈ 10–20% at large x_1, and the discrepancy between the data and current PDFs appears to be larger for the *pd* cross sections than for the *pp* cross sections. The acceptance of these data includes an anti-correlation between x_1 and x_2 [2], so most of the events at large x_1 also have $x_2 < 0.1$. The NLO calculations describe the average x_2 dependence of the cross sections quite well for $x_2 < 0.1$, so the discrepancy in Fig. 2 is isolated to the subset of the low-x_2 events that also have large x_1. Further comparisons between the measured cross sections and the NLO calculations, in which the x_1 and x_2 dependences are examined for limited ranges of the other variable [10], reinforce this conclusion.

Figure 2 shows that the present results fall well outside the experimental uncertainty [9] for the valence quarks in the MRST2001 PDF fit. A similar comparison shows that the present results track the lower limit on the experimental uncertainty [4] in the CTEQ6 PDF fit. These discrepancies reflect additional uncertainties in the PDF fits [4, 9], and they imply that future PDF fits will see a substantial correction to the u and d quark distributions at large x.

The approved FNAL experiment numbered 906 will repeat many of the measurements of FNAL E866/NuSea using the 120 GeV main injector. This will extend the current Drell-Yan measurements to higher x values. Figure 2 shows the expected coverage in x_1. The coverage in x_2 will reach to almost $x_2 \sim 0.5$.

We thank Fermilab for their assistance in performing this experiment. We also thank W.K. Tung of the CTEQ Collaboration for providing the code to calculate the next-to-leading-order Drell-Yan cross sections and U. Baur for providing the code to calculate the radiative corrections to the Drell-Yan cross section. This work was supported in part by the U.S. Department of Energy.

REFERENCES

1. E.A. Hawker *et al.* (FNAL E866/NuSea Collaboration), Phys. Rev. Lett. **80**, 3715 (1998).
2. R.S. Towell *et al.* (FNAL E866/NuSea Collaboration), Phys. Rev. D **64**, 052002 (2001).
3. H.L. Lai *et al.*, Eur. Phys. J. C **12**, 375 (2000).
4. J. Pumplin *et al.*, JHEP **0207**, 012 (2002).
5. A.D. Martin, R.G. Roberts, W.J. Stirling, and R.S. Thorne, Eur. Phys. J. C **4**, 463 (1998).
6. A.D. Martin, R.G. Roberts, W.J. Stirling, and R.S. Thorne, Eur. Phys. J. C **23**, 73 (2002).
7. M. Gluck, E. Reya, and A. Vogt, Eur. Phys. J. C **5**, 461 (1998).
8. G. Moreno *et al.*, (FNAL E605 Collaboration), Phys. Rev. D **43**, 2815 (1991).
9. A.D. Martin, R.G. Roberts, W.J. Stirling, and R.S. Thorne, hep-ph/0211080.
10. J.C. Webb, Ph.D. thesis, New Mexico State University (2002); hep-ex/0301031.
11. U. Baur, S. Keller, W.K. Sakumoto, Phys. Rev. D **63** 033007 (2002).
12. P.L. McGaughey *et al.*, (FNAL E772), Phys. Rev. D **50**, 3038 (1994); **60**, 119903 (1999).
13. D.M. Alde *et al.*, (FNAL E772 Collaboration), Phys. Rev. Lett. **64**, 2479 (1990).

Q^2 Dependence of the Azimuthal Asymmetry in Unpolarized Drell-Yan

E. Di Salvo

Dipartimento di Fisica ad INFN- sez. Genova
Via Dodecaneso 33 - 16146 Genova - Italy

Abstract. I study the azimuthal asymmetry of the unpolarized Drell-Yan in the framework of the T-odd functions. I find, on the basis of quite general arguments, that for $|\mathbf{q}_\perp| << Q$ such an asymmetry decreases as Q^{-2}, where $\mathbf{q}_\perp$ and Q are respectively the transverse momentum and the center-of-mass energy of the muon pair. Data support this conclusion.

INTRODUCTION

The existence of an azimuthal asymmetry in unpolarized Drell-Yan (DY) has been known for a long time[1-3]. There are rather strong indications[4] that the asymmetry could be due to an effect of the so-called T-odd distribution functions[5]. However, according to the traditional parametrization of such functions[5,6], the Q^2 dependence of that asymmetry cannot be reproduced[4], Q being the center-of-mass energy of the muon pair. Here I propose to modify that parametrization, on the basis of quite general arguments. As we shall establish, this modification considerably improves the agreement between theory and experiment.

My talk is organized as follows. Firstly, I recall the parametrization of the angular differential cross section for unpolarized DY. Secondly, I introduce the formalism of the correlation matrix. Thirdly, I determine a scale parameter concerning the T-odd functions. Lastly, I deduce the Q^2 dependence of the asymmetry in unpolarized DY.

THE UNPOLARIZED DRELL-YAN CROSS SECTION

The DY angular differential cross section is conventionally parametrized as

$$\frac{1}{\sigma}\frac{d\sigma}{d\Omega} = \frac{3}{4\pi}\frac{1}{\lambda+3}(1+\lambda cos^2\theta + \mu sin2\theta cos\phi + \frac{1}{2}\nu sin^2\theta cos2\phi). \tag{1}$$

Here $\Omega \equiv (\theta, \phi)$, where θ and ϕ are respectively the polar and azimuthal angle of the momentum of the positive muon in a center-of-mass frame of the muon pair; moreover λ, μ and ν are parameters. In the naive DY model, where the parton transverse momentum and QCD corrections are neglected, one has λ = 1, μ = ν = 0. Experiments[1-3] have

CP747, *Structure of the Nucleon at Large Bjorken x*, edited by C. Bourrely, Z.-E. Meziani, and J. Soffer

shown deviations from these predictions, in particular an azimuthal asymmetry, due to nonvanishing values of μ and ν. Since ν has a considerably larger size than μ[1,2], the asymmetry is almost entirely described by the last term of eq. (1). Such a behavior can be reproduced theoretically by taking into account the intrinsic transverse momentum of a quark inside a hadron[4], which causes, thanks to a coherence effect, a quark in an unpolarized hadron to acquire a transverse polarization, described by a suitable T-odd term[6,4]. Therefore it appears most natural to attribute the azimuthal asymmetry essentially to this coherence effect[4]. In particular, for $Q >> |\mathbf{q}_\perp|$[4] - $\mathbf{q}_\perp$ being the transverse momentum of the muon pair -, the asymmetry is described conveniently as a convolutive product[4] involving the T-odd functions $h_1^\perp$ and $\bar{h}_1^\perp$[6] relative to the active partons.

THE QUARK CORRELATION MATRIX

At leading order in QCD, the DY differential cross section has the following expression:

$$\frac{d\sigma}{d\Omega} = \frac{1}{(x_a + x_b)s} \frac{\alpha^2}{Q^4} L^{\mu\nu} W_{\mu\nu}. \tag{2}$$

Here α is the fine structure constant, s the squared energy in the overall center-of-mass system and x_a and x_b the longitudinal fractional momenta of the active partons. Moreover $L^{\mu\nu}$ and $W^{\mu\nu}$ are respectively the leptonic and hadronic tensor, *i. e.*,

$$L^{\mu\nu} = k^\mu \bar{k}^\nu + k^\nu \bar{k}^\mu - g^{\mu\nu} k \cdot \bar{k}, \tag{3}$$

k and $\bar{k}$ being the four-momenta of the two muons, and

$$W^{\mu\nu} = \frac{1}{6} \int d^2 p_\perp Tr \left[\Phi_A(x_a, \mathbf{p}_\perp) \gamma^\mu \Phi_B(x_b, \mathbf{q}_\perp - \mathbf{p}_\perp) \gamma^\nu\right]. \tag{4}$$

Φ_A and Φ_B are the transverse momentum dependent (tmd) correlation matrices of the active partons inside the two initial hadrons, denoted here as A and B. A sum over quark and antiquark flavors is understood. The tmd correlation matrix can be split into a T-even and a T-odd term, *i. e.*[6,7],

$$\Phi = \Phi_E + \Phi_O, \tag{5}$$

which are parametrized according to the Dirac algebra, taking into account Lorentz and parity invariance[6-9]. In particular, Φ_O reads

$$\Phi_O = \frac{i}{4\mu_0} h_1^\perp [\not{p}_\perp, \not{n}_+] + ... \tag{6}$$

Here ellipses indicate spin dependent terms, if any. Moreover, $p_\perp = p - (P \cdot p)P/M^2$, p and P being respectively the quark and the hadron four-momentum. n_+ is a lightlike four-vector, whose space components are directed along the hadron momentum, and such that $n_+ \cdot n_- = 1$; n_- is lightlike as well, with space components opposite to those of n_+. Lastly, μ_0 is a scale parameter, contained also in the parametrization of Φ_E[7,9,10].

This parameter *must* be the same for all the functions involved, in order to normalize them appropriately, so as to interpret them as probability densities. As I am going to show, the original choice[11,8], $\mu_0 = m_h$, where m_h is the hadron mass, is not the most suitable one.

The term (6) describes the quark polarization in an unpolarized or spinless hadron, due to a coherence effect, for example to an interference between two different partial waves[12-15,4,7,9]. Notice, however, that this effect is described correctly by eq. (6) only if $|\mathbf{q}_\perp| << Q$[4], a necessary condition for assuming factorization[16].

DETERMINING THE PARAMETER μ_0

In order to determine μ_0, I observe that, in the case of an on-shell, noninteracting quark, the T-even part of the correlation matrix turns out to coincide with the density matrix. This latter reads

$$\rho = \frac{1}{2}(\not{p}+m)(f_1 + g_{1L}\gamma_5 \not{S}_{q\parallel} + h_{1T}\gamma_5 \not{S}_{q\perp}). \tag{7}$$

Here

$$S_{q\parallel} = (S_q \cdot n_+)n_- + (S_q \cdot n_-)n_+, \tag{8}$$
$$S_{q\perp} = S_q - S_{q\parallel} \tag{9}$$

and S_q is the Pauli-Lubanski vector of the active quark. Comparing the density matrix with the general parametrization of Φ_E[9,10] yields

$$\mu_0 = |\mathbf{p}|, \tag{10}$$

where $\mathbf{p}$ is the quark momentum. This result is independent of the quark dynamics inside the hadron, since μ_0 is a kinematic parameter. As regards DY, μ_0 can be conveniently calculated in the center-of-mass system of the muon pair, resulting in

$$\mu_0 \simeq \frac{Q}{2}. \tag{11}$$

AZIMUTHAL ASYMMETRY

Inserting the parametrization of the correlation matrix[6-10] into eqs. (4) and (2), the DY differential cross section reads[4,9]

$$\frac{d\sigma}{d\Omega} = \frac{1}{(x_a+x_b)s}\frac{\alpha^2}{12Q^2}\sum_f e_f^2\left[(1+cos^2\theta)F_f + sin^2\theta cos2\phi H_f/\mu_0^2\right]. \tag{12}$$

Here the sum runs over all the light quark and antiquark flavors, $u, d, s, \bar{u}, \bar{d}, \bar{s}$, whose fractional charges are denoted by e_f. Moreover one has (omitting the flavor indices)

$$F = \int d^2p_\perp f_{1A}(x_a, \mathbf{p}_\perp^2) f_{1B}[x_a, (\mathbf{q}_\perp - \mathbf{p}_\perp)^2] \tag{13}$$

and

$$H = \int d^2 p_\perp h_{1A}^\perp(x_a, \mathbf{p}_\perp^2) h_{1B}^\perp[x_a, (\mathbf{q}_\perp - \mathbf{p}_\perp)^2] \mathcal{S}. \tag{14}$$

Here f_{1A} and f_{1B} are the unpolarized densities of the active partons,

$$\mathcal{S} = [2\mathbf{p}_\perp \cdot \mathbf{n}(\mathbf{q}_\perp - \mathbf{p}_\perp) \cdot \mathbf{n} - \mathbf{p}_\perp \cdot (\mathbf{q}_\perp - \mathbf{p}_\perp)] \tag{15}$$

and $\mathbf{n} = \mathbf{q}_\perp/|\mathbf{q}_\perp|$. As explained above, eq. (12) holds true for $|\mathbf{q}_\perp| << Q$.

Eq. (14) implies that, for $\mathbf{q}_\perp \to 0$, $H \to H_0 \mathbf{q}_\perp^2$, where $H_0 = H_0(x_a, x_b, \mathbf{q}_\perp^2)$ assumes a finite value for $\mathbf{q}_\perp = 0$. Now I compare eq. (12) with eq. (1), taking into account eqs. (2) to (6), eq. (11) and the above limit for H. As a result I get

$$\nu = A_0 \frac{\mathbf{q}_\perp^2}{Q^2} = A_0 \rho^2, \tag{16}$$

A_0 being a proportionality constant and $\rho = |\mathbf{q}_\perp|/Q$. My prediction is compared with the experimental results of ν[1,2], both as a function of Q^2 at fixed $|\mathbf{q}_\perp|$ and as a function of ρ, at different energies. This comparison[9] confirms that the value of μ_0 deduced in sect. 4 (see eqs. (10) and (11)) describes satisfactorily the Q^2 dependence of the asymmetry parameter ν, whereas the assumption $\mu_0 = m_h$[8,11] leads to a bad disagreement with experiment[4,9]. In particular, formula (16) provides a very simple interpretation of the behavior of ν versus ρ[1,2], which would be quite difficult to reproduce otherwise.

REFERENCES

1. NA10 Coll., S. Falciano et al.: Z. Phys. C - Particles and Fields **31** (1986) 513
2. NA10 Coll., M. Guanziroli et al.: Z. Phys. C - Particles and Fields **37** (1988) 545
3. J.S. Conway et al.: Phys. Rev. D **39** (1989) 92
4. D. Boer, S. Brodsky and D.-S. Huang: Phys. Rev. D **67** (2003) 054003
5. D. Boer: Phys. Rev. D **60** (1999) 014012
6. D. Boer and P.J. Mulders: Phys. Rev. D **57** (1998) 5780
7. E. Di Salvo: AIP Conf. Proc. **675** (2003) 494
8. P.J. Mulders and R.D. Tangerman: Nucl. Phys. B **461** (1996) 197
9. E. Di Salvo: hep-ph/0407208
10. E. Di Salvo: Int. J. Mod. Phys. A **18** (2003) 5277
11. J. Ralston and D.E. Soper: Nucl. Phys B **152** (1979) 109
12. S.J. Brodsky, D.S. Hwang and I. Schmidt: Nucl. Phys B **642** (2002) 344
13. X. Ji, J.-P. Ma and F. Yuan: Nucl. Phys. B **652** (2003) 383
14. X. Ji and F. Yuan: Phys. Lett. B **543** (2002) 66
15. A.V. Belitsky, X. Ji and F. Yuan: Nucl. Phys. B **656** (2003) 165
16. J.C. Collins: Phys. Lett. B **536** (2002) 43

LATTICE QCD AND NUCLEAR DEEP INELASTIC SCATTERING

Large x Behaviour and the Non-Perturbative Structure of Hadronic Systems

Anthony W. Thomas

Jefferson Lab, 12000 Jefferson Ave., Newport News VA 23606, USA

Abstract. While the traditional interest in structure functions has been the confirmation of the predictions of perturbative QCD, this data also contains a wealth of information on how QCD works in the infrared, or confinement, region. As the challenge of the strong force now turns to the study of QCD in the non-perturbative region, such information is extremely valuable. We outline some of the key issues for both nucleon and nuclear structure functions.

INTRODUCTION

While considerable progress has been made towards understanding hadron structure, it is fair to remark that there are far more questions which remain unanswered than resolved. Amongst the open questions which go to the heart of how QCD is realized we mention:

- Does pQCD describe the behaviour of the d/u ratio as $x \to 1$ [1, 2, 3]?
- What role do di-quark correlations play in the spin and flavour dependence of the parton distribution functions (PDFs) [4]?
- What is the behaviour of the ratio $\Delta d/\Delta u$ as $x \to 1$?
- Is charge symmetry satisfied by the valence PDFs [5, 6]? If not, how much of the NuTeV anomaly [7] does this explain [8]?
- Is the excess of $\bar{d}$ over $\bar{u}$ quarks in the proton [9] entirely explained by the spontaneous breaking of chiral symmetry [10, 11]?
- Is there a similar violation [12] of the naive expectation that $\Delta\bar{u} = \Delta\bar{d}$? What is its source?
- Does the strange sea satisfy $\bar{s} = s$, or is this symmetry also violated by spontaneous chiral symmetry breaking [13] or some other mechanism?
- What is the role of heavier flavours in the nucleon – is there significant intrinsic charm/beauty [14]?
- Can the modern, rigorous version of the Ellis-Jaffe sum rule [15] be tested with precision?
- Does the Gottfried sum rule exist [16]?
- Can modern experiments improve the precision with which fundamental sum rules such as Gross-Llewellyn-Smith (currently 10-15% precision) and Adler (currently 20-30% precision) are tested [17]?

CP747, *Structure of the Nucleon at Large Bjorken x,* edited by C. Bourrely, Z.-E. Meziani, and J. Soffer

- To what extent is SU(3) symmetry satisfied in the PDFs of the other octet baryons [18]?
- Can we relate baryon fragmentation functions to the PDFs of the baryons [19] – in the valence region.... in the sea region?
- What is the origin of the famous, nuclear EMC effect [20] in the valence region?
- Can we differentiate between the effects of short-range correlations, multi-quark states and other potential contributions to nuclear structure functions in the kinematic region $x > 1$?
- What is the contribution of polarized glue in the proton spin structure function [21]?
- Are we able to reliably extract higher twist contributions to free nucleon structure functions [22]?
- What is the role of shadowing in existing nuclear structure functions? Is there a significant higher twist component and what role do differences between higher twist components in electromagnetic, charged and neutral current neutrino scattering [23] play in the NuTeV anomaly [24]?
- What is the role of Lorentz scalar and vector mean fields in atomic nuclei in modifying the nuclear PDFs [25, 26]? Are nuclear PDFs related in any simple way, as usually assumed, to the PDFs of the free nucleon?
- Given recent advances in lattice QCD are we able to contribute anything helpful from that source?

To be honest, after compiling this list of questions I was astonished. All are real and all are important and there are a lot of them! Resolving them is an outstanding challenge in hadronic physics for the next decade, experimentally and theoretically. It is certainly a major part of the mission of Jefferson Lab at 12 GeV and beyond. Apart from listing them here with minimal references for those interested to at least get an introduction to the issues, we cannot explore more than couple of topics.

LATTICE QCD

The problems faced in performing lattice QCD simulations at light quark mass are relatively well known, with the time for a calculation scaling as roughly m_π^7 or worse [27]. In addition, because lattice simulations are performed in Euclidean space, one can only access information on PDFs through their moments. Furthermore, problems of operator mixing associated with the H(4) symmetry of a lattice, mean that only low moments are currently accessible - up to x^3 for unpolarised moments and x^2 in the spin dependent case. Since the large-x behaviour of the PDFs is determined by the higher moments, it is not likely that one will unambiguously fix their behaviour in that region from lattice QCD. Nevertheless it is of great interest to see exactly what one can learn from such calculations.

The outstanding problem in the field for some years was a continuing failure of the lattice simulations to reproduce the momentum fraction of the nucleon carried by its valence quarks. In particular, the integral of the non-singlet combination $u_V - d_V$, weighted by x, was typically a factor of two larger than experiment. For a very fundamental quan-

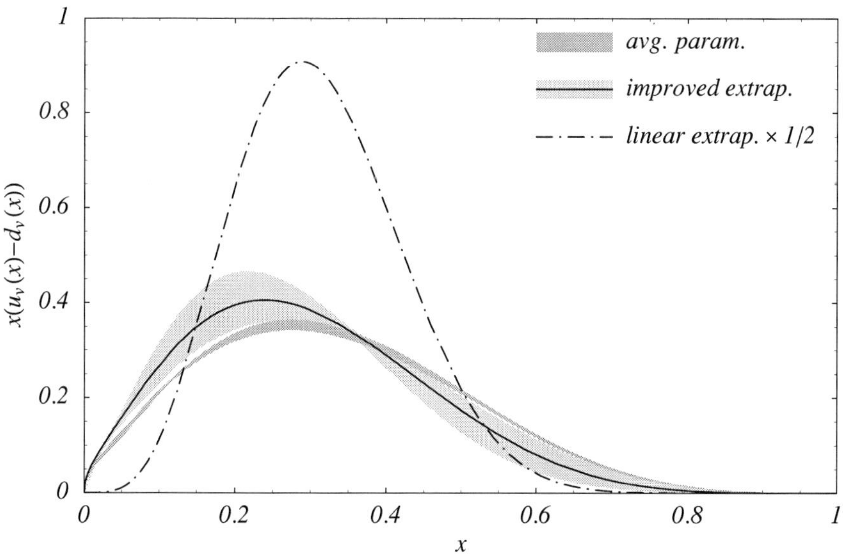

FIGURE 1. Comparison between the isovector valence quark distribution of the proton reconstructed from the lattice moments after a correct chiral extrapolation (solid curve and light shaded area) with the commonly used parametrizations of world data (dark shaded area). The dash-dot curve shows the result if the extrapolation of the moments is made with a simple linear fit – from Ref. [29].

tity which characterizes the sharing of momentum between quarks and gluons in the proton, this was an intriguing problem. Only in 2001 was it realized that because the calculation of PDFs involves a forward matrix element, chiral corrections can be extremely important and indeed the chiral logarithm associated with this problematic moment is expected to lead to a rapid variation as m_π approaches its physical value. Taking this into account removed the long-standing discrepancy [28].

Having realized the importance of a correct chiral extrapolation of the moments it is possible to parametrize all the moments as a function of pion mass. Then by assuming a particular functional form for the PDF one can determine the parameters of the phenomenological function which provide a best fit to the lattice data at each pion mass. This approach necessarily involves a degree of model dependence in any conclusions drawn, but this seems unavoidable. Nevertheless, as we see in Fig. 1, the level of agreement between the PDF extracted from such a fitting process, using moments extrapolated with the correct chiral behaviour to the physical mass, and experiment seems very good [29]. The lack of higher moments, as well as the statistical errors in the lattice moments and, of course, uncertainty in the chiral extrapolation, lead to a relatively large error band on the theoretical curve. A similar study has been carried out for the spin dependent PDFs, with necessarily larger errors as a consequence of having one less moment in that case [36].

Another feature of Fig. 1 that requires attention is the reconstructed valence PDF at a current quark mass above 50 MeV. While this does not seem so much larger than

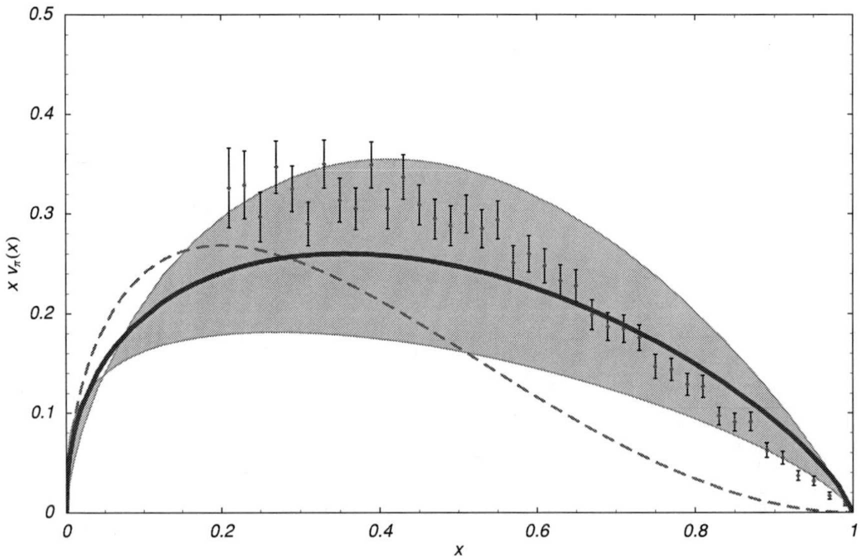

FIGURE 2. Comparison between Drell-Yan data for the valence PDF of the pion and the reconstruction based on a chiral extrapolation of lattice moments (solid line and shaded area). For comparsion the commonly used from based on pQCD is shown as the dashed curve – from Ref. [36].

the physical quark mass the physical difference is enormous. At this point the pion mass is sufficiently large that its Compton wavelength is smaller than the source of the pion field and, as a consequence, pion loops are almost totally suppressed [30]. As a corollary the constituent quark model seems to suddenly describe the structure of the nucleon extremely well. Indeed, we see that the PDF clearly peaks around 1/3, with the three valence quarks sharing the momentum of the nucleon. This strongly supports other evidence based on the behaviour of masses, magnetic moments and charge radii calculated on the lattice, which suggests that this slightly unphysical mass regime is the best place to build a constituent quark model [31].

While much attention has been focussed on the nucleon, perhaps because of the amount of experimental information we have on its PDFs, the pion is also of great interest. The Drell-Yan data, which have yielded much of our information on its large-x structure [32, 33], seem to contradict the perturbative QCD (pQCD) expectation that the valence distribution should go as $(1-x)^2$ as $x \to 1$ [34]. It is therefore of great interest to see whether lattice QCD can give any information [35]. Using the same technique described earlier for the nucleon, to reconstruct the valence PDF of the pion is even more challenging – there are less moments and they alternate (for even and odd moments) between singlet and non-singlet. However, using the fact that chiral loops are suppressed at pion masses above 0.4–0.5 GeV, Detmold *et al.* were able to obtain the distribution shown in Fig. 2 [36]. The errors are clearly large and without more modern data (and ideally more moments) one cannot make a firm conclusion. Nevertheless, it is clear that the calculated PDF does agree with the experimental data and furthermore that it does tend to (slightly) favour a harder distribution than pQCD would suggest. Further progress awaits new and more precise data.

CONCLUDING REMARKS

We have seen that in terms of fundamental calculations based on non-perturbative QCD, there has been considerable progress in extracting PDFs for the nucleon and pion. For the moment this process is model dependent and the accuracy is limited by the small number of moments that can be calculated and the relatively large quark masses for which data is available. In the near future we can expect the rapidly improving computer power to yield partially quenched or, better, full QCD data at masses nearer the physical regime. In addition, group theoretical work as well as improved algorithms will enable us to calculate more moments and hence better constrain the large-x behaviour.

On the other hand, when it comes to the important issues surrounding nuclear PDFs, where in fact much of our information about PDFs has been obtained, lattice QCD is unlikely to be of any help in the forseeable future. We have no choice but to develop more sophisticated models of nuclear structure starting at the quark and gluon level, which respect the constraints of special relativity [37]. In particular, we are going to have to rely on such models, as well as new experimental data and good phenomenology, in order to answer the fundamental question whether there is indeed any simple relation between nucleon and nuclear structure functions. Certainly the recent studies of the different roles of scalar and vector mean fields in models of the QMC type [25, 26], suggest that there may well be significant differences and this would be of considerable importance.

ACKNOWLEDGMENTS

I would like to particularly acknowledge the contributions of Wolfgang Bentz, Will Detmold, Wally Melnitchouk, Fernando Steffens and Kazuo Tsushima to the main topics discussed here. This work was supported by DOE contract DE-AC05-84ER40150, under which SURA operates Jefferson Lab.

REFERENCES

1. G. R. Farrar and D. R. Jackson, Phys. Rev. Lett. **35**, 1416 (1975).
2. W. Melnitchouk and A. W. Thomas, Phys. Lett. B **377**, 11 (1996) [arXiv:nucl-th/9602038].
3. S. J. Brodsky, M. Burkardt and I. Schmidt, Nucl. Phys. B **441**, 197 (1995) [arXiv:hep-ph/9401328].
4. F. E. Close and A. W. Thomas, Phys. Lett. B **212**, 227 (1988).
5. E. Sather, Phys. Lett. B **274**, 433 (1992).
6. E. N. Rodionov, A. W. Thomas and J. T. Londergan, Mod. Phys. Lett. A **9**, 1799 (1994).
7. G. P. Zeller *et al.* [NuTeV Collaboration], Phys. Rev. Lett. **88**, 091802 (2002) [Erratum-ibid. **90**, 239902 (2003)] [arXiv:hep-ex/0110059].
8. J. T. Londergan and A. W. Thomas, Phys. Rev. D **67**, 111901 (2003) [arXiv:hep-ph/0303155].
9. P. Amaudruz *et al.* [New Muon Collaboration], Phys. Rev. Lett. **66**, 2712 (1991).

10. A. W. Thomas, Phys. Lett. B **126**, 97 (1983).
11. D. B. Leinweber *et al.*, Phys. Rev. Lett. **86**, 5011 (2001) [arXiv:hep-ph/0101211].
12. A. W. Schreiber, A. I. Signal and A. W. Thomas, Phys. Rev. D **44**, 2653 (1991).
13. A. I. Signal and A. W. Thomas, Phys. Lett. B **191**, 205 (1987).
14. S. J. Brodsky, P. Hoyer, C. Peterson and N. Sakai, Phys. Lett. B **93**, 451 (1980).
15. S. D. Bass *et al.*, Phys. Rev. D **66**, 031901 (2002) [arXiv:hep-ph/0207071].
16. V. Guzey *et al.*, Phys. Rev. D **64**, 054503 (2001) [arXiv:hep-ph/0102133].
17. A. W. Thomas and W. Weise, "The Structure of the Nucleon," Wiley-VCH, Berlin, 2001.
18. M. Alberg *et al.*, Phys. Lett. B **389**, 367 (1996) [arXiv:hep-ph/9609498].
19. C. Boros *et al.*, Phys. Rev. D **62**, 014021 (2000) [arXiv:hep-ph/0002081].
20. J. J. Aubert *et al.* [European Muon Collaboration], Phys. Lett. B **123**, 275 (1983).
21. J. Ashman *et al.* [European Muon Collaboration], Phys. Lett. B **206**, 364 (1988).
22. Z. E. Meziani *et al.*, arXiv:hep-ph/0404066.
23. C. Boros *et al.*, Phys. Rev. D **59**, 074021 (1999) [arXiv:hep-ph/9810220].
24. G. A. Miller and A. W. Thomas, arXiv:hep-ex/0204007.
25. H. Mineo *et al.*, Nucl. Phys. A **735**, 482 (2004) [arXiv:nucl-th/0312097].
26. F. M. Steffens *et al.*, Phys. Lett. B **595**, 237 (2004) [arXiv:hep-ph/0405096].
27. T. Lippert, S. Gusken and K. Schilling, Nucl. Phys. Proc. Suppl. **83**, 182 (2000).
28. W. Detmold *et al.*, Phys. Rev. Lett. **87**, 172001 (2001) [arXiv:hep-lat/0103006].
29. W. Detmold *et al.*, Mod. Phys. Lett. A **18**, 2681 (2003) [arXiv:hep-lat/0310003].
30. W. Detmold *et al.*, Pramana **57**, 251 (2001) [arXiv:nucl-th/0104043].
31. I. C. Cloet *et al.*, Phys. Rev. C **65**, 062201 (2002) [arXiv:hep-ph/0203023].
32. J. Badier *et al.* [NA3 Collaboration], Z. Phys. C **18**, 281 (1983).
33. B. Betev *et al.* [NA10 Collaboration], Z. Phys. C **28**, 15 (1985).
34. G. P. Lepage and S. J. Brodsky, Phys. Rev. D **22**, 2157 (1980).
35. G. Martinelli and C. T. Sachrajda, Nucl. Phys. B **306**, 865 (1988).
36. W. Detmold *et al.*, Phys. Rev. D **68**, 034025 (2003) [arXiv:hep-lat/0303015].
37. P. A. M. Guichon and A. W. Thomas, Phys. Rev. Lett. **93**, 132502 (2004) [arXiv:nucl-th/0402064].

Generalized Parton Distributions and Nucleon Structure at Large x from Lattice QCD

G. Schierholz

John von Neumann-Institut für Computing NIC,
Deutsches Elektronen-Synchrotron DESY,
15738 Zeuthen, Germany
and
Deutsches Elektronen-Synchrotron DESY,
22603 Hamburg, Germany

E-mail: Gerrit.Schierholz@desy.de

Abstract. Recent progress in lattice calculations of generalized parton distributions is discussed, with emphasis on the large-x behavior of nucleon structure functions.

Keywords: QCD, lattice, form factors, parton distributions
PACS: 11.15.Ha,12.38.Gc,13.40.Gp,13.60.Hb

INTRODUCTION

Understanding the internal structure of hadrons in terms of quarks and gluons (partons), and in particular how quarks and gluons provide the binding and spin of the nucleon, is one of the outstanding problems in particle physics. The region of large fractional parton momentum x, being at the borderline between inclusive and exclusive processes, is of particular interest in this respect, as it provides valuable information about the dynamics of quark confinement.

Generalized parton distributions [1] (GPDs) are a powerful tool to address these questions. While ordinary parton distributions measure the probability of finding a parton with fractional longitudinal momentum x in the fast moving nucleon at a given resolution $1/Q$, and no information on the transverse distribution of partons is provided, GPDs describe the coherence of two different hadron wave functions, one where the parton carries fractional momentum $x+\xi$ and one where this fraction is $x-\xi$, from which information on the transverse motion and binding of partons can be gathered [2].

Moments of GPDs are amenable to lattice calculations [3, 4]. In this talk I will present some selected results obtained by the QCDSF collaboration [5].

GENERALIZED PARTON DISTRIBUTIONS

Let me introduce some notation first and briefly comment on the lattice calculation. Consider the deeply virtual Compton scattering process depicted in Fig. 2, where $p = p_1 + p_2$, $\Delta = p_1 - p_2$, $q = (q_1 + q_2)/2$ and $\xi = q\Delta/qp$. At high energies and large $Q^2 = -q^2$ an operator product expansion (OPE) becomes effective. For $\xi = 0$ the OPE

CP747, *Structure of the Nucleon at Large Bjorken x*, edited by C. Bourrely, Z.-E. Meziani, and J. Soffer

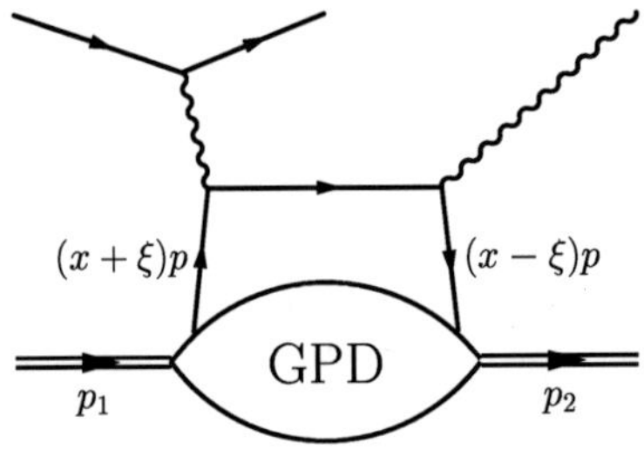

FIGURE 1. The leading diagram for deeply virtual Compton scattering.

can be mapped onto a parton model description [6]. In this case the momentum transfer of the struck quark is purely transverse, i.e. $\Delta = (0, \mathbf{\Delta}_\perp)$. Throughout this talk I shall consider the case $\xi = 0$ only.

The soft part of this process can be parameterized by unpolarized and polarized GPDs [1], $H_q(x,\Delta^2,Q^2)$, $E_q(x,\Delta^2,Q^2)$ and $\tilde{H}_q(x,\Delta^2,Q^2)$, $\tilde{E}_q(x,\Delta^2,Q^2)$, respectively. At zero momentum transfer, $\Delta = 0$, they coincide with the ordinary parton distributions. For example

$$\begin{aligned} H_q(x,0,Q^2) &= q(x,Q^2), \\ \tilde{H}_q(x,0,Q^2) &= \Delta q(x,Q^2). \end{aligned} \tag{1}$$

The GPDs are not directly calculable, but only their moments:

$$\begin{aligned} \int_0^1 dx x^n H_q(x,\Delta^2,Q^2) &= A^q_{n+1}(\Delta^2), \\ \int_0^1 dx x^n E_q(x,\Delta^2,Q^2) &= B^q_{n+1}(\Delta^2), \\ \int_0^1 dx x^n \tilde{H}_q(x,\Delta^2,Q^2) &= \tilde{A}^q_{n+1}(\Delta^2), \\ \int_0^1 dx x^n \tilde{E}_q(x,\Delta^2,Q^2) &= \tilde{B}^q_{n+1}(\Delta^2), \end{aligned} \tag{2}$$

where A^q_n, B^q_n, $\tilde{A}^q_n$ and $\tilde{B}^q_n$ are generalized form factors, which can be related to (off-forward) nucleon matrix elements of certain operators. The lowest form factors (for $n = 0$) are

$$\begin{aligned} A^q_1(\Delta^2) &= F^q_1(\Delta^2), \\ B^q_1(\Delta^2) &= F^q_2(\Delta^2), \\ \tilde{A}^q_1(\Delta^2) &= G^q_A(\Delta^2), \\ \tilde{B}^q_1(\Delta^2) &= G^q_P(\Delta^2), \end{aligned} \tag{3}$$

where F_1^q, F_2^q are the Dirac and Pauli form factors and G_A^q, G_P^q the axial vector and pseudoscalar form factors of the nucleon. At $\Delta^2 = 0$ we have

$$A_{n+1}^q(0) = \langle x_q^n \rangle , \quad \tilde{A}_{n+1}^q(0) = \langle \Delta x_q^n \rangle . \tag{4}$$

When Fourier transformed to impact parameter space, both

$$\begin{aligned} H_q(x, \mathbf{b}_\perp^2, Q^2) &= \int \frac{d^2\mathbf{b}_\perp}{(2\pi)^2} e^{i\mathbf{b}_\perp \Delta_\perp} H_q(x, \boldsymbol{\Delta}_\perp^2, Q^2), \\ \tilde{H}_q(x, \mathbf{b}_\perp^2, Q^2) &= \int \frac{d^2\mathbf{b}_\perp}{(2\pi)^2} e^{i\mathbf{b}_\perp \Delta_\perp} \tilde{H}_q(x, \boldsymbol{\Delta}_\perp^2, Q^2), \end{aligned} \tag{5}$$

admit a probability interpretation. While $H_q(x, \mathbf{b}_\perp^2, Q^2)$ describes the probability of finding a quark of flavor q with fractional momentum x at impact parameter $\mathbf{b}_\perp$, $\tilde{H}_q(x, \mathbf{b}_\perp^2, Q^2)$ measures the fraction of the nucleon's spin carried by the quark q with fractional momentum x and at impact parameter $\mathbf{b}_\perp$.

The generalized form factors are obtained from nucleon matrix elements of local operators,

$$\begin{aligned} &\langle p_1 | \bar{\mathcal{O}}^q_{\{\mu_1 \cdots \mu_n\}}(a) | p_2 \rangle \\ &= \left(\frac{i}{2}\right)^{n-1} \langle p_1 | \bar{q} \gamma_{\{\mu_1} \overleftrightarrow{D}_{\mu_2} \cdots \overleftrightarrow{D}_{\mu_n\}} q | p_2 \rangle \\ &= \left(\frac{1}{2}\right)^{n-1} \bar{u}(p_1) \big[\bar{A}_n^q(\Delta^2) \gamma_{\{\mu_1} \\ &+ \frac{\mathrm{i}\Delta^\alpha}{2m_N} \bar{B}_n^q(\Delta^2) \sigma_{\alpha\{\mu_1} \big] \bar{p}_{\mu_2} \cdots \bar{p}_{\mu_n\}} u(p_2) + \cdots , \end{aligned} \tag{6}$$

which we compute on the lattice. Similar formulae hold for the polarized case. To match the Wilson coefficients, the (bare) lattice operators must be renormalized:

$$\mathcal{O}_{\{\mu_1 \cdots \mu_n\}}(\mu) = Z_{\mathcal{O}}(a\mu) \bar{\mathcal{O}}_{\{\mu_1 \cdots \mu_n\}}(a). \tag{7}$$

Due to bad convergence of lattice perturbation theory, this is gradually being done nonperturbatively [7].

The calculations I am going to present have been performed using Wilson fermions with $N_f = 2$ flavors of light dynamical quarks. The lightest quark mass that has been reached so far corresponds to a pion mass of $O(500)$ MeV. Our smallest lattice spacing is $a \approx 0.07$ fm, which translates into a cut off of ≈ 2.8 GeV.

LATTICE RESULTS

Once we know the generalized form factors, we can reconstruct the GPDs. Consider $H_q(x, \Delta^2, Q^2)$. The generalized form factor can be written

$$\int_0^1 dx x^n C_q(x, \Delta^2) = A_{n+1}^q(\Delta^2)/A_{n+1}^q(0). \tag{8}$$

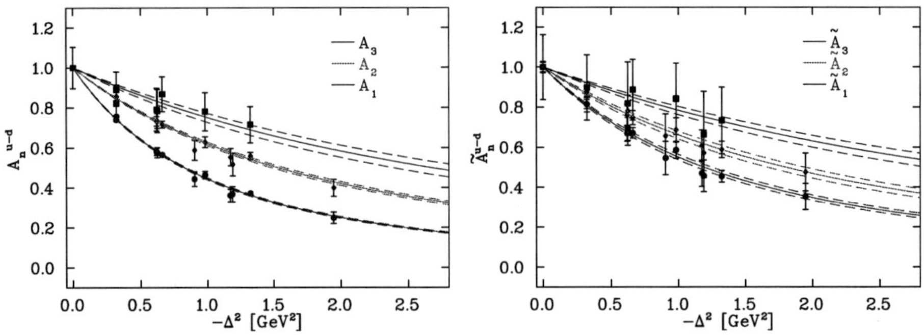

FIGURE 2. The nonsinglet generalized form factors A_n^{u-d} and $\tilde{A}_n^{u-d}$, together with a dipole fit.

All we have to do is find $C_q(x,\Delta^2)$ by an inverse Mellin transform. The corresponding GPD is then given by the convolution integral

$$H_q(x,\Delta^2,Q^2) = \int_x^1 \frac{dy}{y} C_q(\frac{x}{y},\Delta^2)\, q(x,Q^2). \tag{9}$$

In Fig. 2 I show the Δ^2 dependence of the first three nonsinglet generalized form factors A_n^{u-d} and $\tilde{A}_n^{u-d}$, respectively, for some particular coupling and quark mass. The form factors are well described by the dipole ansatz

$$A_n^q(\Delta^2) = \frac{A_n^q(0)}{(1-\Delta^2/M_n^2)^2}. \tag{10}$$

The solid lines show a dipole fit, and the dashed lines indicate the error. We note here that the various form factors are well separated, and that their slopes decrease with increasing n. This does not come unexpectedly. For $x \to 1$ we expect $H_q(x,\mathbf{b}_\perp^2,Q^2)$ to approach a δ-like function in $\mathbf{b}_\perp$. This requires that $A_n(\Delta^2) \to const.$ as $n \to \infty$. Fitting the form factors with a dipole is purely phenomenological. It provides us with a means to measure the change in slope by monitoring the extracted dipole masses. In Fig. 3 I show the dipole masses M_n and (for the polarized case) $\tilde{M}_n$ as a function of n. Our results obtained so far support a Regge-like behavior

$$M_n^2 = c + n/\alpha'. \tag{11}$$

An important issue is the extrapolation to the chiral limit. In lack of any guidance from chiral perturbation theory, we have to rely on more or less phenomenological fits. At the physical point we find $c \approx -0.5$ GeV^2 and $1/\alpha' \approx 1.1$ GeV^2. The singlet form factors give generally different c's and α''s.

Assuming that the Regge behavior continues to hold for the higher moments as well, we can compute $C_q(x,\Delta^2)$. We find

$$C_q(x,\Delta^2) = \delta(1-x) + \alpha'\Delta^2\left(2+\alpha'\Delta^2 \ln x\right) x^{\alpha'(c-\Delta^2)}. \tag{12}$$

Taking $q(x,Q^2)$ from the literature and Fourier transforming (9) to impact parameter space, we finally obtain $H_q(x,\mathbf{b}_\perp^2,Q^2)$. In Fig. 4 I show the distribution of the valence u

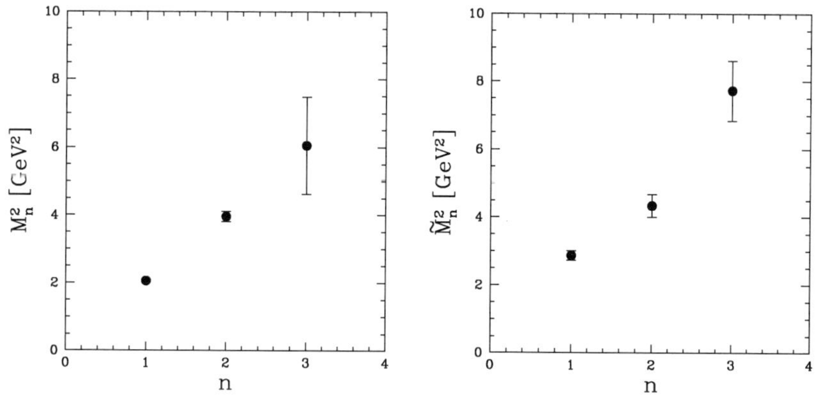

FIGURE 3. The dipole masses M_n and $\tilde{M}_n$ extracted from the generalized form factors shown in Fig. 2

quark in the proton at the scale $Q^2 = 4\ \mathrm{GeV}^2$ as a function of x and the impact parameter $\mathbf{b}_\perp$. It appears that at small x the proton looks like a doughnut, being largely empty in the middle, while at larger values of x the hole closes and the distribution becomes increasingly narrow. Similar results are found for $\tilde{H}_q(x, \mathbf{b}_\perp^2, Q^2)$.

Let us discuss the behavior at large x in some more detail now. For $x \to 1$ we find

$$H_q(x, \mathbf{b}_\perp^2, Q^2) \simeq const.\, \mathrm{e}^{-\mathbf{b}_\perp^2/4\alpha'(1-x)}\, q'(x, Q^2). \tag{13}$$

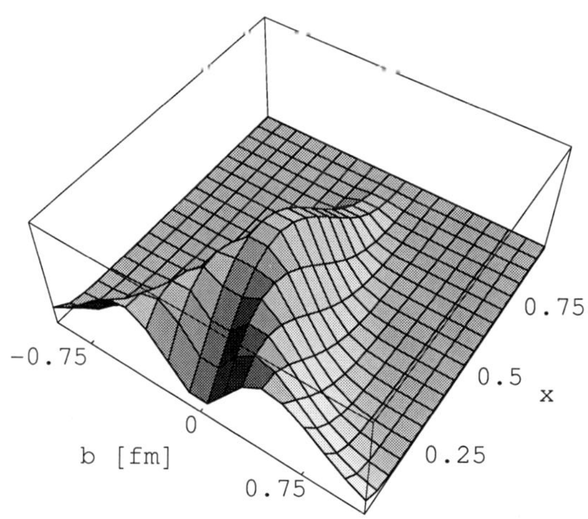

FIGURE 4. The generalized parton distribution $H_u(x, \mathbf{b}_\perp^2, Q^2)$ of the proton as a function of x and impact parameter $b = |\mathbf{b}_\perp|$ at $Q^2 = 4\ \mathrm{GeV}^2$.

Assuming that $q'(x,Q^2)$ remains finite as $x \longrightarrow 1$, we are then led to

$$\langle \mathbf{b}_\perp^2 \rangle \simeq 4\,const.\,(1-x). \tag{14}$$

While the impact parameter is the transverse distance between the struck parton and the center of momentum of the nucleon, an estimate of the transverse size $\mathbf{r}_\perp$ of the nucleon is provided by the distance between the struck quark and the system of spectators. Its average square is given by [8]

$$\langle \mathbf{r}_\perp^2 \rangle = \frac{\langle \mathbf{b}_\perp^2 \rangle}{(1-x)^2}. \tag{15}$$

Hence, the average transverse size, $r = \sqrt{\langle \mathbf{r}_\perp^2 \rangle}$, of the nucleon diverges like $(1-x)^{-1/2}$ in the limit $x \longrightarrow 1$. This behavior can certainly not be reconciled with confinement. Perhaps our dipole ansatz for the generalized form factors is too crude. Or it was premature to assume a Regge-like behavior of the dipole masses. Future calculations will have to tell.

CONCLUSIONS

I have presented first lattice results for the lowest three moments of GPDs for $N_f = 2$ flavors of dynamical quarks. Our findings, though very crude yet, kill already many models of the spatial structure of the nucleon in the literature. There are still many problems to overcome. The most immediate challenges are to extend these calculations to more realistic quark masses, so that controlled extrapolations to the chiral limit can be performed, and to higher moments.

ACKNOWLEDGMENTS

I wish to thank my collaborators M. Göckeler, P. Hägler, R. Horsley, D. Pleiter, P. Rakow, W. Schroers and J. Zanotti for a most pleasant and profitable collaboration. Furthermore, I like to thank the organizers of HiX2004, and in particular Jacques Soffer, for a most productive and pleasant workshop.

REFERENCES

1. D. Müller et al., Fortsch. Phys. 42, 101 (1994); X. Ji, Phys. Rev. Lett. 78, 610 (1997); A.V. Radyushkin, Phys. Rev. D56, 5524 (1997); M. Diehl et al., Phys. Lett. B411, 193 (1997); X. Ji, J. Phys. G24, 1181 (1998); J. Blümlein, B. Geyer and D. Robaschik, Nucl. Phys. B560, 283 (1999).
2. M. Diehl, Eur. Phys. J. C25, 223 (2002); M. Burkardt, Int. J. Mod. Phys. A18, 173 (2003).
3. M. Göckeler et al., hep-ph/0304249.
4. P. Hägler et al., Phys. Rev. D68 (2003) 034505.
5. M. Göckeler et al., Nucl. Phys. B (Proc. Suppl.) 128, 203 (2004); M. Göckeler et al., hep-lat/0409162; M. Göckeler et al., hep-lat/0410023.
6. M. Burkardt, Ref. [2].
7. G. Martinelli et al., Nucl. Phys. B445, 81 (1995); M. Göckeler et al., Nucl. Phys. B544, 699 (1999).
8. M. Burkardt, Phys. Lett. B595, 245 (2004).

Nucleon Structure Functions and Nuclear DIS

Alexander Molochkov

Institut für Theoretische Physik, Justus-Liebig Universität, Giessen, Germany
Far Eastern National University, Vladivostok, Russia

Abstract. The nucleon structure study in nuclear deep inelastic scattering is considered. It is shown that nuclear data provide a new source of information about dynamics of parton distributions in the nucleon. An exmple of the neutron structure function extraction from the deuteron and proton data is considered. The limit $x \rightarrow 1$ of the neutron to proton structure functions ratio is studied. A link between the deep inelastic scattering off the nucleon at high x and elastic scattering off nuclei in high Q^2 region is discussed.

NUCLEON STRUCTURE CHANGE IN NUCLEI

The nuclear data is one of the important sources of information about nucleon structure. First of all, it is connected with absence of a free neutron target for deep inelastic scattering (DIS) experiments, what makes nuclear data the only source of information about neutron structure functions. Another important advantage of the nuclear DIS is access to the kinematic region of large Bjorken x that is unreachable in DIS off the nucleon. For example, the average nuclear Bjorken x ($x_A = 1/A$) corresponds to the very large nucleon x ($x_{\rm N} = M_A/m_{\rm N}x_A \simeq 1$). However, such study cannot be performed without consistent separation of the nuclear and nucleon hard structure. The European Muon Collaboration demonstrated in the nuclear DIS experiments that connection between the nuclear and nucleon hard structure cannot be trivially explained by the Fermi-motion of bound nucleons [1], what was called the EMC-effect. Study of the A-dependence of the effect [2] and nuclear Drell-Yan scattering [3] together with the theoretical investigations performed within the QMC [4] and Chiral soliton [5] models show that the EMC-effect cannot be understood without introducing nucleon structure change in nuclei, which is strongly model dependent. Recently, the nuclear effects in DIS were studied within a fully covariant approach [6] based on the Bethe-Salpeter equation. In the paper [7] it was found that the EMC-effect results from the Fermi-motion of the bound nucleon in the four-dimensional space, which leads to the space- and time-smearing of the bound nucleon. The time-smearing leads to increase of the nucleon localization radius in the four-dimensional space, what corresponds to the explanation of the effect provided by the Q^2–rescaling model [8]. In a 3D projection of the relativistic Fermi-motion the time-smearing of the bound nucleon reduces to the dynamical distortions of the nucleon structure, which correspond to the nucleon structure change obtained within the QMC [4] and Chiral soliton models [5].

Consistent analytical calculations within this approach in the approximations of high Q^2 and small binding energy give the following expression for the nuclear structure

CP747, *Structure of the Nucleon at Large Bjorken x,* edited by C. Bourrely, Z.-E. Meziani, and J. Soffer

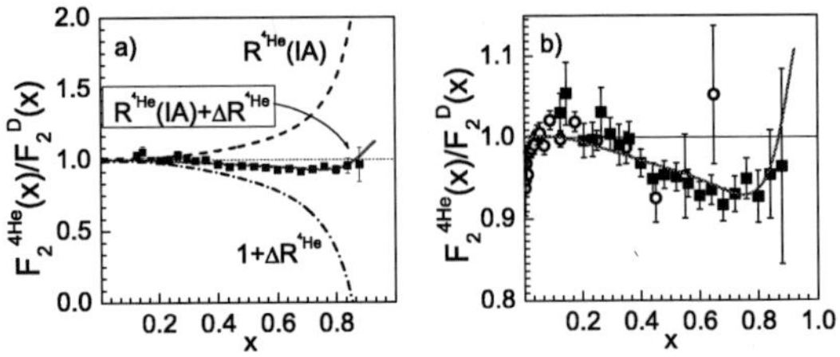

FIGURE 1. Contribution of the impulse approximation (dashed curve $R^{^4He}(IA)$) and nucleon structure function derivative (dashed-dot curve $\Delta R^{^4He}$) to the ratio of the ^{4}He and D structure functions. The experimental values are shown by the dark squares [2].

function:

$$F_2^A(x) = \int \frac{d^3p}{(2\pi)^3} \left[\frac{M_A - E_{A-1} - p_3}{E_{\rm N}} F_2^{\rm N}(x_{\rm N}) - \frac{\Delta_A^{\rm N}}{E_{\rm N}} x_{\rm N} \frac{dF_2^{\rm N}(x_{\rm N})}{dx_{\rm N}} \right] \frac{f^{{\rm N}/A}(M_A, \mathbf{p})}{8M_A E_{\rm N} E_{A-1} \Delta_A^{{\rm N}2}}, \quad (1)$$

where $f^{{\rm N}/A}(M_A,\mathbf{p})$ is a function defined by the solutions of the Bethe-Salpeter equation, $E_{\rm N}$ is the nucleon on-shell energy, $\Delta_A^{\rm N} = M_A - E_{\rm N} - E_{A-1} \simeq -T_{kin} + \varepsilon$ is the nucleon energy change due to the Fermi motion ($T_{kin} > 0$ – kinetic energy of the Fermi motion) and binding ($\varepsilon < 0$ – binding energy of the nucleon).

Numerical calculations of the contributions of the different terms in Eq.(1) to the ratio $R^{^4\rm He} = 2\sigma^{^4\rm He}/(A\sigma^{\rm D}) = F_2^{^4\rm He}/F_2^{\rm D}$ are presented at Fig.1 a). The first term ($R^{^4\rm He}(IA)$), which results from the Fermi motion in 3D-space, is presented by the monotone increasing dashed curve. The second term ($\Delta R^{^4\rm He}$) depicted by the point-dashed curve results from the time-smearing of the bound nucleon. Since $M_A < E_{\rm N} + E_{A-1}$ and $F_2^{\rm N}(x_{\rm N})/dx_{\rm N} < 0$ this term puts the ratio $R^{^4\rm He}$ bellow unity in the middle x region and, therefore, produces the EMC-effect resulted from cancellation of the two large contributions. This result is presented by the solid curve, which fits to the experimental data within the experimental errors with rather good accuracy (see Fig.1 b)). Due to the contribution of $dF_2^{\rm N}(x)/dx$ the expression (1) establishes a connection between the nuclear data and dynamics of parton distributions in the nucleon expressed by the nucleon structure function derivative.

It is worth noting that according to (1) the functional behavior of the EMC-ratio is defined by the nucleon structure function properties, since then the duality observed in $F_2^{\rm N}$ can lead to the dual behavior of the R^A in the kinematic region of high and low Q^2. This conclusion is consistent with the experimental results obtained recently in [9].

NEUTRON STRUCTURE FUNCTION

Let's consider Eq.(1) as an integral equation with the unknown function $F_2^{\rm n}$. The functions $F_2^{\rm p}$ and $F_2^{\rm D}$ are defined by the existing experimental data. The boundary conditions

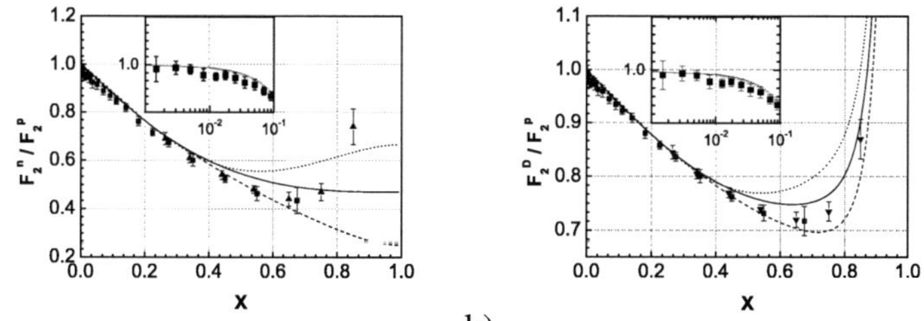

FIGURE 2. Comparison of the obtained $F_2^{\rm n}(x)$ with experimental data for the a) $F_2^{\rm n}/F_2^{\rm p}$ and b) $F_2^{\rm D}/F_2^{\rm p}$ ratios. Experimental data are taken from SLAC (triangles) and NMC (squares).

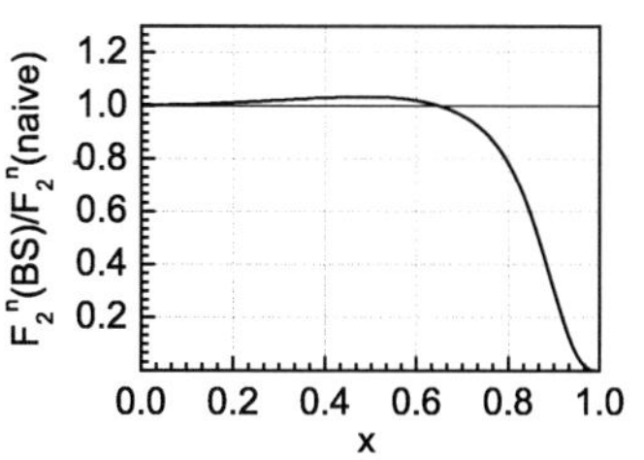

FIGURE 3. The ratio of the $F_2^{\rm n}(x)$ extracted within the present approach to the naive approximation $F_2^{\rm n}(x) = 2F_2^{\rm D}(x) - F_2^{\rm p}(x)$.

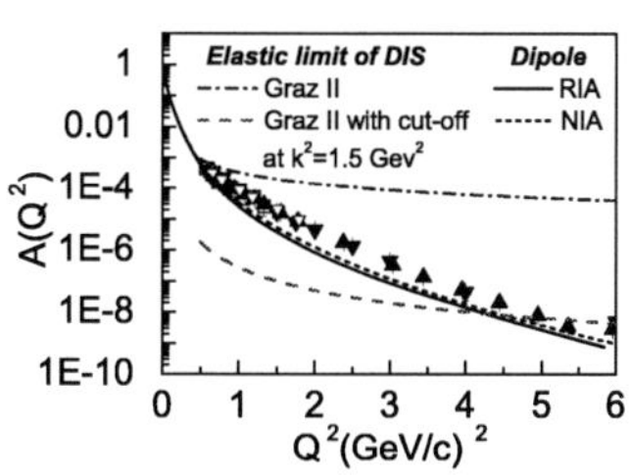

FIGURE 4. Comparison of the elastic limit of $F_2^{\rm D}(x,Q^2)$ $(x = 2)$with the deuteron structure function $A(Q^2)$.

for the equation are fixed with the help of asymptotics of the nucleon structure functions at $x = 0$ and $x = 1$. Introducing the neutron structure function $F_2^{\rm n}(x)$ as follows:

$$F_2^{\rm n}(x,Q^2) = r^{\rm n/p}(x)F_2^{\rm p}(x,Q^2); \quad r^{\rm n/p}(x) = a_1(1-x)^{\alpha_1} + a_2x^{\alpha_2} + b_1x^{\beta_1}(1-x)^{\beta_2}(1+c_1x^{\gamma_1}), \tag{2}$$

we fix the parameters a_1, a_2, α_1, b_1, β_2 by the boundary conditions. According to the limit $F_2^{\rm p}(0) = F_2^{\rm n}(0)$, the parameter $a_1 = 1$. The parameter a_2 is fixed according to $lim_{x\to 1}F_2^{\rm n}(x)/F_2^{\rm p}(x)$, which is model dependent. We study the three limits predicted by the following three different models. The quark model with SU(6) symmetry gives $r^{\rm n/p}_{x=1} = 2/3$. The elastic limit gives $r^{\rm n/p}_{x=1} = \sigma^{\rm n}_{\rm elastic}/\sigma^{\rm p}_{\rm elastic} \simeq 0.47$. The minimal value of the ratio has been predicted by the model with SU(6) symmetry breaking with scalar diquark dominance — $r^{\rm n/p}_{x=1} = 1/4$. Assuming same asymptotic behavior for $F_2^{\rm p}(x)$ and $F_2^{\rm n}(x)$ at the limit $x \to 1$: $\lim\limits_{x\to 1} F_2^{\rm p}(x) \simeq C_{\rm p}(1-x)^3$, $\lim\limits_{x\to 1} F_2^{\rm n}(x) \simeq C_{\rm n}(1-x)^3$; we get that the derivative of $r^{\rm n/p}$ at $x = 1$ is zero. It constrains the following parameters of Eq. (2): $\alpha_1 = 1$, $\beta_2 = 1$, $b_1 = (\alpha_2 a_2 - 1)/(1+c_1)$. The remaining four parameters (α_2, β_1, c_1, γ_1) are used to fit Eq. (1) to the data for $F_2^{\rm D}(x,Q^2)$ in the range $10^{-3} < x < 0.6$ and $0.5GeV^2 < Q^2 < 40GeV^2$. The result of the $F_2^{\rm n}$ extraction is presented at Fig.2 a). The experimental data are obtained from the ratio $F_2^{\rm D}/F_2^{\rm p}$ within the naive approximation that neglect all nuclear effects in the deuteron. Due to the cancellation of the Fermi-motion and time-smearing discussed in the previous section the naive approximation works well up to $x \simeq 0.7$, what explains the good agreement of the obtained here $F_2^{\rm n}$ with the experimental data. This agreement is illustrated on Fig.3, which shows that at $x > 0.7$ the correct accounting of the Fermi motion in 4D space becomes very important.

The obtained nucleon structure function that contains inelastic channels only allows to study contribution of these channels to the elastic lepton scattering off the deuteron. Calculation of the deuteron structure function at the point $x = 2$, which corresponds to the elastic scattering off the deuteron, we get the result presented at the Fig.4. Thus, the nucleon resonances, contribution of which is presented by the dashed curve, can give sizable contribution to the lepton deuteron elastic scattering at high Q^2.

SUMMARY

The nuclear short-range structure can be expressed in terms of the nucleon structure and four-dimensional Fermi motion of the nucleons. The time-smearing broadens the bound nucleon localization area, what lead to the observation of the nucleon structure change in nuclei – EMC-effect. The pattern of the EMC-effect is defined by cancellation of the two large contributions that are the 3D Fermi motion and the time-smearing of the bound nculeons, which are important at the average and large Bjorken x. Due to the time-smearing the nuclear data enable direct access to the information about nucleon structure dynamics expressed by the nucleon structure function derivative, $dF_2^{\rm N}(x)/dx$, that is difficult to detect in DIS off the nucleon. Better accuracy in $F_2^{\rm D}/F_2^{\rm p}$ data at $x \simeq 0.7-0.8$ can give information about the ratio $F_2^{\rm n}/F_2^{\rm p}$ at $x \to 1$ (see Fig.2 b)). Study of the elastic limit of $F_2^{\rm D}$ shows that nucleon structure excitations give sizable contribution to lepton deuteron elastic scattering at $Q^2 > 4GeV^2$.

At the end I would like to thank Alexander von Humboldt Foundation for financial support of my participation in this meeting. I would like to thank also G.I. Smirnov and U. Mosel for useful discussions. I am grateful to the organizers of the Workshop for support and for warm hospitality during the meeting.

REFERENCES

1. EMC, J.J. Aubert et al., Phys. Lett. **B123**, 275 (1983).
2. SLAC, J. Gomez et al., Phys. Rev. **D49**, 4348 (1994).
3. D.M. Alde, et al., Phys. Rev. Lett. **64** (1990) 2479.
4. K. Saito and A.W. Thomas, Nucl. Phys. **A574**, 659 (1994); K. Tsushima et al., nucl-th/0301078.
5. G.A. Miller, J.R. Smith, Phys. Rev. **C65**, 015211 (2001);
6. V.V. Burov, A.V. Molochkov, G.I. Smirnov, Phys. Lett **B466**, 1 (1999).
7. A. Molochkov, Nucl. Phys. **A666**, 169 (2000); A. Molochkov, e-print, nucl-th/0407077.
8. F.E. Close, R.L. Jaffe, R.G. Roberts, G.G. Ross, Phys.Rev. **D31**, 1004, (1985).
9. J. Arrington, et al., e-print, nucl-ex/0307012.

Mapping High x Structure of the Nucleon: which data is missing?

G.I. Smirnov

Joint Institute for Nuclear Research, Dubna, Russia
and
Université Blaise Pascal, Clermont-Ferrand, France

Abstract. The analysis of numerous experiments on lepton-nucleon and lepton-nucleus deep inelastic scattering reveals several non-trivial features of the nucleon structure, which are particulary interesting in the range of large Bjorken x ($x > 0.8$). It is shown that new data form the lightest nuclei can provide desisive information needed for the understanding of the nucleon structure.

INTRODUCTION

Understanding of the physical processes related to the nucleon parton distributions in the high x regime remains a topical problem for the last 30 years. It is of importence for the comprehension both of the nucleon structure and of the high-energy nucleon-nucleon interections that might lead to the formation of the quark-gluon plasma. This is why the first suggestion for the meausurements of the nucleon structure function $F_2(x,Q^2)$ on nuclear targets in the region of $x > 1$ has been put forward even before precise data on $F_2(x,Q^2)$ became available in the medium x range [1].

Two important effects were discovered by means of the deep inelastic scattering (DIS) of leptons on nuclei since that time: (a) appreciable difference between the distributions of u and d quarks in the nucleon and (b) modification of $F_2(x,Q^2)$ in the nuclear environment (EMC effect). The two discovered effects involved the dynamics of the parton distributions in nucleon and nuclei, which is very hard to study experimentally. One of the problem is related with a fast decrease in the DIS cross section when $x \to 1.0$, whereas the second one is more fundumantal: experiments with targets containig free neutrons are impossible. As a result, the measurements of the parton distributons in the high x region become strongly model dependent. There also exists a common prejudice that modifications of the nucleon (neutron) structure in nuclear environment can be related to the variations in nuclear density. This unjustified assumption always leads to uncontrolable systematic errors in the analysis of the nucleon structure. In what follows, I show that it is the physics of a few nucleon interactions which can provide full understanding of the relations between the structure of free and bound nucleons. This will require new data from DIS on the lightest nuclei and an accurate check of consistency in theoretical description of the data from several nuclear targets.

CP747, *Structure of the Nucleon at Large Bjorken x*, edited by C. Bourrely, Z.-E. Meziani, and J. Soffer

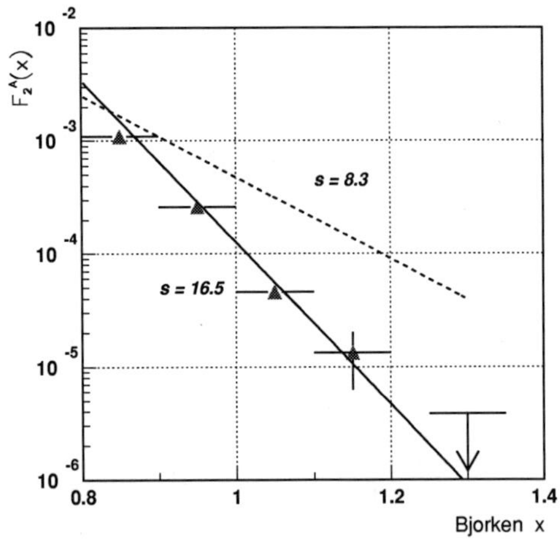

FIGURE 1. Nucleon structure function measured by the BCDMS [2] on a carbon target (data points) in the range of large x. Full and dashed lines show approximation of $F_2^A(x)$ with $\exp(-sx)$ and parameters $s = 16.5$ (BCDMS) and 8.3 (CCFR), respectively.

STRUCTURE FUNCTIONS FROM NUCLEAR TARGETS DATA

Measurements of $F_2^A(x)$ beyond $x = 1$. Measurements of the nucleon structure function in the domain of large x and high Q^2 became feasible in the muon beam of CERN SPS owing to high luminosity of the BCDMS set-up. The results obtained on a 40 m length carbon target at $Q^2 = 150$ GeV^2 are displayed in Fig. (1) [2]. Another experiment has been realyzed in the Fermilab neutrino beam by the CCFR [3]. The two experiments are in disagreement, which can be partialy due to systematic errors caused by substantial smearing corrections. In the case of the BCDMS, such corrections have been checked by replacing carbon taget with the hydrogen one, which was unfeasible for the CCFR. In principle, the resolution of the BCDMS-type set-up can be further improved by employing a target calorimeter [4].

Measurements of $F_2^A(x)$ at $x < 0.9$. Experiments on the lepton-nucleus scattering [5] yielded valuable data, which allowed to establish a pattern [6] of the high x behavior of nuclear structure functions $F_2^A(x)$ with respect to the deuteron structure function $F_2^{\mathrm{D}}(x)$. It is well to bear in mind, that the first nucleus with reasonable statistics, as it is illustrated in Fig. (2), was ^{9}Be. Therefore, a crucial link with the ligthest nuclei, which should carry most important information about modifications of $F_2^p(x)$ and $F_2^n(x)$ in nuclear environment was missing.

One can try to describe the entire set of data using models of the modificaton of the nucleon structure, which was in fact done in numerous publications, but with a

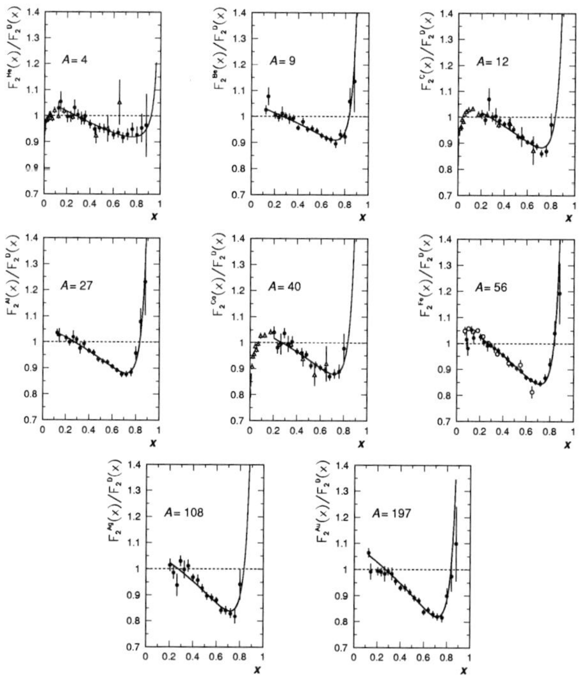

FIGURE 2. High x behavior of F_2^A/F_2^D, measured by SLAC – •, BCDMS – ○ and NMC – △. Full lines correspond to theoretical calculations [7] for ^{4}He (first panel) and for ^{3}He scaled according to nuclear density effects given by the Woods-Saxon potential ($9 \leq A \leq 197$).

little success if we require a consistent description of the high x range. On the other hand, the data accumulated on nuclei with $A > 4$ clearly demonstrate that the pattern of $F_2^A(x)/F_2^D(x)$ does not change as a function of A copying that obtained theoretically for $A = 3$ [6]. This means that the deviation of the ratio from unity is due to nuclear density effects, whereas the physics of the observed x dependence is fully determined by the topology of a few-nucleon system, and the 3N system is the most preferable candidate.

EXPERIMENTS THAT MUST BE DONE

Hence there remains a white spot in the region of nuclei, which is of the particular interest — $A \leq 4$. The theoretical points [7] for the high x cross-over shown in Fig. (3) for A= 3 and 4 should be confronted with future data. The points evaluated from available data are virtually independent of A as it should be if only nuclear density effects determine modifications, whereas the theory predicts $1 - x_3 \approx 1/6$ and $1/12$ for $A = 3$ and 4, respectively. Thus, new experiments should be realized in order to obtain data on:

- ^{3}He (for the fist time);
- ^{4}He (with better statistics, in particular, around $x = 0.9$);
- Heavy nuclei (with improved accuracy at $x \sim 0.84$ and, if feasible, up to $x \sim 1$).

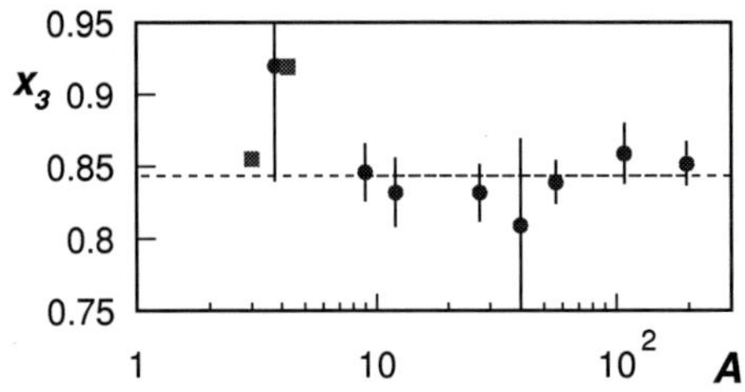

FIGURE 3. The coordinate of the cross-over of the ratio F_2^A/F_2^D in the high x region (closed circles) [6] and theoretical prediction [7] (closed squares). The average value $x_3 = 0.84$ is shown with the dashed line.

SUMMARY

High statistics data which can be collected on D, ^{3}He and ^{4}He targets at $x \sim 0.8$ and beyond are of special interest because they can confirm existance of two different types of x dependence of the nucleon structure function $F_2(x)$ measured in 3N and 4N systems. This will prove, in turn, that the modification of the nucleon structure in nuclear environment developes in two stages, the first of which is limited by formation of the dense 4N system — the helium nucleus. The practical merit which follows from this finding will be utilisation of the measurements on heavy nuclei, where the second stage of modifications developes through nuclear density effects, in mapping $F_2(x)$ of the 3N system in the high x region.

ACKNOWLEDGMENTS

I am grateful for the financial support provided by the Organizing Committee of the HiX2004 Workshop. I whish to thank V. V. Burov and A. M. Molochkov with whom modifications of the nucleon structure function in ^{3}He and ^{4}He have been obtained.

REFERENCES

1. A.M. Baldin et al., Sov. J. Nucl. Phys., **18**, 41–44 (1974).
2. BCDMS, A. Benvenuti et al., Zeit. Phys. C, **63**, 29–36 (1994).
3. CCFR, M. Vakili et al., Phys. Rev. D, **61**, 052003 (2000).
4. S.P. Baranov et al., Dubna, JINR, E1-524 (1985).
5. M. Arneodo, Phys. Rep., **240**, 301–393 (1994).
6. G.I. Smirnov, Eur. Phys. J. C, **10**, 239–247 (1999).
7. V. Burov, A. Molochkov and G. Smirnov, Phys. Lett. B, **466**, 1–10 (1999).

FUTURE AND SUMMARY

Probing correlations in the large x proton wave function at pp and pA colliders

Mark Strikman

104 Davey lab, Penn State University, University Park, PA16802, U.S.A.

Abstract.
We discuss several examples of how the transverse spatial distribution of partons in the nucleon, as well as multiparton correlations, can be probed by observing hard processes (dijets) in high–energy pp ($\bar{p}p$) and pA (dA) collisions. Such studies can complement the information gained from measurements of hard exclusive processes in ep scattering. The transverse spatial distribution of partons determines the distribution over pp impact parameters of events with hard dijet production. Correlations in the transverse positions of partons can be studied in multiple dijet production. We find that the correlation cross section measured by the CDF Collaboration, $\sigma_{\text{eff}} = 14.5 \pm 1.7^{+1.7}_{-2.3}$ mb, can be explained by "constituent quark" type quark–gluon correlations with $r_q \approx r_N/3$, as suggested by the instanton liquid model of the QCD vacuum. Longitudinal and transverse multiparton correlations can be separated in a model–independent way by comparing multiple dijet production in pp and pA collisions.

So far major successes of QCD in describing high energy hadron hadron collisions were for hard inclusive processes - collision of two partons. To describe these processes it is sufficient to know only longitudinal single parton densities.

At the same time the knowledge of the transverse spread of partons, longitudinal and transverse correlations of partons, their dependence on flavor, x polarization of the parton, are necessary for building a realistic description of the global structure of the final states in pp collisions, understanding microscopic structure of nucleon bound state.

Hard exclusive processes in ep scattering allow to probe not only the distribution of partons with respect to longitudinal momentum, but also their spatial distribution in the transverse plane. Examples include the hard electroproduction of light mesons and real photons (deeply virtual Compton scattering), as well as the photoproduction of heavy quarkonia (J/ψ, ψ', Υ). Thanks to QCD factorization theorems the amplitudes for these processes can be separated into a "hard" part, calculable in perturbative QCD, and "soft" parts characterizing the non-perturbative structure of the involved hadrons. The information about the nucleon is contained in so-called generalized parton distributions (GPD's). These are functions of the parton momentum fractions, x and x', as well as of the invariant momentum transfer to the nucleon, t, and thus combine aspects of the usual parton distributions, measured in inclusive deep–inelastic scattering, with those of the elastic nucleon form factors. For $x = x'$, their Fourier transform with respect to t describes the spatial distribution of partons in the transverse plane. The GPD's thus, in a sense, provide us with a "3D parton image" of the nucleon.

Extensive data on VM production from HERA support dominance of the pQCD dynamics. Numerical calculations including transverse size of the wave function of vector

CP747, *Structure of the Nucleon at Large Bjorken x*, edited by C. Bourrely, Z.-E. Meziani, and J. Soffer

meson explain key elements of high Q^2 data. In particular, they confirm prediction of [1] that t-slope of convergence of the slopes of the ρ-meson diffractive electroproduction at large Q^2 to the slope of the J/ψ meson photo(electro) production and describe the Q^2 dependence of the difference of the slopes [2]. Hence the t-dependence of J/ψ photo/electro production is given (up to small corrections) by the square of two - gluon form factor of the nucleon, which is related to the transverse distribution of the gluons at given x as:

$$F_g(x,t) = \int d^2\rho \, e^{i(\vec{\Delta}_\perp \vec{\rho})} \, F_g(x,\rho), \qquad (t = -\vec{\Delta}_\perp^2). \tag{1}$$

The function $F_g(x,\rho)$ is normalized as $\int d^2\rho \, F_g(x,\rho) = 1$. Concerning the shape of the two-gluon form factor, it has been argued that at $x \geq 10^{-1}$ (where pion cloud contributions are absent) the two–gluon form factor should follow the axial form factor of the nucleon, and thus be described by the dipole parameterization

$$F_g(x,t) = (1 - t/m_g^2)^{-2}, \qquad m_g^2 = 1.1\,\text{GeV}^2 \qquad (x \geq 10^{-1}). \tag{2}$$

This form indeed describes well the t–dependence of the fixed–target J/ψ photoproduction experiments. Increase of the slope between $x \sim 0.1$ and $x \sim 0.01$ appears to be mostly due to the scattering off the pion field [3] Based on Eq. (2), we have suggested in Ref. [5] a generalized dipole parametrization valid also at small x, which incorporates the observed increase in the gluonic transverse size (as well as the effects of DGLAP evolution) by way of an x– and Q^2–dependent of the dipole mass parameter, $m_g^2(x, Q^2)$. This simple parametrization conveniently summarizes our knowledge of the transverse spatial distribution of gluons from ep scattering. For details, see Refs. [5, 6].

Exclusive processes in ep scattering, however, are not the only reactions which probe the "3D" parton distributions. In fact, a lot more information about the longitudinal momentum and transverse spatial distribution of partons, as well as about multiparton correlations, can be obtained from the study of selected hard processes in (not necessarily exclusive) pp and pA scattering. Comparative studies of ep and pp/pA induced hard processes will help to improve the quantitative description of both classes of processes and offer many new, fascinating insights into the partonic structure of the nucleon.

Turning now to pp collisions, an immediate application of the transverse spatial distribution of partons is in the description of the impact parameter dependence of the cross section for hard dijet production [5]. In a pp collision with c.m. energy $\sqrt{s}$, a hard dijet with transverse momentum $q_\perp$ at zero rapidity is produced in the collision of two partons carrying momentum fractions x_1,x_2. The probability for such a parton–parton collision, as function of the impact parameter of the pp system, b, is given by the convolution of the spatial distributions of the partons; for gluons

$$P_2(b) \equiv \int d^2\rho_1 \int d^2\rho_2 \, \delta^{(2)}(\vec{b} - \vec{\rho}_1 + \vec{\rho}_2) \, F_g(x_1,\rho_1) \, F_g(x_2,\rho_2). \tag{3}$$

The scale of the parton distributions here is $q_\perp^2$.

While exclusive processes in ep scattering provide in principle the cleanest way to access the transverse spatial distribution of partons, there are several instances in which

pp scattering is more effective. One is the study of large x, where the cross sections for exclusive processes in ep are small. Besides, in this kinematics, when $x \sim 1$ are probed, one is mostly sensitive to the Feynman mechanism; see the discussion above. In this case the global transverse distribution of matter can be measured more directly using various reactions combining a soft and hard trigger, in particular in connection with pA collisions [7]. New opportunities for such studies will emerge at LHC, where the high luminosity will allow, for example, to compare the characteristics of W^+ and W^- production at the same forward rapidities, corresponding to relatively high x where the d/u ratio deviates strongly from the naive value 1/2. By studying the accompanying production of hadrons one can learn which configurations in the nucleon have larger transverse size — those with a leading u–quark or with a leading d-quark. One suitable observable is, for example, the distribution of the number of events over the number of the produced soft particles. A larger transverse size corresponds to a larger probability of soft interactions, and hence to a larger probability of events with large multiplicity. It is interesting to note that the studies of the associated soft hadron multiplicity in the production of $W^\pm$ and Z bosons in $\bar{p}p$ collisions by the CDF collaboration at Fermilab find an increase of this multiplicity by a factor of two as compared to generic inelastic events [8]. This appears natural if one takes into account that the hard quarks producing the weak bosons have a narrower transverse spatial distribution than the soft partons. As a result, the average impact parameters in events with weak boson production are much smaller than in generic inelastic collisions, leading to an enhancement of multiple soft and semi-hard interactions [5].

Single parton densities and GPDs do not carry information about longitudinal and transverse correlations of partons in the hadron wave function. Such information can be extracted from high energy pp and pA collisions where two (or more) pairs of partons can collide to produce multiple dijets, with a kinematics distinguishable from those produced in $2 \to 4$ parton processes. Since the momentum scale of the hard interaction, p_t, corresponds to much smaller transverse distances in coordinate space than the hadronic radius, in a double parton collision the two interaction regions are well separated in transverse space. Experimentally, one measures the ratio

$$\frac{\frac{d\sigma(p+\bar{p}\to jet_1+jet_2+jet_3+\gamma)}{d\Omega_{1,2,3,4}}}{\frac{d\sigma(p+\bar{p}\to jet_1+jet_2)}{d\Omega_{1,2}} \cdot \frac{d\sigma(p+\bar{p}\to jet_3+\gamma)}{d\Omega_{3,4}}} = \frac{f(x_1,x_3)f(x_2,x_4)}{\sigma_{eff} f(x_1)f(x_2)f(x_3)f(x_4)}, \tag{4}$$

where $f(x_1,x_3), f(x_2,x_4)$ are the longitudinal light-cone double parton densities at the hard scale μ^2 (we assume for simplicity that the virtuality in both hard processes is comparable; in the following equations we suppress dependence on μ^2), and the quantity σ_{eff} can be interpreted as the "transverse correlation area". The variables Ω_i characterize the observed jets (or photons) — their transverse momenta, rapidities, cuts on the opening angle, etc.

Parton correlations can emerge due to nonperturbative effects at a low resolution scale, or due to the effects of QCD evolution. One possible nonperturbative mechanism is the existence of "constituent quarks" within the nucleon, which appear due to the interaction of current quarks with localized non-perturbative gluon fields, resulting in local short–range correlations in the transverse spatial distribution of gluons. The instanton model of the QCD vacuum suggests a constituent quark radius of about $1/3$ the nucleon radius, $r_q \approx r_N/3$. Another nonperturbative mechanism, relevant at small x, are fluctuations of the color field in the nucleon due to the fluctuations of the transverse size of the quark distribution. Perturbative correlations emerge due to small transverse distances in the emission process in the perturbative partonic ladder in DGLAP evolution. Of all the mentioned mechanisms, only the first one is effective at $x \geq 0.05$, where the data of the CDF experiment were collected.

The CDF experiment observed correlation effects in a restricted x range (two balanced jets, and jet plus photon) and found $\sigma_{\text{eff}} = 14.5 \pm 1.7^{+1.7}_{-2.3}$ mb. This value is significantly smaller than the naive estimate obtained by taking a uniform distribution of partons of a transverse size determined by the e.m. form factor of the nucleon, which gives $\sigma_{\text{eff}} \approx 53$ mb, indicating strong correlations between the transverse positions of partons in the transverse plane. The longitudinal correlation between partons in the measured kinematics due to energy conservation is likely to be small, as $x_1 + x_2$ and $x_3 + x_4$ are much smaller than 1. If this effect were important it would likely lead to a suppression of the double parton collision cross section, and hence to an increase of σ_{eff}. However, no dependence of σ_{eff} on x_i was observed in the experiment.

For a more quantitative analysis of the CDF data, we can make use of the information about the transverse spatial distribution of gluons gained from J/ψ photoproduction, as summarized above. Since the x values of the partons probed were reasonably small compared to 1, the simple "geometric" picture of the $\bar{p}p$ collision in transverse position in the spirit of Eq. (3) is justified, and one has

$$\sigma_{\text{eff}} = \left[\int d^2 b \, P_2^2(b) \right]^{-1} . \tag{5}$$

Evaluating this with the dipole parametrization of the two–gluon form factor (2), this comes to $\sigma_{\text{eff}} = 28\pi/m_g^2 \approx 34$ mb. Thus, about 50% of the enhancement compared to the naive estimate of the previous paragraph is due to smaller actual transverse radius of the gluon distribution. Still, our value indicates significant correlations in the transverse positions of the partons. In the kinematics discussed here the relevant partons are both quarks and gluons. We can estimate the effect of correlations assuming that most of the partons are concentrated in a small transverse area associated with the "constituent quarks", as implied by the instanton liquid model of Diakonov and Petrov [9]. Assuming a constituent quark radius of $r_q \sim r_N/3$, we obtain an enhancement factor due to transverse spatial correlations of partons of

$$\frac{8}{9} + \frac{1}{9} \frac{r_N^2}{r_q^2} \sim 1.6 \div 2. \tag{6}$$

This is roughly the value needed to explain the remaining discrepancy with the CDF data. Thus, the combination of the relatively small transverse size of the distribution of large–x gluons and the quark–gluon correlations implied by "constituent quarks" with $r_q \approx r_N/3$ is sufficient to explain the trend of the CDF data. Further studies of multijet events at hadron–hadron colliders, with a broader range of final states, would in principle allow to measure separately quark–quark, quark–gluon, and gluon–gluon correlations for different x.

However, studies based on $\bar{p}p$ or pp collisions alone do not allow for a model–independent separation of transverse and longitudinal correlations. This is possible only in pA collisions at RHIC and LHC. The reason is that the nucleus, having a thickness which practically does not change on the nucleon transverse scale, provides an important contribution which is sensitive only to the longitudinal correlations of hadrons [10]. This is the contribution when two partons of the incident nucleon interact with partons belonging to two different nucleons in the nucleus, σ_2,

$$\sigma_2 \;=\; \sigma^{NN}_{\text{double}} \frac{A-1}{A} \int d^2b\, T^2(b)\, \frac{f(x_1)f(x_2)}{f(x_1,x_2)}. \tag{7}$$

The other term is the impulse approximation — two partons of the incoming nucleon interact with two partons of the same nucleon in the nucleus, σ_1, which is simply equal to A times the cross section of double scattering in pp collisions. Thus, by measuring the ratio of pA and pp double scattering cross sections we can determine $1/\sigma_{\text{eff}}$. For the ratio of double to single scattering terms we find, for $A \geq 12$,

$$R \;=\; \frac{\sigma_2}{\sigma_1} \;\approx\; 0.68 \left(\frac{A}{12}\right)^{0.38} \frac{\sigma_{\text{eff}}}{14\text{ mb}}. \tag{8}$$

Taking the CDF value of $\sigma_{\text{eff}} \sim 14$ mb, we obtain $R \sim 3$ for $A \sim 200$. Thus, the separation of the two terms will be quite straightforward. Even in the case of deuteron–nucleus scattering, which was studied at RHIC recently, the contributions from two partons of one nucleon of the deuteron interacting with two different nucleons in the nucleus remains significant.It constitutes about 50% of the cross section for $A \sim 200$. Hence we conclude that both measurements of pA and dA collisions will allow to measure of σ_{eff} if it is ≥ 5 mb, with pA being a better option. Finally, if σ_{eff} will have been measured in pA collisions, it will be possible to extract the longitudinal two–parton distributions in a model independent way.

To summarize, we have demonstrated that future experiments will be able to measure independently the longitudinal and transverse two–parton distributions in the nucleon. With a detector of sufficiently large acceptance it would be possible to extend these studies even to the case of three parton correlations.

REFERENCES

1. S. J. Brodsky, L. Frankfurt, J. F. Gunion, A. H. Mueller and M. Strikman, Phys. Rev. D **50**, 3134 (1994)
2. L. Frankfurt, W. Koepf, and M. Strikman, Phys. Rev. D**57**, 512 (1998).
3. M. Strikman and C. Weiss, Phys. Rev. D **69**, 054012 (2004)
4. L. Frankfurt and M. Strikman, Phys. Rev. D **66**, 031502 (2002).
5. L. Frankfurt, M. Strikman and C. Weiss, Phys. Rev. D **69**, 114010 (2004).
6. M. Strikman and C. Weiss, arXiv:hep-ph/0408345.
7. L. L. Frankfurt and M. I. Strikman, Nucl. Phys. B **250**, 143 (1985).
8. T. Affolder *et al.* [CDF Collaboration], Phys. Rev. D **65**, 092002 (2002).
9. For a review, see D. Diakonov, Prog. Part. Nucl. Phys. **51**, 173 (2003).
10. M. Strikman and D. Treleani, Phys. Rev. Lett. **88**, 031801 (2002).

Research Perspectives at Jefferson Lab: 12 GeV and Beyond

Kees de Jager

Jefferson Laboratory, Newport News, VA 23606, USA

Abstract. The plans for upgrading the CEBAF accelerator at Jefferson Lab to 12 GeV are presented. The research program supporting that upgrade is illustrated with a few selected examples. The instrumentation under design to carry out that research program is discussed. Finally, a conceptual design of a future upgrade which combines a 25 GeV fixed-target facility and an electron-ion collider facility at a luminosity of up to 10^{35} cm^{-2}s^{-1} and a CM energy of up to 65 GeV.

INTRODUCTION

The design parameters of the Continuous Electron Beam Accelerator Facility (CEBAF) at the Thomas Jefferson National Accelerator Facility (JLab) were defined over two decades ago. Since then our understanding of the behaviour of strongly interacting matter has evolved significantly, providing important new classes of experimental questions which can be optimally adressed by a CEBAF-type accelerator at higher energy. The original design of the facility, coupled to developments in superconducting RF technology, makes it feasible to triple the initial design value of CEBAF's beam energy to 12 GeV in a cost-effective manner (at about 15% of the cost of the initial facility).

The research program with the 12 GeV upgrade will provide breakthroughs in two key areas: (1) gluonic excitations of mesons and the origin of quark confinement and (2) a direct exploration of the quark-gluon structure of the nucleon. The upgrade will provide important advances in two additional areas: (3) a unified description of high-density cold nuclear matter and (4) measurements which test the Standard Model. An overview of the upgrade research program is given in its pre-Conceptual Design Report [1].

ACCELERATOR

At present CEBAF accelerates electrons to 6 GeV by recirculating the beam four times through two superconducting linacs, each producing an energy gain of 600 MeV per pass. Both linac tunnels provide sufficient space to install five additional newly designed cryomodules. The new cryomodules will each provide over 100 MV (compared to the 28 MV from the existing ones), by increasing the gradient to 20 MV/m and the number of cavity cells from five to seven. This will result in a maximum energy gain per pass of 2.2 GeV, providing a maximum beam energy to Halls A, B and C of 11 GeV. The

CP747, *Structure of the Nucleon at Large Bjorken x*, edited by C. Bourrely, Z.-E. Meziani, and J. Soffer

new Hall D will be provided with the desired maximum energy of 12 GeV by adding a tenth arc and recirculating the beam a fifth time through one linac. A total of 90 μA of CW beam can be provided at the maximum beam energy. Further modifications required are changing the dipoles in the arcs from C-type to H-type magnets, replacing a large number of power supplies and doubling the central helium liquifier capacity to 10 kW. An overview of the upgrade of the accelerator is shown in Fig. 1.

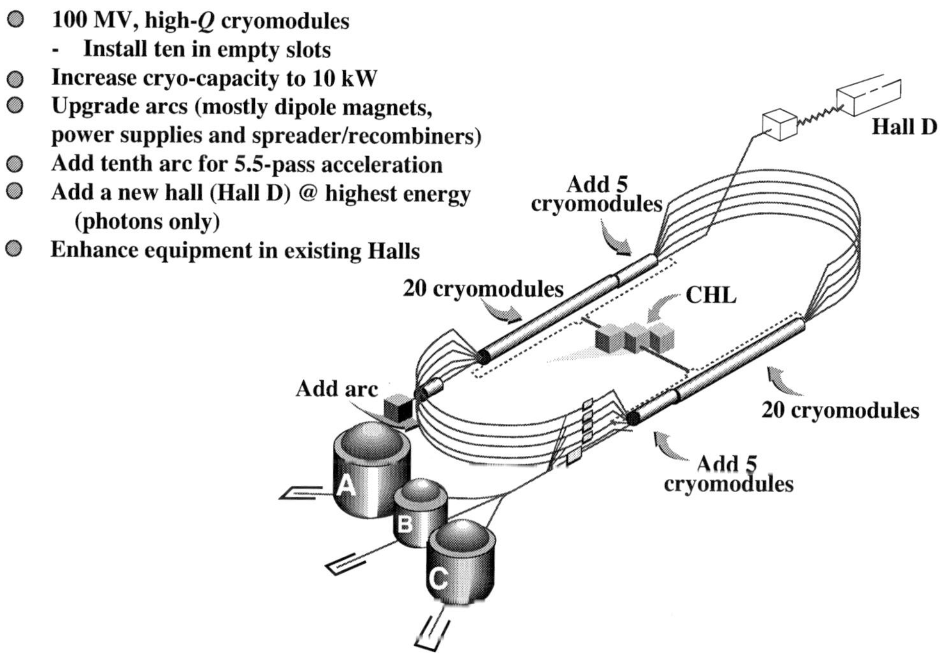

FIGURE 1. Overview of the accelerator upgrade to 12 GeV.

HALL UPGRADES

A main focus of the research program will be the Generalized Parton Distributions (GPD) through the study of exclusive processes at large momentum transfer. The GPDs can be considered as overlap integrals between different components of the hadronic wave function[2], governed by the selection of the final state. Measurements of these GPDs will thus make it possible to map out quark and gluon wave functions. One way to access the orbital angular momentum contribution to the nucleon spin is through GPDs. Factorization is an essential ingredient in the extraction of GPDs. For Deeply Virtual Compton Scattering (DVCS) scaling will have been achieved at 11 GeV, but this has to be established experimentally for other processes.

The 12 GeV upgrade will also make it possible to map out with high precision the quark distribution functions in the high-x (valence) region. Figure 2 shows two examples of such measurements. On the left is shown how the neutron polarization asymmetry A_1^n can be measured up to x-values close to 0.8 outside the nucleon resonance region, on the right how polarized valence and sea quark distributions can be extracted from semi-inclusive deep inelastic scattering by detecting the leading $\pi^{\pm}$ and $K^{\pm}$ hadrons.

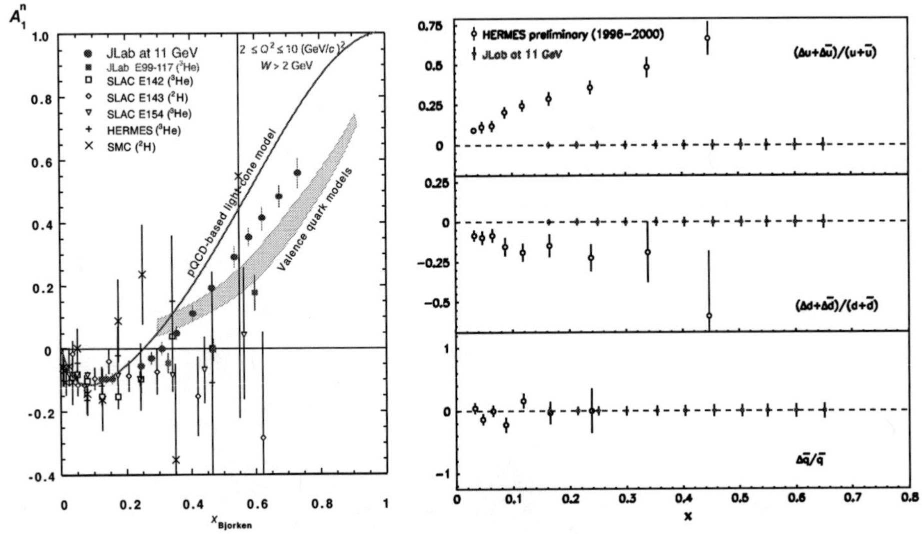

FIGURE 2. On the left is shown the projected measurement of A_1^n, on the right the projected determination of various combinations of polarized valence and sea quark distributions from semi-inclusive deep inelastic scattering.

The present base instrumentation in Hall A has been used with great success for experiments which require high luminosity and high resolution in momentum and/or angle of at least one of the reaction products. The central elements are the two High Resolution Spectrometers (HRS). Both of these devices provide a momentum resolution of better than 2×10^{-4} and an angular resolution of better than 1 mrad with a design maximum central momentum of 4 GeV/c. Two instrumentation upgrades are proposed to allow an optimal study of the intended experiments: a magnetic spectrometer, dubbed MAD (Medium Acceptance Detector), and an electro-magnetic calorimeter. The proposed MAD device is an extremely versatile magnetic spectrometer built from two combined-function (quadrupole and dipole) superconducting magnets. In its standard configuration the most forward scattering angle is 12°. The design provides a momentum acceptance (resolution) of ± 15 (0.1)% and an angular acceptance (resolution) of 6-23 msr (1 mrad). The maximum central momentum is 7.5 GeV/c at a total bend angle of 20°. With a septum magnet a forward angle of 5° can be reached with an angular acceptance of 5 msr. With the quadrupole component in the magnets turned off the central momentum can be increased to over 10 GeV/c, while the momentum resolution only deteriorates to 0.4% with an angular acceptance(resolution) of 2-5 msr(1 mrad). The basic detector package for the MAD spectrometer, covering the full momentum and angular acceptance, includes: fast high-resolution tracking chambers, a hydrogen-gas Čerenkov counter, trigger scintillator counters and a lead-glass hadron rejector. For the detection of hadrons a variable-pressure gas Čerenkov counter, two diffuse-reflective aerogel counters, a Ring Imaging Čerenkov counter and a Focal Plane Polarimeter will also be available.

The CEBAF Large Acceptance Spectrometer (CLAS) in Hall B is used for experiments that require the detection of several, loosely correlated particles in the hadronic final state at a limited luminosity. CLAS is a magnetic toroidal multi-gap spectrometer. Its magnetic field is generated by six superconducting coils. The detection system consists of drift chambers to determine the track of a charged particle, gas Čerenkov counters for particle identification, scintillation counters for the trigger and for measuring time-of-flight and electromagnetic calorimeters to detect showering particles like electrons and photons. CLAS presently operates at a luminosity of 10^{34} $cm^{-2}s^{-1}$. The CLAS upgrade incorporates two major improvements: (1) increasing the luminosity by an order of magnitude to account for the lower cross section values, (2) providing a more complete detection of the hadronic final state. The use of major components (torus magnet, scintillators, Čerenkov counters and EM calorimeters) will be retained. The tracking chambers will be replaced and a new central detector added.

The Hall C facility has generally been used for experiments which require high luminosity at moderate resolution. The core spectrometers are the High Momentum Spectrometer (HMS) and the Short Orbit Spectrometer (SOS). These two devices have been used flexibly as either electron or hadron arms, at times in coincidence with each other, at times in coincidence with a third experiment-specific arm. The HMS has a maximum momentum of 7.6 GeV/c, the SOS a value limited to only 1.7 GeV/c. For the upgrade the SOS spectrometer will be replaced by the so-called Super High Momentum Spectrometer (SHMS). The SHMS design consists of two superconducting quadrupoles and one combined-function magnet. It will have a maximum momentum of 11 GeV/c, a minimum scattering angle of 5.5°, a momentum acceptance (resolution) of ± 20 (0.2)% and an angular acceptance (resolution) of 2 msr (2 mrad). The basic configuration of the detector stack would consist of DC tracking chambers, trigger scintillator hodoscopes and a lead glass calorimeter.

Hall D

The Hall D research program will be focused on a definitive measurement of the spectrum of exotic hybrid mesons, which are expected in a mass range from 1 to 2.5 GeV/c^2. Lattice QCD calculations have convincingly illustrated[3] the linear quark-quark potential necessary for confinement. However, very little is still known about the direct excitation of the flux tube. The observation of such direct manifestations of gluonic degrees of freedom will provide understanding of confinement[4]. The quantum numbers of the flux tube, added to those of a $q\bar{q}$ meson, can produce exotic hybrids with unique J^{PC} quantum numbers. These excitations can be probed far more effectively with photons than with π- or K-mesons, because the quark spins are aligned in the virtual vector-meson component of the photon. For a full partial-wave analysis of such excitations linearly polarized photons are a requisite.

The optimum photon energy for production of exotic hybrids in its expected mass range is between 8 and 9 GeV. Linearly polarized photons in this energy range are optimally produced by coherent bremsstrahlung with 12 GeV electrons. The Hall D detector provides a nearly hermetic acceptance for both charged and neutral particles and includes several particle identification systems. Momentum analysis of charged particles is achieved with a superconducting solenoid and tracking chambers. The final planned photon flux is 10^8 photons/s. At this flux the experiment will acumulate in one year of running a factor of 100 more meson data than are presently available even from pion production.

ELECTRON-ION COLLIDER

Conceptual design studies[5] have indicated that the JLab facility can be further upgraded to provide a high-luminosity Electron Light-Ion Collider (ELIC) in the 20-65 GeV center-of-mass energy range. Such a facility would uniquely address the following research goals: (1) to complete our understanding of how quarks and gluons provide the binding and the spin of the nucleon, (2) to understand how quarks and gluons evolve into hadrons via the dynamics of confinement and (3) to refine our understanding of how the nuclear binding arises from QCD. A schematic lay-out of the ELIC collider is shown in Fig. 3. Longitudinally polarized electrons are injected into the CEBAF accelerator and accelerated to 5-7 GeV in a single pass (after all of the remaining 20 MV cryo-modules have been replaced by new 100 MV modules). The electrons are then injected into a circulator ring, in which they collide with the ions. After appr. 100 revolutions they are extracted and transported back for deceleration and energy recovery.

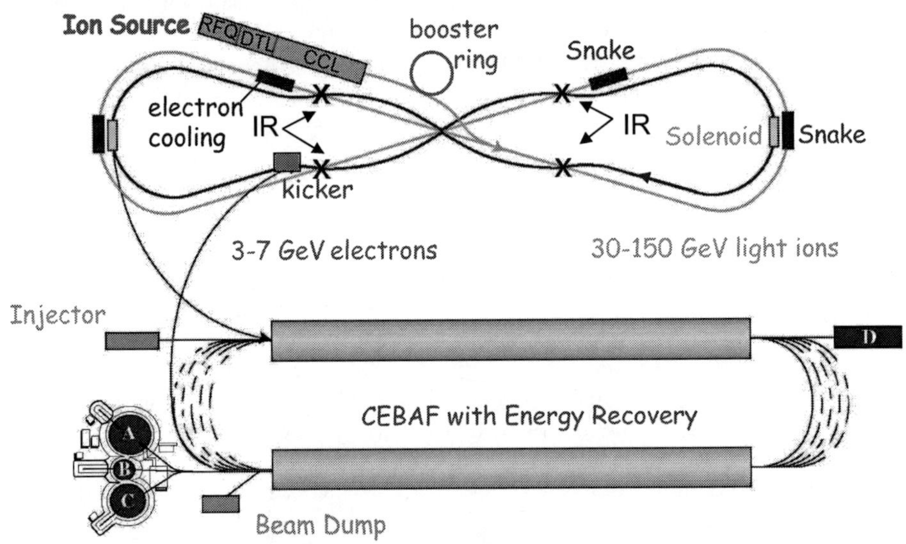

FIGURE 3. Schematic view of ELIC.

^{1}H, ^{2}H and ^{3}He ions are injected longitudinally polarized and accelerated in a conventional RF linac to a maximum energy of 200 MeV. After pre-boosting to 2-3 GeV the ions are stacked and accelerated in a "figure 8" ring to a maximum of 150 GeV. The "figure 8" ring provides flexible spin manipulation and preservation. This design results in a maximum CM energy of appr. 65 GeV. A luminosity of close to 10^{35} $cm^{-2}s^{-1}$ appears feasible through the use of electron cooling and crab crossings. Ions of up to mass 40, albeit unpolarized, will also be available. This design will also provide a 25 GeV beam in fixed-target mode.

SUMMARY

In April of 2004 the US Department of Energy (DOE) signed CD-0 approval for the 12 GeV Upgrade project. With this Critical Decision #0 DOE acknowledges the mission need for this project. Further review processes in increasing level of detail, spaced 12 to 18 months apart, have to be successfully passed before funding for the project will be allocated. A Conceptual Design Report is being prepared for the first of such reviews which is expected to take place in July of 2005.

ACKNOWLEDGMENTS

This work was supported by DOE contract DE-AC05-84ER40150 Modification No. M175, under which the Southeastern Universities Research Association (SURA) operates the Thomas Jefferson National Accelerator Facility.

REFERENCES

1. *Pre-Conceptual Design Report for the Science and Experimental Equipment for the 12 GeV Upgrade of CEBAF*, 2003, L.S. Cardman *et al.*, editors, http://www.jlab.org/div_dept/physics_division/pCDR_public/pCDR_final.
2. X. Ji, Phys. Rev. Lett. 78 (1997) 610 ; A. Radyushkin, Phys. Lett. B 380 (1996) 417.
3. G.S. Bali *et al.*, Proceedings of Int. Conf. on Quark Confinement and the Hadron spectrum, World Scientific, 1995, p. 225.
4. N. Isgur, R. Kokoski and J. Paton, Phys. Rev. Lett. 54 (1985) 869;
   S. Godfrey and J. Napolitano, Rev. Mod. Phys. 71 (1999) 1411.
5. L. Merminga, D.R. Douglas and G.A. Krafft, Annu. Rev. Nucl. Part. Sci. 53 (2003) 387.

Overview on the Future Electron-Ion Collider Projects

Emlyn W. Hughes

W.K. Kellogg Radiation Laboratory
California Institute of Technology
Pasadena, CA 91125

Abstract. One of the large new projects envisioned for the future US nuclear physics program is the construction of a high-energy electron ion collider. The motivation for such a machine would be primary directed at the study of the nucleon spin structure using polarized light ions and the study of low x gluon distributions in heavy nuclei. Two candidates coming from the two large DOE nuclear physics facilities, one at Brookhaven called eRHIC, and one at Jefferson Laboratory called ELIC are the primary candidates today.

Introduction

In November of 2003, the Secretary of Energy, Spencer Abraham, gave a speech to the US Congress outlining twenty-eight new large projects that represent the future science priorities for the DOE over the next 20 years [1]. Included on the list was an electron-ion collider (EIC). That was the good news. Unfortunately, the EIC only ranked #23 and fell into the category of "long-term" projects. However, there is some substantial reality to this decision. Today we have two concepts of what an EIC might be. One is based upon an upgrade of an interaction region with a new electron accelerator in the existing polarized proton program at the Relativistic Heavy Ion Collider (RHIC) at Brookhaven. This upgrade is called eRHIC (electron collider at RHIC) [2]. The second involves the construction of a new polarized hadron accelerator that would be used to collide with the electron beam from CEBAF at Jefferson Laboratory. This project is called ELIC for Electron Light Ion Collider [3]. However, both Jefferson Lab and Brookhaven have two additional projects that appear on the DOE list ahead of the EIC. At CEBAF, a 12 GeV upgrade of the beam energy is approved in spot #7 and at Brookhaven, an upgrade of the RHIC heavy ion program is being proposed and sits at spot #18. If one takes the DOE list seriously, then both the Brookhaven and Jefferson Lab "mid-term" projects will have to be done before an EIC is seriously considered. This places the EIC project well into the next decade.

In this paper, I will review briefly some of the global considerations for embarking on an EIC. The construction of a new collider in the US for nuclear physics would be based upon the interests of a broad community of physicists with a variety of topics that somehow unify into a single proposal. Now comes the most important sentence in the article. *The strategy and opinions given in this paper are my own personal view of the projects and present-day situation and absolutely do not represent any community or group*. So, let's start.

CP747, *Structure of the Nucleon at Large Bjorken x*, edited by C. Bourrely, Z.-E. Meziani, and J. Soffer

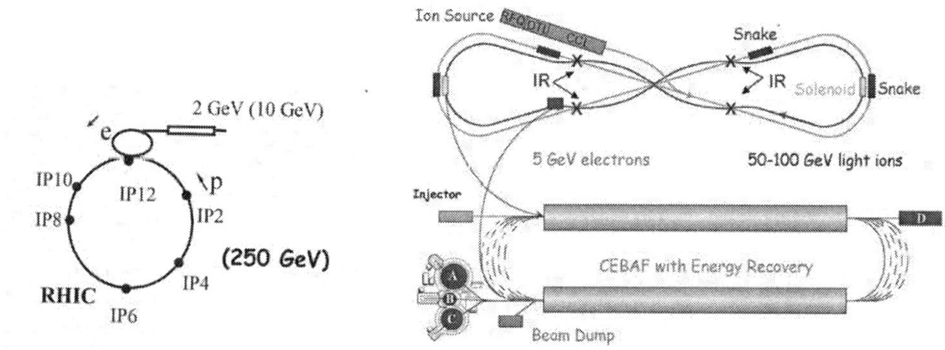

(a) (b)

FIGURE 1. (a) Schematic of the eRHIC collider layout. (b) Schematic of the ELIC collider at Jefferson Lab.

The Two EICs

No matter where it is built, the EIC will cost one billion dollars.

The eRHIC proposal at present envisions a single interaction region in which a circular or possibly a linear electron accelerator is constructed to collide with the polarized proton or heavy ion beam produced at RHIC (Figure 1a). The project will require a large 4π detector for studying the interactions. At a glance, the general idea of this proposal, to build a single ring and a single detector, should cost closer to \$500 million. The problem with the strawman eRHIC proposal is that (1) it assumes today's dollars and (2) it basically ignores the needs of the physics community interested in the study of low x structure functions of the proton and heavy ions. Although eRHIC is the highest energy electron-proton collider on the table today, its energy compared to HERA, for example, is limited (see Table 1). To continue the study of low x structure functions beyond HERA requires a dedicated forward-angle detector to access the low x region. A conventional 4π detector will simply not fit into the same interaction region as a detector needed for low x structure function studies. The solution to this problem is to build two interaction regions at RHIC, one per program. But 2 x \$500 million is \$1 billion ☺

To construct a new hadron machine at Jefferson Laboratory for ELIC (Figure 1b) simply costs \$1 billion; hadron machines are expensive.

TABLE 1. EIC Parameters

Collider	Luminosity (10^{31})	Center of Mass Energy (GeV)	Q^2	Bjorken x range
HERA	7	320	1	10^{-5}
			5	10^{-4}
eRHIC	50	100	1	10^{-4}
			5	10^{-3}
ELIC	100-10,000	31	1	10^{-3}
			5	10^{-2}

The ELIC design has the nice feature of multiple interaction regions. Although this seems extravagant, multiple interaction regions give a wonderful boost to the nuclear community, similar to having multiple experimental halls. The freedom that multiple interaction regions provide should not be underestimated as a primary advantage of this idea.

However, ELIC operates at substantially lower center of mass energy. The machine will be able to perform detailed studies of the nucleon structure only in the mid x range. An important consideration for the ELIC machine will be the study of energy upgrades. It is fine to start the collider at low energy and high luminosity, but without an upgrade to 6 GeV electrons on, for example, 150 GeV protons, the nucleon spin community will be limited. In particular, the kinematic coverage of ELIC may not extend out knowledge significantly compared to the fixed target programs that already have studied nucleon spin structure functions in the 1990s [4].

The subsequent four sections discuss how the two collider ideas stack up in addressing four mainline physics topics.

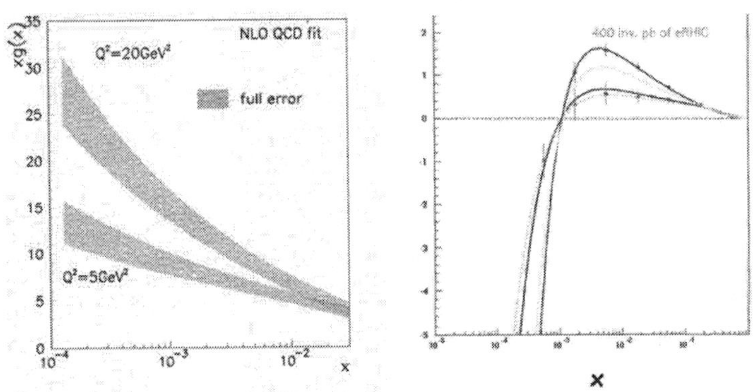

(a) (b)

FIGURE 2. (a) Results on the unpolarized gluon distribution as measured by the H1 collaboration at HERA. (b) Simulation of the proton spin structure function results at low x from eRHIC for different polarized gluon distributions.

Polarized Gluon Distribution

The proton spin structure can be decomposed into three pieces,

$$\text{Proton spin} = 1/2 = 1/2\ \Delta q + \Delta G + L_Z,$$

where Δq corresponds to the quark contribution to the proton spin, ΔG is the gluon contribution and L_Z is the contribution from orbital angular momentum of the constituents. One of the greatest challenges today is to try to measure ΔG for the first time. The expectation is that ΔG should be quite large and positive, since the quark contribution only accounts for about 30% of the total spin. RHIC Spin [5] and COMPASS [6] will make important first measurements of ΔG over the next five years, but the kinematic range of these measurements will be limited ($x > 0.01$), and theoretical uncertainties arising from the determination of structure functions in proton-proton scattering, for example, are likely to be large.

Measurements of G(x) from HERA represent the state-of-the art for determining the gluon distribution of the proton. Figure 2a presents results for G(x) coming from the H1 collaboration [7] evolved to an average Q^2 of 5 GeV^2. The dream for the spin community is to produce exactly such a plot for $\Delta G(x)$ coming from a polarized collider measurement. Figure 2b shows a simulation of the precision on the proton spin structure function coming from the eRHIC program. The ability to make precision measurements of spin structure functions down to x of 10^{-3} is one of the main goals of the eRHIC program.

At present, with the limited center-of-mass energy, ELIC could not make a competitive measurement at low x.

Orbital Angular Momentum

If the integral over the polarized gluon distribution gives a result that accounts for the remaining proton constituent contribution to its total spin, then the field of study will be quiet for awhile, and certainly any vestige of the "Proton Spin Puzzle" will disappear. However, if ΔG is small or negative, there will still be a lively hunt to extract information on the quark or gluon orbital angular momentum contribution to the proton's spin.

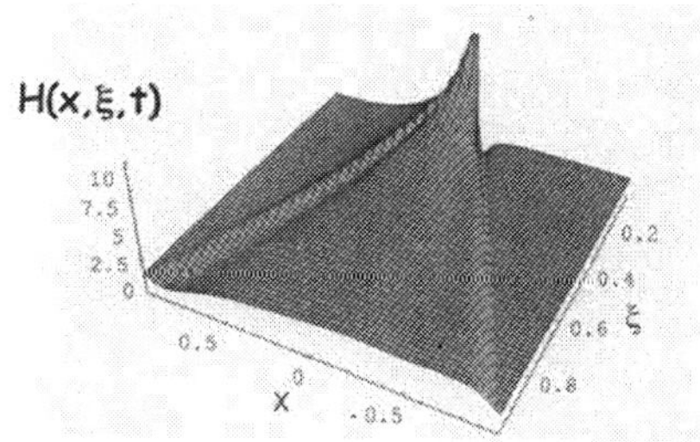

FIGURE 3. Sample GPD distribution as predicted from a theoretical model.

Theoretically, the best hope today is that one may be able to find information on the quark orbital angular momentum contribution through deeply virtual Compton scattering [8] or, more generally, through the study of Generalized Parton Distributions (GPDs) [9]. Unfortunately, the measurements of these functions is very difficult and covers an enormous parameter space. Figure 3 presents a theorist vision of what one of these functions might look like. A primary goal of the Jefferson Laboratory 12 GeV upgrade will be to make first extensive measurements of GPDs and try to evaluate whether this is a viable field of study.

Certainly, with the extended kinematic reach of an ELIC compared to the 12 GeV program, the GPD studies, if fruitful, could be continued with further precision at higher energies. For this idea to become a reality, a very high luminosity collider will be needed, and this makes ELIC a natural candidate, with it potential to reach 10^{35} cm^{-2} per second. This is to be contrasted to eRHIC which will not have a high luminosity compared to the 12 GeV fixed-target program.

In fact, at present, it is probably reasonable to think of eRHIC as a collider to for measuring ΔG and ELIC as a potential collider to study orbital angular momentum.

Gluon Shadowing

There has been a significant interest recently in the study of low x structure functions in the nucleon and nuclei and their implication for understanding unpolarized low x gluon distributions [10]. With limited space, I will just reference the analysis of HERA data [11] and the recent results from Brahms [12]. Results from these experiments indicate that the gluon distribution may be saturating at low x compared to the predictions of the DGLAP evolution equations [13-15]. It is intriguing that a heavy ion measurement can play a role in this study. Measurements made at the LHC will certainly be of enormous interest to continue probing at low x.

If the study outlasts the heavy ion program at RHIC and the proton and heavy ion program at LHC, then eRHIC could make some important "clean" measurements of low x gluon distributions, especially in nuclei. Since ELIC is a low energy collider and only aimed at very light nuclei, this field of study is an eRHIC project.

But, will it still be interesting a decade from now, post LHC?

Precision Electroweak

I cannot resist, but end on a humorous note. First pointed out by Rolf Ent, ELIC has a luminosity that could, in principle, perform precision electroweak measurements via parity violation in polarized electron scattering from unpolarized protons. Figure 4 presents measurements of the electroweak mixing angle versus energy comparing what exists today and what an ELIC could do.

There is still a long list of difficulties that one would have to address even if the full luminosity ELIC collider worked flawlessly. Just to mention one, super precise polarimetry of the electron beam will be needed (also for an NLC, by the way!) Another problem is that many competitive precision parity-violation measurements will come earlier from the existing Jefferson Laboratory fixed target program plus experiments at the 12 GeV upgrade.

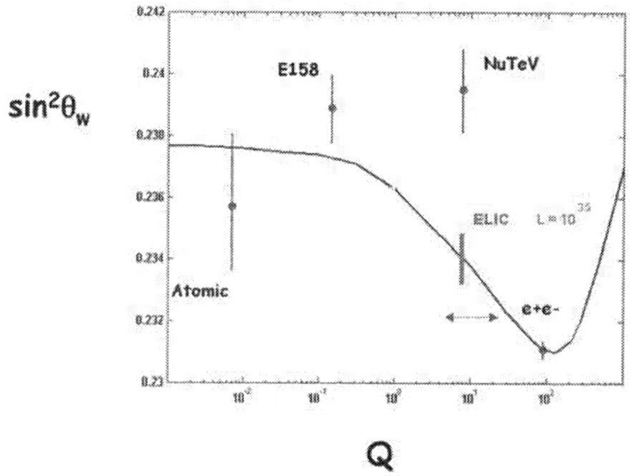

Figure 4. Measurements of the electroweak mixing parameter versus Q compared to a simulation from a possible measurement at the ELIC collider.

But, it is interesting and a testament to the new ideas possible at a very high luminosity collider. However, like the last section, after the LHC has finished its new particle searches, will this field of physics still be interesting?

Summary

The existence of an EIC in the US would certainly bring a significant community to collider physics for the study of strong interactions, taking up the flag where HERA will soon leave off. The biggest challenge faced by the proponents of an EIC today is to convince the world that this machine must be built in a reasonable timeframe.

ACKNOWLEDGMENTS

I would like to thank the conference organizers for the invitation to come to beautiful Marseille. A special thanks goes to both Zein-Eddine Meziani and Jacques Soffer for their enormous efforts. Also, I apologize that this collider discussion had nothing to do with high x! After the 12 GeV program at CEBAF and the high energy measurements from HERA, I do not believe that high x play a large role in the EIC business. I could certainly be wrong. HERA may still find leptoquarks at high x, in which case all bets are off on the reason for building a US EIC. In that case, just shred this article.

REFERENCES

1. Abraham, Spencer, talk given to US Congress, Nov. 10, 2003, *Facilities for the Future of Science: A Twenty-year Oulook*, http://www.er.doe.gov/Sub/Facilities_for_future/facilities_future.htm
2. R. Holt et al., *The Electron-Ion Collider: A High Luminosity Probe of the Partonic Substructure of Nucleons and Nuclei*, BNL-68933, Feb. 2002.
2. L. Merminga et al., *ELIC: An Electron-Light Ion Collider Based at CEBAF*, Proc. of EPAC, p. 203, Paris, France, 2002.
3. E.W. Hughes and R. Voss, *Spin Structure Functions*, Ann. Rev. Nucl. Part. Sci, **49**, 303 (1999).
5. G. Bunce et al., Prospects for Spin Physics at RHIC, Ann. Rev. Nucl. Part. Sci, **50**, 525 (2000).
6. G. Baum et al. (COMPASS Collaboration), *Common Muon and Proton Apparatus for Structure and Spectroscopy*, CERN/SPSLC 96-14, March 1996.
7. S. Aid et al. (H1 Collaboration), Nucl. Phys. **B470**, 3 (1996).
8. X. Ji, Int. J. Mod. Phys. **A18**, 1303 (2003).
9. X. Ji, Phys. Rev. Lett. **78**, 610 (1997); Phys. Rev. **D55**, 7114 (1997).
10. J.-P. Blaizot & F. Gelis, hep-ph/0405305, August 2004
11. L. Lindemann et al. (H1 and ZEUS Collaborations), Nucl. Phys. Proc. Suppl., 64, 179 (1998).
12. I. Arsene et al. (BRAHMS Collaboration), nucl-ex/0403005, March 2004.
13. G. Altarelli and G. Parisi, Nucl. Phys. **B126**, 298 (1977).
14. Yu, Dokshitzer, Sov. Phys. JETP **46**, 641 (1977).
15. V.N. Gribov, L.N. Lipatov, Sov. J. Nucl. Phys. **15**, 438 (1972); **15**, 675 (1972).

APPENDIX

HiX 2004 Workshop Agenda

Monday July 26, 2004

08:00-09:00	Registration at CPPM	
09:00-09:10	J. Soffer (Marseille)	Welcome and introduction

Session 1: Overview (Chair: A. Thomas)

09:10-09:50	X. Ji (Maryland)	QCD theory at large x
09:50-10:20	W. Vogelsang (BNL)	Large-x resummations in QCD
10:20-10:50	J. Stirling (Durham)	Parton distributions from global fits
10:50-11:10	Coffee Break	

Session 2: Unpolarized Structure Functions (Chair: K. de Jager)

11:10-11:40	W.K. Tung (Michigan)	Global QCD analyses of parton distributions at large x and small
11:40-12:10	G. Petratos (Kent State)	Measuring F_2^n/F_2^p and d/u: Past and Future
12:10-12:40	Z. Zhang (LAL)	Prospects for and status of the Measurements of High x Parton Distributions at HERA
12:40-13:00	S. Bültmann (ODU)	The BoNuS experiment at Jefferson Lab
13:00-14:00	Lunch (Cafeteria on campus)	
14:00-14:20	H. Böettcher (DESY)	Non-singlet QCD analysis of the structure function F_2 in 3-loops

Session 3: Polarized Structure Functions (I) (Chair: W. Vogelsang)

14:20-14:50	C. Bourrely (Marseille)	Statistical polarized parton distributions at large x
14:50-15:20	L. Bland (BNL)	Status of the RHIC-spin program
15:20-15:50	A. Magnon (CEA)	The Compass experiment at CERN
15:50-16:10	Coffee Break	

Session 4: Polarized Structure Functions (II) (Chair: Z.E. Meziani)

16:10-16:40	J.-P. Chen (JLab)	Neutron spin structure measurements in Jlab Hall A
16:40-17:10	P. Bosted (JLab)	Spin structure of the proton and deuteron with CLAS at Jefferson Lab
17:10-17:40	O. Rondon (UVA)	Polarized structure functions: proton/deuteron measurements in Hall C
17:40-18:00	J. Franklin (Temple)	How high is high x ?
19:00	Welcome reception at CIRM	

Tuesday July 27, 2004

Session 5: Duality and Higher Twists (Chair: P. Souder)

09:00-09:30	W. Melnitchouk (JLab)	Quark-hadron duality : higher twists and large-x structure functions
09:30-09:55	E. Christy (Hampton)	Quark-hadron duality: experimental status and prospects
09:55-10:20	N. Liyanage (UVA) (presented by J.-P. Chen)	Spin duality in the nucleon: measurements at Jefferson Lab Hall A
10:20-10:45	W. Korsch (Kentucky)	Higher twists and the measurements of g_2
10:45-11:05	Coffee Break	
11:05-11:25	A. Fantoni (Frascati)	Quark-hadron duality and higher twist contributions in structure functions

Session 6 : Semi-Inclusive DIS (Chair: O. Rondon)

11:25-11:55	P.G. Ratcliffe (Insubria)	Semi-Inclusive DIS and transversity
11:55-12:25	J.-C. Peng (Illinois)	Transversity and SIDIS at Jefferson Lab
12:25-12:55	M. Stratmann (Regensburg)	Semi-Inclusive DIS in NLO QCD
12:55-14:00	Lunch (Cafeteria on campus)	
14:00-14:30	L. De Nardo (Alberta)	Helicity and transversity distributions at Hermes
14:30-15:00	L. Gamberg (Penn State Berks)	Novel transversity properties in SIDIS

Session 7: Exclusive Reactions (Chair: W. Melnitchouk)

15:00-15:30	A Schäfer (Regensburg)	GPDs from small to large x
15:30-16:00	M. Garçon (CEA)	DVCS in Hall A and B at Jefferson Lab
16:00-16:20	Coffee Break	
16:20-16:50	H. Gao (Duke)	Pion photoproduction on the nucleon
16:50-17:20	M. Traini (Trento)	Generalized parton distributions and constituent quark model
17:20-17:50	M. Sargsian (Florida)	Nucleon DIS structure functions from semi-inclusive $ed \rightarrow enX$ reactions
20:00	Conference dinner in Cassis	

Wednesday July 28, 2004

Session 8: Parity Violating Experiments and Drell-Yan (Chair: J. Stirling)

09:00-09:30	P. Souder (Syracuse)	Parity at high x
09:30-10:00	T. Londergan (Indiana)	Current status of the NuTeV anomaly
10:00-10:20	R. Towell (ACU)	The past and future of Drell-Yan physics at Fermilab
10:20-10:40	E. Di Salvo (Genova)	Q2 dependence of the azimuthal asymmetry in unpolarized Drell-Yan
10:40-11:00	Coffee Break	

Session 9: Lattice QCD and Nuclear DIS (Chair: X. Ji)

11:00-11:30	A. Thomas (JLab)	Large x behaviour and the non-perturbative structure of hadronic systems
11:30-12:00	G. Schierholz (DESY)	Recent Results from the lattice
12:00-12:20	A. Molochkov (Giessen)	Nucleon structure functions and nuclear DIS
12:20-12:40	G. Smirnov (Dubna)	Mapping high x structure of the nucleon : which data is missing?
13:00-14:00	Lunch (Cafeteria on campus)	

Session 10: Future and Summary (Chair: J.P. Chen)

14:00-14:20	M. Strikman (PSU)	Probing correlations in the large x proton wave function at pp and pA colliders
14:20-14:50	K. de Jager (JLab)	Research perspective at Jefferson Lab: 12GeV and beyond
14:50-15:20	E. Hughes (Caltech)	Overview on the future electron-ion collider projects
15:20-15:50	Coffee Break	
15:50-16:20	Z.-E. Meziani (Temple)	Summary (Experimental)
16:20-16:50	J. Soffer (Marseille)	Summary (Theory)
16:50-17:20	Concluding remarks and Adjourn	

Workshop Web site http://www.cpt.univ-mrs.fr/hix2004

HiX 2004 List of Participants

Les BLAND
Phys. Dept.
Brookhaven National Lab.
U.S.A.
bland@bnl.gov

Helmut BÖETTCHER
DESY Zeuthen
Germany
hboett@ifh.de

Peter E. BOSTED
Jefferson Lab MS 12H
U.S.A.
bosted@jlab.org

Claude BOURRELY
CPT – Marseille
France
bourrely@cpt.univ-mrs.fr

Franco BUCCELLA
Phys. Dept. Napoli Univ.
Italy
Franco.Buccella@na.infn.it

Stephen BÜLTMANN
Phys. Dept.
Old Dominion Univ.
U.S.A.
stephenb@physics.odu.edu

Jian-Ping CHEN
Jefferson Lab.
U.S.A.
jpchenc@jlab.org

M. Eric CHRISTY
Phys. Dept.
Hampton Univ.
U.S.A.
christy@jlab.org

Lara DE NARDO
Desy-Hermes
Germany
ldenardo@hermes.desy.de

Elvio DI SALVO
Phys. Dept.
Genova Univ.
Italy
disalvo@ge.infn.it

Alessandra FANTONI
INFN Frascati
Italy
fantoni@lnf.infn.it

Jerrold FRANKLIN
Phys. Dept.
Temple Univ.
U.S.A.
jerry.f@temple.edu

Leonard P. GAMBERG
Dept. Phys.
Penn State Berks
U.S.A.
lpg10@psu.edu

Haiyan GAO
Phys. Dept.
Duke Univ. (Durham)
U.S.A.
gao@phy.duke.edu

Michel GARÇON
DAPNIA / SPhN
CEA-Saclay
France
mgarcon@cea.fr

Emlyn W. HUGHES
Kellogg Radiation Lab.
Pasadena
U.S.A.
emlyn@caltech.edu

Kees de JAGER
Jefferson Lab.
U.S.A.
kees@jlab.org

Xiandong JI
Phys. Dept.
Maryland Univ.
U.S.A.
xji@physics.umd.edu

Mina KATRAMATOU
Phys. Dept.
Kent State Univ.
U.S.A.
akatrama@kent.edu

Wolfgang KORSCH
Phys. Dept.
Kentucky Univ.
U.S.A.
korsch@pa.uky.edu

John Timothy LONDERGAN
Wells Scholars
Indiana Univ.
U.S.A.
tlonderg@indiana.edu

Alain MAGNON
CEN Saclay
France
Alain.Magnon@cern.ch

Wally MELNITCHOUK
Jefferson Lab.
U.S.A.
wmelnitc@jlab.org

Zein-Eddine MEZIANI
Temple Univ. Philadelphia
U.S.A.
meziani@jlab.org

Alexander MOLOCHKOV
Theoretische Physik Giessen
Germany
Alexander.Molochkov@
theo.physik.uni-giessen.de

Jen-Chieh PENG
Department of Physics
UIUC
U.S.A.
jcpeng@uiuc.edu

Gerassimos G. PETRATOS
Phys. Dept.
Kent State Univ.
U.S.A.
gpetrato@kent.edu

Philip RATCLIFFE
Dip. di Fisica e Mathematica
Como
Italy
philip.ratcliffe@uninsubria.it

Oscar RONDON-ARAMAYO
Phys. Dept.
Virginia Univ.
U.S.A.
or@virginia.edu

Misak SARGSIAN
Phys. Dept.
Florida Int. Univ.
U.S.A.
sargsian@fiu.edu

Andreas SCHÄFER
Inst. für Theoretische Physik
Regensburg Univ.
Germany
andreas.schaefer@
physik.uni-regensburg.de

Gerrit SCHIERHOLZ
DESY
Germany
Gerrit.Schierholz@desy.de

George I. SMIRNOV
LPC
Clermont-Ferrand
France
george@clermont.in2p3.fr

Jacques SOFFER
CPT – Marseille
France
soffer@cpt.univ-mrs.fr

Paul A. SOUDER
Phys. Dept.
Syracuse Univ.
U.S.A.
souder@physics.syr.edu

James STIRLING
IPPP Durham Univ.
U.K.
w.j.stirling@durham.ac.uk

Marco STRATMANN
Institute für Physik
Regensburg Univ.
Germany
marco.stratmann@
physik.uni-regensburg.de

Mark STRIKMAN
Davey Lab.
Park Univ.
U.S.A.
strikman@phys.psu.edu

Anthony W. THOMAS
Jefferson Lab.
U.S.A.
awthomas@jlab.org

Rusty S. TOWELL
Abilene Christian Univ.
U.S.A.
rusty.towell@acu.edu

Marco TRAINI
Phys. Dept.
di Trento Univ.
Italy
traini@science.unitn.it

Wu-Ki TUNG
Phys. Dept.
Michigan State Univ.
U.S.A.
tung@pa.msu.edu

Jean-Marc VIREY
CPT – Marseille
France
virey@cpt.univ-mrs.fr

Werner VOGELSANG
RIKEN-BNL
U.S.A.
wvogelsang@bnl.gov

Zhiqing ZHANG
LAL Orsay
France
zhangzq@lal.in2p3.fr

Author Index

B

Bianchi, N., 126
Bland, L. C., 63
Blümlein, J., 50
Boffi, S., 185
Bosted, P. E., 89
Bourrely, C., 57
Böttcher, H., 50
Bültmann, S., 46

C

Chen, J.-P., 83, 141
Christy, M. E., 114

D

de Jager, K., 248
De Nardo, L., 153
Di Salvo, E., 215

F

Fantoni, A., 126
Franklin, J., 101

G

Gamberg, L. P., 159
Gao, H., 179
Garçon, M., 173
Goldstein, G. R., 159
Guffanti, A. 50

H

Hughes, E. W., 254

J

Ji, X., 3
Jiang, X., 141

K

Korsch, W., 122

L

Liuti, S., 126
Liyanage, N., 118
Londergan, J. T., 205

M

Magnon, A., 72
Martin, A. D., 16
Melnitchouk, W., 107
Molochkov, A., 233

O

Oganessyan, K. A., 159

P

Pasquini, B., 185
Peng, J.-C., 141
Petratos, G. G., 32

R

Ratcliffe, P. G., 133
Roberts, R. G., 16
Rondon, O. A., 95

S

Sargsian, M., 191
Schäfer, A., 167
Schierholz, G., 227
Smirnov, G. I., 237
Souder, P. A., 199
Stirling, W. J., 16
Stratmann, M., 147
Strikman, M., 243

T

Thomas, A. W., 221
Thorne, R. S., 16
Towell, R. S., 211
Traini, M., 185
Tung, W.-K., 25

V

Vogelsang, W., 9

Z

Zhang, Z., 40
Zhu, L., 141, 179